BAYESIAN INFERENCE

Edited by **Javier Prieto Tejedor**

Bayesian Inference
http://dx.doi.org/10.5772/66264
Edited by Javier Prieto Tejedor

Contributors

Claudia Ceci, Katia Colaneri, Sherif Sherif, Ishan Maduranga Wickramasingha, Michael Sobhy, Anamaria Berea, Daniel Maxwell, Hafedh Ben Zaabza, Abderrahmen Ben Gara, Boulbaba Rekik, Solomon Tesfamicael, Steven Kim, Tai VoVan, Poom Kumam, Plern Saipara, Cristiano Premebida, Diego Faria, Francisco Souza, Loc Nguyen, Alonso Ortega, Gorka Navarrete, Xin Tong, Dingjing Shi, Mengxi Chen, Sien Chen, Jie Cheng, Junjiang Zhong, Wenqiang Huang, Hamid El Maroufy, El Houcine Hibbah, Abdelmajid Zyad, Taib Ziad, Naveen Bansal, Jørund Gasemyr, Bent Natvig, Valeria Sambucini, Pablo Emilio Verde

Published by InTech
Janeza Trdine 9, 51000 Rijeka, Croatia

Publishing Process Manager Mirena Calmic
Technical Editor SPi Global
Cover InTech Design team

First published October, 2017
Printed in Croatia
Legal deposit, Croatia: National and University Library in Zagreb

Additional hard copies can be obtained from orders@intechopen.com

Bayesian Inference, Edited by Javier Prieto Tejedor
p. cm.
Print ISBN 978-953-51-3577-7
Online ISBN 978-953-51-3578-4

For EU product safety concerns:
IN TECH d.o.o., Prolaz Marije Krucifikse Kozulić 3, 51000 Rijeka, Croatia,
info@intechopen.com or visit our website at intechopen.com.

Meet the editor

Javier Prieto received his PhD degree in Information and Communications Technology and the Extraordinary Performance Award for Doctorate Studies in 2012 from the University of Valladolid (Spain). Since 2015, he is a lecturer and researcher at the University of Salamanca (Spain). Previously, he was with the University of Valladolid from 2009 to 2014 and with a Spanish technological center from 2007 to 2009. In 2010, he was a visiting researcher at the Massachusetts Institute of Technology (MIT), USA. Dr. Prieto serves as an associate editor for various journals. His research interests include artificial intelligence for smart cities, navigation for indoor environments, and Bayesian inference for dynamic systems.

Contents

Preface

The range of Bayesian inference algorithms and their different applications has been greatly expanded since the first implementation of a Kalman filter by Stanley F. Schmidt for the Apollo program. Extended Kalman filters, unscented Kalman filters, particle filters, and belief condensation filters are just some examples of these algorithms that have been applied to logistics, medical services, search and rescue operations, or automotive safety, among others. The essence of the mentioned algorithms is to explain how we should update our existing beliefs in the light of new evidence. Stephen Senn defined a Bayesian as "one who, vaguely expecting a horse and catching a glimpse of a donkey, strongly concludes he has seen a mule."

From the Bayesian perspective, both the parameter to estimate and the observations are random variables, against the frequentist approach where the parameter to estimate is an unknown deterministic value. This Bayesian point of view leads to a common resolution framework where what we infer is a density function of the parameter conditioned on the observation. Under this context, the task is to determine the posterior distribution of the desired state, from the knowledge of the prior and the likelihood, by using Bayes' rule. This setting can be modeled with a hidden Markov model (HMM), where the sequence of variables of interest is called hidden states and the sequence from which one can obtain realizations is called observations.

Achilles' heel of Bayesian inference is the implementation of algorithms for solving nonlinear and/or non-Gaussian system models, where the traditional Kalman approach is inaccurate. Optimal algorithms exist in some restricted cases, while researchers have proposed many practical implementations of Bayesian inference that rely on the use of approximations. In addition, the inclusion of relevant prior information has been deeply addressed in the literature where there is no consensus about the existence of a truly non-informative prior. The election of appropriate prior information has extensively nourished the research on Bayesian inference, and it is still today an element of intense discussion.

This book takes a look at both theoretical foundations of Bayesian inference and practical implementations in the fields of life sciences, engineering, and economics. The book is organized into four sections according to this twofold perspective.

Section 1 is dedicated to the theoretical foundations of Bayesian inference.

In Chapter 1, the authors make a review of Bayesian inference components and its application to game theory. The chapter includes definitions of the former (prior, likelihood, and posterior distributions) and comprises several examples of the latter (Bayesian and fuzzy games).

In Chapter 2, the authors review methods for checking the modeling assumptions at each node of a directed acyclic graph that represents the understanding of the underlying structure of a problem. The chapter shows how nodes in a graph correspond to data or parameters and directed edges between parameters correspond to conditional distributions.

In Chapter 3, the author presents a Bayesian classification method, the associated Bayes error, and its relationship with other measures and proposes an algorithm to determine prior probability that can make to reduce Bayes error is proposed. The chapter applies the proposed algorithm in three domains, biology, medicine, and economics, through specific problems.

In Chapter 4, the author introduces the problem of high-dimensional multiple hypothesis testing by using the Bayesian approach. The chapter demonstrates the practical application of the Bayesian decision theoretic approach by means of a real example of directional hypothesis testing with skewed alternatives that use gene expression data.

In Chapter 5, the author provides formal criteria to determine the adequate sample size in the design of experiments based on combined frequentist-Bayesian or fully Bayesian approaches. The chapter defines four power functions for sample size calculations where the Bayesian predictive power is the one that allows to add more flexibility.

In Chapter 6, the author addresses the problem of converting graphic relationships into conditional probabilities in order to construct a simple Bayesian network from a graph. The chapter applies this research in learning context in which a Bayesian network is used to assess students' knowledge.

Section 2 is dedicated to the applications of Bayesian inference in life sciences.

In Chapter 7, the authors propose a first-order autoregressive hidden Markov model as a suitable model to characterize a marker of breast cancer disease progression. The chapter shows how this model captures the complexity and the dynamics of the evolution of breast cancer by introducing the latent states and permits to evaluate the efficacy of a treatment by their transition probabilities.

In Chapter 8, the author deals with the limitations of the Bayesian framework applied to dose-response studies relying on small samples and sparse data. The chapter addresses three practical issues in small-sample dose-response studies: model sensitivity, disagreement in prior knowledge, and conflicting perspective in decision rules.

In Chapter 9, the author illustrates the application of Bayesian inference to two meta-analyses in medical research when indirect evidence is available for analysis. The chapter presents the hierarchical meta-regression method for meta-analysis, which is an integrated approach for evidence synthesis when a multiplicity of bias, coming from indirect and disparate evidence, has to be incorporated.

In Chapter 10, the authors provide a review of statistical methods applied to animal and plant breeding programs with a particular focus on Bayesian methods. The chapter illustrates the flexibility of the Bayesian approaches and their high accuracy of prediction of the breeding values.

In Chapter 11, the authors propose four types of two-stage least squares models with instrumental variables to model normal and non-normal data in causal inference research. The chapter evaluates the performance of the robust method using Student's t-distributions in the four distributional 2SLS models by means of Monte Carlo simulation.

In Chapter 12, the authors present the Bayesian hypothesis testing and its application to psychology and social sciences as an alternative to traditional frequentist null hypothesis significance testing (NHST). The chapter shows the advantages of this Bayesian approach over frequentist NHST, providing examples that support its use.

Section 3 is dedicated to the applications of Bayesian inference in engineering.

In Chapter 13, the authors motivate the use of Bayesian inference in compressed sensing signal processing method. The chapter provides three use cases of its applicability, such as magnetic resonance images, remote sensing, and wireless communication systems, specifically on multiple-input multiple-output (MIMO) systems.

In Chapter 14, the authors present different methods to estimate an unknown signal from its linear measurements where the number of measurements is less than the dimension of the unknown signal. The chapter introduces the concept of signal sparsity and describes how it could be used as prior information for either regularized least squares or Bayesian signal estimation.

In Chapter 15, the authors explore the use of dynamic Bayesian networks (DBN) for time-dependent classification problems in mobile robotics and present some experiments in semantic place recognition and daily activity classification. The chapter formulates the DBN as a time-dependent classification problem and gives a general expression for a DBN in terms of classifier priors and likelihoods through the time steps.

Section 4 is dedicated to the applications of Bayesian inference in economics.

In Chapter 16, the authors propose a new Bayesian model to aid the investment decision in early-stage start-ups and ventures informed by previous academic literature on entrepreneurship and venture capital investment practices. The chapter assesses this model in an anonymized experiment where reviewers with previous experience in entrepreneurship and/or investment scored a list of 20 anonymous real companies.

In Chapter 17, the authors present some results about nonlinear filtering for Markovian partially observable systems where the state and the observation processes are described by jump diffusions with correlated Brownian motions and common jump times. The chapter applies this theory to the financial problem of derivatives hedging for a trader who has limited information on the market.

In Chapter 18, the authors illustrate how a Bayesian belief network (BBN) can enable airlines to optimize the loyalty based on dynamic recommendations of relevant contents from predictions of passengers' choice. The chapter establishes BBN models by using the use case of China Southern Airlines with real transaction data, including passengers' basic information, history decision options, and purchase characteristics.

Overall, this book is intended as an introductory guide for the application of Bayesian inference in the fields of life sciences, engineering, and economics, as well as a source document of fundamentals for intermediate Bayesian readers.

Dr. Javier Prieto Tejedor
University of Salamanca,
Salamanca, Spain

Section 1

Theoretical Foundations of Bayesian Inference

Chapter 1

Bayesian Inference Application

Wiyada Kumam, Plern Saipara and Poom Kumam

Additional information is available at the end of the chapter

http://dx.doi.org/10.5772/intechopen.70530

Abstract

In this chapter, we were introduced the concept of Bayesian inference and application to the real world problems such as game theory (Bayesian Game) etc. This chapter was organized as follows. In Sections 2 and 3, we present Model-based Bayesian inference and the components of Bayesian inference, respectively. The last section contains some applications of Bayesian inference.

Keywords: statistical inference, Frequentist inference, Bayesian inference

1. Introduction

In statistical inference, there are two ways for interpretations of probability include Frequentist (or Classical) inference and Bayesian inference. It usually is unlike with each other in the classical nature of probability. Classical inference defines probability as the limit of an event's relative frequency for a large number of experiments and only in the sense of random experiments which are well defined. Other side, Bayesian inference can to impose probabilities to each statement when a random process is not associated. In the sense of Bayesian, probability is a way to show an individual's degree of believes in a statement. Bayesian inferences are different interpretations of probability, and also different approaches depend on those interpretations. Bayes' theorem presents the relativity about two conditional probabilities that are the reverse of anything other. The initials of the term Bayes' theorem is in honor of Reverend Thomas Bayes, and is referred to as Bayes' law (see [1]). This theorem shows the conditional probability or *posterior probability* of an event A after B is observed in terms of the *prior probability* of A, prior probability of B and the conditional probability of B given A. It is valid in all interpretations of probability. Bayes' formula is how to revise probability statements using data. The Bayes' law (or Bayes' rule) is

$$P(A|B) = \frac{P(B|A)P(A)}{P(B)}. \tag{1}$$

The conditional probability definition is defined as follows

$$P(A \cap B) = P(A|B)P(B) = P(B|A)P(A). \tag{2}$$

For example, let a dice is thrown under a dice-box. From the standard model, all of outcomes have probability equal to 1/6. Now, the dice is lifted a bit and a random corner of the upper side is able to visible which it contains a dot. The new probability distribution of the outcomes shows as follows. Let A_i is the outcome of the throw, for $i = 1, 2, 3, 4, 5, 6$ and B is the randomly chosen corner contains a dot. So, we get $P(A_i) = 1/6$ and $P(B) = 2/3$. We get the following table:

A_i	$P(A_i)$	$P(B \mid A_i)$	$P(A_i \cap B)$	$P(A_i \mid B)$
A_1	1/6	0	0	0
A_2	1/6	1/2	1/12	1/8
A_3	1/6	1/2	1/12	1/8
A_4	1/6	1	1/6	1/4
A_5	1/6	1	1/6	1/4
A_6	1/6	1	1/6	1/4

The simplest way to construct the fourth column is to multiply. For any A_i, $P(A_i \mid B)$ and $P(B \mid A_i)$, to sum these values and divide by this sum. This final term is said to be scaling and corresponds to the formula as

$$\sum_{i=1}^{6} P(B|A_i)P(A_i) = \sum_{i=1}^{6} P(A_i \cap B) = P(B).$$

An simpler argument is that $P(A_i \mid B)$ has to be a probability distribution, thus sum to unity. As the scaling operation is trivial, Bayes' rule is also shown as

$$P(A|B) \propto P(A)P(B|A)$$

where $P(A)$ the prior (distribution), $P(B \mid A)$ is the likelihood and $P(A \mid B)$ is the posterior (distribution).

The main result of Bayesians statistics is that statistical inference may depend on the simple device $posterior \propto prior * likelihood$. By dice-throwing example is not of controversial. The disputations about the possibility of using Bay's rule as

$$P(Truth|Data) = \frac{P(Data|Truth)P(Truth)}{P(Data)}. \tag{3}$$

So, we get

$$P(Truth) = the\ prior. \tag{4}$$

The second ingredient we need is data, plus a how the data associate to the truth which is nothing but the classical concept of specifying a random relationship

$$P(Data|Truth) = the\ likelihood \tag{5}$$

for all associated values of *Truth*. Note that *P*(*Data*| *Truth*) is not applied as probability distribution for different data, but as the probability of the given data for different values of *Truth*. Various authors do apply *P*(*Data*| *Truth*) for likelihood to sheer this misconstrue.

Now, noting that (replace *Truth* with *T*), probability of Data (*P*(*Data*)) can be written as

$$P(Data) = \int P(T)P(Data|T)dT \tag{6}$$

that is as a function of *P*(*T*) and *P*(*Data*| *T*), it is obvious that the prior and likelihood enable, using 1 to construct a new probability statement about *T* given the data as follows

$$P(Truth|Data) = the\ posterior. \tag{7}$$

The purpose of this chapter was to introduce the concept of Bayesian inference and application to the real world problem such as game theory (Bayesian Game). In this chapter was organized as follows. In Sections 2 and 3, we present Model-based Bayesian inference and the components of Bayesian inference, respectively. The last section contains some applications of Bayesian inference.

2. Model-based Bayesian inference

The basic of Bayesian inference is continued by Bayes' theorem. From (1), replacement *B* with observations *y*, *A* with the set of parameter Θ, and probabilities *P* with densities *p*, results as the following

$$p(\Theta|y) = \frac{p(y|\Theta)p(\Theta)}{p(y)} \tag{8}$$

which *p*(*y*) is the marginal likelihood of *y*, *p*(Θ) is the set prior distributions of the set of parameter Θ before *y* is observed, *p*(*y*|Θ) is the likelihood of *y* underneath a model and *p*(Θ|*y*) is the joint posterior distribution of Θ that expresses uncertainty about parameter set Θ after taking both the prior and data into system. Because there are often multiple parameters, Θ presents a set of *j* parameters, we have

$$\Theta = \theta_1, \theta_2, \ldots, \theta_j.$$

The term

$$p(y) = \int p(y|\Theta)p(\Theta)d\Theta \tag{9}$$

determines the *marginal likelihood* (or the *prior predictive distribution*) of y which it was introduced by Jeffreys [2], and may be set to c where c is an unknown constant. This distribution shows what y should be similar to given the model, before y has been observed. Only the prior probabilities and the model's likelihood function are applied for $p(y)$. The presence of $p(y)$ normalizes the joint posterior distribution, $p(\Theta\,|\,y)$ guarantee it is a proper distribution and integrates to 1. From replacement $p(y)$ with a *constant of proportionality c*, the Bayes' theorem becomes to

$$p(\Theta|y) = \frac{p(y|\Theta)p(\Theta)}{c}. \tag{10}$$

We get

$$p(\Theta|y) \propto p(y|\Theta)p(\Theta) \tag{11}$$

when $\propto$ is *proportional to*.

This formulation (11) be shown as the unnormalized joint posterior being proportional to the likelihood multiply with the prior. Howsoever, the aim of this model is often not to concluding the non-normalized joint posterior distribution, however to concluding the marginal distributions of the parameters. The set of all Θ can partitioned as

$$\Theta = \{\Phi, \Lambda\} \tag{12}$$

when the interest sub-vector denote by Φ and the complementary sub-vector of Θ denoted by Λ, usually called to as a vector of nuisance parameters. For a Bayesian scope, the presence of nuisance parameters does not pose any formal, theoretical problems. A nuisance parameter is a parameter that exists in the joint posterior distribution of a model, though it is not a interest parameter. The marginal posterior distribution of ϕ, the interest parameter, can be shown as

$$p(\phi|y) = \int p(\phi, \Lambda|y)d\Lambda. \tag{13}$$

In model-based Bayesian inference, Bayes' theorem is applied to approximate the non-normalized joint posterior distribution, and lastly the user can evaluate and make inferences by the marginal posterior distributions.

3. The components of Bayesian inference

In this section, we presents about the components of Bayesian inference which contains the prior distributions, the likelihood or likelihood function and the joint posterior distribution as follows.

1. $p(\Theta)$ is the prior distributions for set of Θ, and uses probability as a methods of quantifying uncertainty about Θ before taking the data into system.

2. $p(y \mid \Theta)$ is the function of likelihood which all variables are associated in a full probability model.

3. $p(\Theta \mid y)$ is the joint posterior distribution that shows uncertainty about Θ after taking both the prior and the data into system. If Θ is partitioned into a single parameter of interest ϕ and the remaining parameters are considered nuisance parameters, then the marginal posterior distribution of ϕ denote by $p(\phi \mid y)$.

3.1. Prior distribution

The prior distribution is a main concept of Bayesian and shows the information about an uncertain Θ that is merged with the probability distribution of new data to yield the posterior distribution which in turn is applied for future inferences and decisions about Θ. The existence of a prior distribution for any problem can justified by some axioms of decision theory; which we now focus for how to set up a prior distribution for every given application. Generally, Θ will be a vector, but for easiness we will point as on $p(\Theta)$.

By well-identified and large sample sizes, suitable alternatives of $p(\Theta)$ will have minor effects on posterior inferences. This definition might look like to be circular, but in practice one can check the dependence on $p(\Theta)$ by a sensitivity analysis: comparing posterior inferences under different suitable alternatives of $p(\Theta)$.

If the sample size is small, or available data provide only indirect information about the parameters of interest, then $p(\Theta)$ becomes more important. In various cases, nevertheless, models can be set up hierarchically, such that clusters of parameters have shared $p(\Theta)$, which can themselves be approximated from data. Prior probability distributions have belonged to one of two kinds as informative and uninformative priors. In this section, four kinds of priors which include informative, weakly informative, least informative, and uninformative, are shown according to information and the aim in the use of the prior.

3.1.1. Informative prior

If prior information is obtainable about Θ, it should be included in $p(\Theta)$. If the current model is homologous to a previous model, and the current model is goal to be an adjusted version dependent on more current data, then the posterior distribution of Θ from the previous model maybe used as $p(\Theta)$ for the current model.

Now, every version of a model is not start from scratch, based only on the current data, but the cumulative effects of all data, past and current, can be taken into system. To sure the current data do not dominate the prior, in 2000, Ibrahim and Chen [3] presented the power prior which it is a class of informative prior distribution that takes early data and results into system. If the current data is very homologous to the previous data, then the precision of the posterior distribution increases when including more information from previous models. If the current data differs tremendously, then the posterior distribution of Θ maybe in the tails of the prior distribution for Θ, therefore $p(\Theta)$ contributes less density in its tails.

Sometimes informative prior is not ready to be applied, for example when it resides in other person, as in an expert. For this way, their human personal beliefs of the probability for the event must be elicited into the form of a suitable probability density function which this process is said to be prior elicitation.

3.1.2. Weakly informative prior

Weakly informative prior (in the short term: WIP) use prior information for regularization and stabilization, providing sufficient prior information to prevent results that contradict our knowledge for example an algorithmic failure to explore the state space. Other aim is for WIPs to use less prior information than is really available. WIPs should provide some of the useful of prior information while avoiding some of the risk from using information which does not exist. WIPs are the most common priors in practice and are liked by subjective Bayesians.

Selecting WIPs may be cumbersome. WIPs distributions should change with the sample size, since a model should have sufficient prior information to learn from the data, but the prior information must also be weak sufficient to learn from the data.

In practice, this is an example of WIPs. It is favor, for well reasons, to center and scale all continuous predictors [4]. Though centering and scaling predictors is not talked about here, but it should be clear that the potential range of the posterior distribution of θ for a centered and scaled predictor should be small. A favor WIPs for a centered and scaled predictor may be $\theta \sim \mathcal{N}(0, 10, 000)$ where θ is normal distribution agreeable to a mean of 0 and a variance of 10,000. Here, the density for θ is nearly flat. Nonetheless, the fact that it is not perfectly at yields well properties for numerical estimation algorithms. In both Bayesian and Frequentist inference, it is possible for numerical estimation algorithms to become stuck in regions of at density which become more common as sample size decreases or model complexity increases. Numerical estimation algorithms in Frequentist inference function as though a at prior were used, thus numerical estimation algorithms in Frequentist inference become stuck more frequently than numerical estimation algorithms in Bayesian inference. Prior distributions that are not completely at allow sufficient information for the numerical estimation algorithm to continue to diagnose the goal density, the posterior distribution.

After updating a model in which WIPs exist, the user should be investigating the posterior. If the posterior contradicts knowledge, then the WIPs must be revised by including information that will make the posterior consistent with knowledge [4]. A favor purpose Bayesian criticism against WIPs is that there is no precise mathematical form to derive the optimal WIPs for a given model and data.

3.1.2.1. Vague priors

A vague prior, is said to be a diffuse prior which difficult to define, after considering WIPs. In 2005, Lambert, Sutton, Burton, Abrams and Jones introduce the first formal move from vague to WIPs. After conjugate priors were introduced by Raiffa and Schlaifer in 1961 which most applied Bayesian has applied vague priors, parameterized to estimate the idea of uninformative priors.

Normally, a vague prior is a conjugate prior together with a large size parameter. Howsoever, if the sample size is small then vague priors may be problems. All most problems about vague priors and small sample size are implicated with scale rather than location. The problem can be particularly acute in random-effects models which it is used rather loosely in this here to imply exchangeable, hierarchical and multilevel structures. A vague prior is defined as commonly being a conjugate prior that is intent to estimate an uninformative prior and without two goals of regularization and stabilization.

3.1.3. Least informative prior

Least informative priors (for short term LIP) is applied here to describe a class of prior in which the aim is to minimize the amount of subjective information content, and to apply a prior that is determined only by the model and observed data. The rationale for using LIPs is often called *to let the data speak for themselves*. LIPs are preferred by objective Bayesians. LIPs are contains Flat Priors [12], Hierarchical Prior [4], Jeffreys Prior [2], MAXENT [5] and Reference Priors [6–8] etc.

3.1.4. Uninformative prior

Traditionally, most of the above descriptions of prior distributions were classified as uninformative priors. However, uninformative priors do not really exist (see in [9]) and all priors are informative in some ways. Moreover, there have been various names associated with uninformative priors including diffuse, minimal, non-informative, objective, reference, uniform, vague, and perhaps weakly informative etc.

3.1.5. Proper and improper priors

It is important for the prior distribution to be proper. A prior distribution, $p(\theta)$, is improper when $\int p(\theta)d\theta = \infty$.

Before, an unbounded uniform prior distribution is an inappropriate prior distribution since $p(\theta) \propto 1$, for $\theta \in (-\infty, \infty)$. An inappropriate prior distribution may be cause an inappropriate posterior distribution. If the posterior distribution is inappropriate, then inferences are invalid.

To determine the propriety of a joint posterior distribution, the marginal likelihood should be finite for any y. Again, the marginal likelihood is $p(y) = \int p(y \mid \Theta)p(\Theta)d\Theta$. Although inappropriate prior distributions can be applied, it is good practice to avoid them.

3.2. Likelihood

To completely for the definition of a Bayesian, both the prior distributions and the likelihood must be estimated or completely specified. The likelihood or $p(y \mid \Theta)$, contains the available information provided by the sample. The likelihood is $p(y|\Theta) = \prod_{i=1}^{n} p(y_i|\Theta)$.

The data y effect to the posterior distribution $p(\Theta | y)$ only through the likelihood $p(\Theta | y)$. In this way, Bayesian inference believes the likelihood principle which states that for a given sample of data, any two probability models $p(\Theta | y)$ that have the same likelihood yield the same inference for Θ.

3.3. Posterior distribution

Recent theoretical and applied overviews of Bayesian statistics, including many examples and uses of posterior distributions, see [10–12]. The posterior distributions for decision-making about home radon exposure are discussed in [13].

The posterior distribution summarizes the current state of knowledge about all the uncertain quantities in a Bayesian analysis. Analytically, the posterior density is the product of the prior density and the likelihood. In a complicated analysis, the joint posterior distribution can be summarized by a set of L simulation draws of the vector of uncertain quantities $w_1, w_2, \ldots, w_J$, as illustrated in the following matrix:

l	w_1	w_2	...	w_J
1	.	.	...	.
2	.	...	...	.
...	...	...	...	.
L	.	.	...	.

The marginal posterior distribution for any unknown quantity w_l can be summarized by its column of L simulation draws. In many examples it is not necessary to construct the entire table ahead of time; rather, one creates the L vectors of posterior simulations for the parameters of the model and then uses these to construct posterior simulations for other unknown quantities of interest, as necessary.

4. Application to games theory

In this section, we present the application of Bayesian inference to the real world problems such as Bayesian Game as follows.

4.1. The classical games

The basic contents of the n-person game was presented by John Forbes Nash [14] in 1950. Also, he first shows the existence of equilibrium for this model when the player's preferences are representable by continuous quasi-concave utilities and the sets of strategy are simplex. The definition of an n-person game can be written as below.

http://dx.doi.org/10.5772/intechopen.70530

Definition 4.1

The normal form of an $n-$ person game is $(X_i, r_i)_{i=1}^n$, where for each $i \in \{1, 2, \ldots, n\}$, the set of individual strategies of player i denoted by X_i which X_i is a non-empty set and r_i is the preference relation on $X := \prod_{i \in I} X_i$ of player i.

The individual preferences r_i are usually represented by utility functions, i.e. for each $i \in \{1, 2, \ldots, n\}$ there exist a real valued function $u_i : X := \prod_{i \in I} X_i \to \mathbb{R}$ such that:

$$x r_i y \Leftrightarrow u_i(x) \geq u_i(y), \forall x, y \in X.$$

Then the normal form of $n-$ person game is transformed to $(X_i, u_i)_{i=1}^n$.

The solution of this game is called Nash equilibrium as below.

Definition 4.2

The Nash equilibrium for the game $(X_i, u_i)_{i=1}^n$ is a point $x^* \in X$ which satisfies for each $i \in \{1, 2, \ldots, n\} : u_i(x^*) \geq u_i(x^*, x_i)$ for each $x_i \in X_i$.

The following theorem offers sufficient conditions for the existence of Nash equilibrium.

Theorem 4.3

Let $\Gamma = (X_i, u_i)_{i=1}^n$ *be a n-person game and denoted by f the real-valued function on* $X \times X$ *defined by* $f(x, y) = \Sigma_{i=1}^n u_i(x_{-i}, y_i)$. *Let us assume that*

1. *for each* $i \in \{1, 2, \ldots, n\}$, X_i *is a non-empty compact convex subset of a Hausdorff linear topological space;*
2. *for each* $i \in \{1, 2, \ldots, n\}$, $u_i(\cdot, x_i)$ *is continuous on* $X_{-i} = \prod_{i \neq j} X_j$ *for each fixed* $x_i \in X_i$*;*
3. $\Sigma_{i=1}^n u_i$ *is continuous on X;*
4. $f(x, \cdot)$ *is quasi-concave on X, for each* $x \in X$.

Then, Γ has an equilibrium.

Proof. See in [34].

Next, we present some examples of Nash equilibrium for two persons game as follows.

Example 4.4

The battle of the sexes game has two Nash equilibrium (MT, FT), (MS, FS) with $(3, 2)$ and $(2, 3)$, where "Male like playing tennis" denoted by MT, "Male like shopping" denoted by MS, "Female like playing tennis" denoted by FT and "Female like shopping" denoted by FS, see in **Figure 1**.

Example 4.5

The oligopoly behavior game is a unique Nash equilibrium (Aa, Ba) where "A coffee shop use a strategy for don't advertising" denoted by Ad, "A coffee shop use a strategy for advertising"

		Female	
		Tennis	Shopping
Male	Tennis	(3, 2)	(1, 1)
	Shopping	(1, 1)	(2, 3)

Figure 1. The battle of the sexes game.

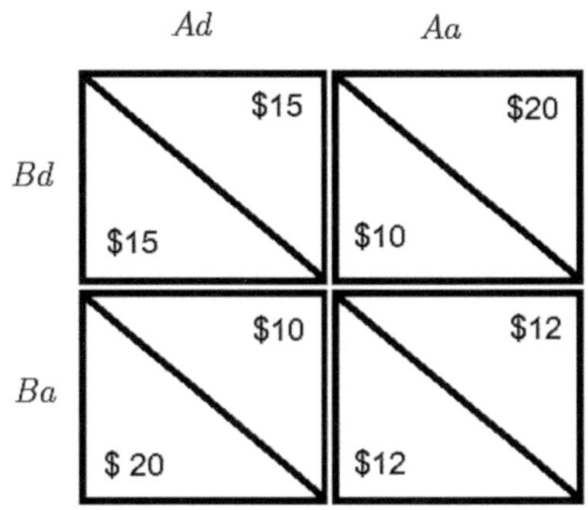

Figure 2. The oligopoly behavior game.

denoted by *Aa*, "A coffee shop use a strategy for do not advertising" denoted by *Bd*, and "A coffee shop use a strategy for advertising" denoted by *Ba*, see in **Figure 2**.

4.2. The Bayesian games

For a long time, we have been supposed that everything in the game was normal knowledge for everyone playing. However, real players may have private information about their own payoffs, their type or preferences, etc. The way to modeling this situation of asymmetrical information is by recurring to the concept was defined by Harsanyi in 1967. The key is to introduce a move by the nature, which changes the uncertainty by converting an asymmetrical information problem into an imperfect information problem. The concept is the nature moves determining players' types, a concept that collects all the private information relevant them (i.e. payoffs, preferences, beliefs of another players, etc.).

Definition 4.6

The normal form of Bayesian games with incomplete information include:

1. the players $i \in \{1, 2, \ldots, I\}$;
2. the set of finite action for each player $a_i \in A_i$;

3. the finite type set for each player $\theta_i \in \Theta_i$;
4. a probability distribution on types $p(\theta)$
5. $u_i : A_1 \times A_2 \times \ldots \times A_I \times \Theta_1 \times \Theta_2 \times \ldots \times \Theta_I \rightarrow \mathbb{R}$, where u_i is utilities function.

It is important to discuss some parts of the definition. Players' types comprise all relevant information about some player's private characteristics. The type of θ_i is only observed by player i who uses this information both to make decisions and to update itself beliefs about the likelihood of opponents' types.

Combining actions and types for each player it is possible to create the strategies. Strategies will be given by $s_i : \Theta_i \rightarrow A_i$, with elements $s_i(\theta_i)$ where Θ_i is the type space and A_i is the action space. A strategy may determine different actions to different types. Lastly, utilities are computed by each player by taking expectations over types using itself own conditional beliefs about opponents' types. Hence, if player i uses the pure strategy s_i, other players use the strategies s_i and player i's type is θ_i, the expected utility can be presented as follows

$$Eu_i(s_i|s_{-i},\theta_i) = \sum_{\theta_{-i}\in\Theta_{-i}} u_i(s_i, s_{-i}(\theta_{-i}), \theta_i, \theta_{-i})p(\theta_{-i}|\theta_i).$$

A Bayesian Nash Equilibrium (for short term: BNE) is basically the same concept than a Nash Equilibrium with the addition that players need to take expectations over opponents' types as follows.

Definition 4.7

A Bayesian Nash Equilibrium is a Nash Equilibrium of a Bayesian Game, i.e. $Eu_i(s_i|s_{-1},\theta_i) \geq Eu_i(s_i'|s_{-i},\theta_i)$ for all $s_i' \in S_i$ and for all types θ_i occurring with positive probability.

The following theorem for the existence of Bayesian Nash Equilibrium.

Theorem 4.8

Every finite Bayesian Games has a Bayesian Nash Equilibrium.

Example 4.9

Consider the Bayesian games as follows:

1. Nature decides that the payoffs are as in matrix I or II, with probabilities;
2. ROW is informed of the choice of nature but COL is not;
3. The choices of ROW include U or D and the choices of COL include L or R where these choices are made simultaneously;
4. Payoffs are as in the matrix chosen from nature.

For any of the Bayesian games, we will find all BNE. All equilibrium in mixed behavioral strategies can be written as.

Matrix I:

	L	R
U	(1, 1)	(0, 0)
D	(0, 0)	(0, 0)

Matrix II:

	L	R
U	(0, 0)	(0, 0)
D	(0, 0)	(2, 2)

4.2.1. Pure strategy BNE

First, we will deflate the case of incomplete information problem as a static extended game with all of possible strategies: $\widehat{\Gamma}$. It can be presented follow Harsanyi, that the Nash Equilibrium of $\widehat{\Gamma}$ is the same equilibrium of the imperfect game presented. The idea is to deflate a game such that all the ways the game can follow is considered in the extended game $\widehat{\Gamma}$.

The first step is to define the strategies for all player.

Since he does not know in which matrix the game is played, so, COL has only two strategies which contain L and R.

ROW knows in which Matrix the game occurs, and the strategies are UU, UD, DU and DD where UD is played U in Matrix I and D in Matrix II.

The probability knowledge, the nature locates the game in any matrix. The new extended game $\widehat{\Gamma}$ can be shown as:

	L	R
UU	$(\frac{1}{2}, \frac{1}{2})$	$(0, 0)$
UD	$(\frac{1}{2}, \frac{1}{2})$	$(1, 1)$
DU	$(0, 0)$	$(0, 0)$
DD	$(0, 0)$	$(1, 1)$

Remember that DU is a dominated strategy for ROW. After displacement that possibility, the game has 3 pure Nash Equilibrium as follows $\{(UU; L); (UD; R); (DD; R)\}$.

4.2.2. Mixed strategy BNE

Sequent to obtain the mixed strategies we will make another kind of analysis and try to repeat the three pure BNE obtained before.

Suppose the probabilities of playing each action are as displayed in the matrices as below, where y is the probability COL plays L, if the game is in Matrix I then x is the probability ROW plays U and if the game is in Matrix II then z is the probability ROW plays U.

4.2.3. Player's best respones

- In Matrix I: we get ROW's best response as follows

  ROW would play U, $x=1$, if $1y+0(1-y)>0$, then $y>0$, which can be concluded as:

  a. if $y>0$, then $x=1$;

  b. if $y=0$, then $x\in[0,1]$.

- In Matrix II: we get ROW's best response as follows

  ROW would play D, $z=0$ if $0<2(1-y)$ then $y<1$ which can be concluded as:

  c. if $y<1$, then $z=0$;

  d. if $y=1$ then $z\in[0,1]$.

- In Matrix I and II: we get COL's best response as follows

  COL would play L, $y=1$ if

$$\frac{1}{2}[1x+0(1-x)]+\frac{1}{2}[0z+0(1-z)]>\frac{1}{2}[0x+0(1-x)]+\frac{1}{2}[0z+2(1-z)]$$

  then $x>2(1-z)$ which can be summarized as:

  e. if $x=2(1-z)$, then $y\in[0,1]$;
  f. if $x>2(1-z)$, then $y=1$;
  g. if $x<2(1-z)$, then $y=0$.

Next, we can check each the possibilities in order to find the Nash Equilibrium, such as those strategies stable for any players. Let us start by checking COL's strategies since there are less combinations.

4.2.4. Mixed equilibrium

Case 1:

If $y=0$, we have *b.* $x\in[0,1]$ and *c.* $z=0$. Here, we want to check this is a equilibrium from COL's point of view. By *g.*, we can see that if $z=0$, then $x<2$ which always hold and that $y=0$.

This Nash equilibrium supports two of the three pure BNE found before: (DD, R), which is the same as $y=0$, $x=0$ and $z=0$ and (UD, R) which is the same as $y=0$, $x=1$ and $z=0$.

Thus, we get Nash equilibrium of $y=0$, $x\in[0,1]$ and $z=0$.

There are many BNE in which column plays R and row plays $xU+(1-x)D$, when $x\in[0,1]$ if Matrix I occurs and D if Matrix II occurs.

Case 2:

If $y=0$, we have *d*. $z\in[0,1]$ and from *a*. $x=1$.

From *f*., we can see that when $x=1$, then it should be the case that $z\geq\frac{1}{2}$ in order to be true that $y=1$. Hence, these BNE are restricted to $y=1$, $z\in\left[\frac{1}{2},1\right]$ and $x=1$.

This BNE supports the third pure Nash Equilibrium found before: (UU, L), which is the same as $y=1$, $x=1$ and $z=1$.

There are many BNE in which column plays L and row plays U if Matrix I occurs and $zU+(1-z)D$, where $z\in\left[\frac{1}{2},1\right]$ if Matrix II occurs.

Case 3:

If $y\in(0,1)$, we have *a*. $x=1$ and *c*. $z=0$. By *e*., we can see that in order y belongs to $[0,1]$ it should be the case that $x=(1-z)$. However it is impossible this equality to hold if both $z=0$ and $x=1$.

Therefore, the case if $y\in(0,1)$ is not a Bayesian Nash equilibrium.

4.3. Abstract economy model

Later, the existence of social equilibrium was proved Debreu [15]. Also Arrow and Debreu [16] proved the existence of Walrasian equilibrium. The classical abstract economy game introduced by Shafer and Sonnenschein [17] or Borglin and Keiding [18] consists of a finite set of agents, each characterized by certain constraints and preferences, explained by correspondences. Following many previous authors ideas, they studied the existence of equilibrium for generalized games (see, for example, [19–27] and the references therein). Now, we show some definitions of an abstract economy model and equilibrium of this model as follows. Let the set of agents be the finite set $\{1,2,\ldots,n\}$. For each $i\in\{1,2,\ldots,n\}$ let X_i be a non-empty set.

Definition 4.10

An abstract economy $\Gamma=(X_i,A_i,P_i)_{i=1}^{n}$ is defined as a family of n ordered triplets (X_i, A_i, P_i), where for each $i\in I$:

1. $A_i:\prod_{i\in I}X_i\rightarrow 2^{X_i}$ is constraint correspondence and
2. $P_i:\prod_{i\in I}X_i\rightarrow 2^{X_i}$ is preference correspondence.

Definition 4.11

An equilibrium for Γ is a point $x^*\in\prod_{i\in I}X_i$ which satisfies for each $i\in\{1,2,\ldots,n\}$:

1. $x^*\in A_i(x^*)$;
2. $A_i(x^*)\cap P_i(x^*)=\varnothing$.

Theorem 4.12

Let $\Gamma=(X_i,A_i,P_i)_{i=1}^{n}$ *be an abstract economy which satisfies, for each* $i\in\{1,2,\ldots,n\}$:

1. *X_i is a non-empty compact convex subset in $\mathbb{R}^l$;*
2. *A_i is a continuous correspondence;*
3. *for each $x \in X$, $A_i(x)$ is non-empty compact and convex;*
4. *P_i has an open graph in $X \times X_i$ and for each $x \in X$, $P_i(x)$ is convex;*
5. *for each $x \in X$, $x_i \notin P_i(x)$.*

Then, Γ has an equilibrium.

Proof. See in [34].

4.4. Fuzzy games

The first concept of a fuzzy set was introduced by Zadeh [28] in 1965. Fuzzy set theory has been shown to be a gainful tool to describe situations in which the data are imprecise or vague. The theory of fuzzy sets has become a well framework for studying results concerning fuzzy equilibrium existence for abstract fuzzy economies. The first study of a fuzzy abstract economy (or a fuzzy game) has been studied by Kim and Lee in [29], they shown the existence of the equilibrium for 1-person fuzzy game. Also Kim and Lee [29] shown the existence of equilibrium for generalized games when the constraints or preferences are vague due to the agent's behavior. In 2009, Patriche [30] studied the Bayesian abstract economy game and proved the existence of equilibrium for an abstract economy game with differential information and a measure space of agents. However, the existence of random fuzzy equilibrium for fuzzy game has not been studied so far. In 2013, Patriche [31] defined the Bayesian abstract economy game and proved the existence of the Bayesian fuzzy equilibrium for this game. Also, Patriche [32] defined the new Bayesian abstract fuzzy economy game and proved the existence of the Bayesian fuzzy equilibrium for this game which it is characterized by a private information set, an action fuzzy mapping, a random fuzzy constraint one and a random fuzzy preference mapping. Recently, Patriche [33] defined the fuzzy games and applications to systems of generalized quasi-variational inequalities problem. The Bayesian fuzzy equilibrium concept is an extension of the deterministic equilibrium. She also generalized and extended the former deterministic models introduced by Debreu [15], Shafer and Sonnenschein [17] and Patriche [34]. Very recently, Saipara and Kumam [35] introduced the model of general Bayesian abstract fuzzy economy for product measurable spaces, and proved the existence for Bayesian fuzzy equilibrium of this model as follows.

For each $i \in I$, let $(\Omega_i, \mathcal{Z}_i)$ be a measurable space, $(\Omega, \mathcal{Z})$ be the product measurable space where $\Omega := \prod_{i \in I} \Omega_i$, $\mathcal{Z} := \otimes_{i \in I} \mathcal{Z}_i$ and μ is a probability measure on $(\Omega, \mathcal{Z})$. Let Y denote the strategy or commodity space, where Y is a separable Banach space.

Let I be a non-empty finite set (the set of agents). For each $i \in I$, let $X_i : \Omega_i \to \mathcal{F}(Y)$ be a fuzzy mapping, and let $z_i \in (0, 1]$.

Let $L_{X_i} = \{x_i \in S(X_i(\cdot))_{z_i} : x_i \text{ is } \Sigma_i\text{-measurable}\}$. Denote by $L_X = \prod_{i \in I} L_{X_i}$ and by the set $\prod_{i \neq j} L_{X_j}$. An element x_i of L_{X_i} is called a strategy for agent i. The typical element of L_{X_i} is denoted by x_i and that of $(X_i(\omega_i))_{z_i}$ by $x_i(\omega_i)$ (or x_i). We can define a general Bayesian abstract fuzzy economy model of product measurable spaces as follow.

Definition 4.13

A general Bayesian abstract fuzzy economy model of product measurable spaces is defined as follows:

$$\Gamma = \left(\left((\Omega_i, \mathcal{Z}_i)_{i\in I}, \mu\right), \left(X_i, \Sigma_i, (A_i, a_i), \left(P_i, p_i\right), z_i\right)_{i\in I}\right),$$

where I is non-empty finite set (the set of agents) and:

a. $X_i : \Omega_i \to \mathcal{F}(Y)$ is a *action (strategy) fuzzy mapping* of agent i;

b. Σ_i is a sub σ-algebra of $\mathcal{Z} = \otimes_{i\in I}\mathcal{Z}_i$, which denotes the *private information* of agent i;

c. for each $\omega_i \in \Omega_i$, $A_i(\omega_i, \cdot) : L_X \to \mathcal{F}(Y)$ is the *random fuzzy constraint mapping* of agent i;

d. for each $\omega_i \in \Omega_i$, $P_i(\omega_i, \cdot) : L_X \to \mathcal{F}(Y)$ is the *random fuzzy preference mapping* of agent i;

e. $a_i : L_X \to (0, 1]$ is a *random fuzzy constraint function*, and $p_i : L_X \to (0, 1]$ is a *random fuzzy preference function* of agent i;

f. $z_i \in (0, 1]$ is such that for all $(\omega_i, x) \in \Omega_i \times L_X$, $(A_i(\omega_i, \tilde{x}))_{a_i(\tilde{x})} \subset (X_i(\omega_i))_{z_i}$ and $(P_i(\omega_i, \tilde{x}))_{p_i(\tilde{x})} \subset (X_i(\omega_i))_{z_i}$.

The Bayesian fuzzy equilibrium for a general Bayesian abstract fuzzy economy model of product measurable spaces is defined as follows.

Definition 4.14

A Bayesian fuzzy equilibrium for Γ is a strategy profile $\tilde{x}^* \in L_X$ such that for all $i \in I$,

i. $\tilde{x}^*(\omega_i) \in cl(A_i(\omega_i, \tilde{x}^*))_{a_i(\tilde{x}^*)}$ $\mu - a.e.$;

ii. $(A_i(\omega_i, \tilde{x}^*))_{a_i(\tilde{x}^*)} \cap (P_i(\omega_i, \tilde{x}^*))_{p_i(\tilde{x}^*)} = \varnothing$ $\mu - a.e.$.

Theorem 4.15

Let I be a non-empty finite set. Let the family

$\Gamma = \left(\left((\Omega_i, \mathcal{Z}_i)_{i\in I}, \mu\right), \left(X_i, \Sigma_i, (A_i, a_i), \left(P_i, p_i\right), z_i\right)_{i\in I}\right)$ *be a general Bayesian abstract economy model of product spaces satisfy (a)-(j). Then, there exists a Bayesian fuzzy equilibrium for Γ.*

For each $i \in I$, the following conditions are sastisfied:

a. *$X_i : \Omega_i \to \mathcal{F}(Y)$ is such that $\omega_i \to X_i(\omega_i)_{z_i} : \Omega_i \to 2^Y$ is a non-empty convex weakly compact-valued and integrably bounded correspondence;*

b. *$X_i : \Omega_i \to \mathcal{F}(Y)$ is such that $\omega_i \to X_i(\omega_i)_{z_i} : \Omega_i \to 2^Y$ is $\Sigma_i -$ lower measurable;*

c. *For each $(\omega_i, \tilde{x}) \in \Omega_i \times L_X$, $(A_i(\omega_i, \tilde{x}))_{a_i(\tilde{x})}$ is convex and has a non-empty interior in the relative norm topology of $(X_i(\omega_i))_{z_i}$;*

d. *the correspondence* $(\omega_i, \tilde{x}) \to (A_i(\omega_i, \tilde{x}))_{a_i(\tilde{x}):\Omega_i \times L_X \to 2^Y}$ *has a measurable graph, i.e.,* $\left\{(\omega_i, \tilde{x}, y) \in \Omega_i \times L_X \times Y : y \in (A_i(\omega_i, \tilde{x}))_{a_i(\tilde{x})}\right\} \in \mathcal{F}_i \otimes \mathfrak{B}(L_X) \otimes \mathfrak{B}(Y)$, *where* $\mathfrak{B}_{\omega_i}(L_X)$ *is the Borel* $\sigma-$ *algebra for the weak topology on* L_X *and* $\mathfrak{B}(Y)$ *is the Borel* $\sigma-$ *algebra for the norm topology on Y;*

e. *the correspondence* $(\omega_i, \tilde{x}) \to (A_i(\omega_i, \tilde{x}))_{a_i(\tilde{x})}$ *has weakly open lower sections, i.e., for each* $\omega_i \in \Omega_i$ *and for each* $y \in Y$, *the set* $\left((A_i(\omega_i, \tilde{x}))_{a_i(\tilde{x})}\right)^{-1}(\omega_i, \tilde{x}) = \left\{\tilde{x} \in L_X : y \in (A_i(\omega_i, \tilde{x}))_{a_i(\tilde{x})}\right\}$ *is weakly open in* L_X*;*

f. *For each* $\omega_i \in \Omega_i$, $\tilde{x} \to cl(A_i(\omega_i, \tilde{x}))_{a_i(\tilde{x})} : L_X \to 2^Y$ *is upper semicontinuous in the sense that the set* $\left\{\tilde{x} \in L_X : cl(A_i(\omega_i, \tilde{x}))_{a_i(\tilde{x})}\right\}$ *is weakly open in* L_X *for every norm open subset V of Y;*

g. *the correspondence* $(\omega_i, \tilde{x}) \to (P_i(\omega_i, \tilde{x}))_{p_i(\tilde{x})} : \Omega_i \times L_X \to 2^Y$ *has open convex values such that* $(P_i(\omega_i, \tilde{x}))_{p_i(\tilde{x})} \subset (X(\omega_i))_{z_i}$ *for each* $(\omega_i, \tilde{x}) \in \Omega_i \times L_X$*;*

h. *the correspondence* $(\omega_i, \tilde{x}) \to (P_i(\omega_i, \tilde{x}))_{p_i(\tilde{x})} : \Omega_i \times L_X \to 2^Y$ *has a measurable graph;*

i. *the correspondence* $(\omega_i, \tilde{x}) \to (P_i(\omega_i, \tilde{x}))_{p_i(\tilde{x})} : \Omega_i \times L_X \to 2^Y$ *has weakly open lower sections, i.e. for each* $\omega_i \in \Omega_i$ *and for each* $y \in Y$, *the set* $((P_i(\omega_i, \tilde{x}))_{p_i(\tilde{x})})^{-1}(\omega_i, y) = \{\tilde{x} \in L_X : y \in (P_i(\omega_i, \tilde{x}))_{p_i(\tilde{x})}\}$ *is weakly open in* L_X*;*

j. *For each* $\tilde{x}_i \in L_{X_i}$, *for each* $\omega_i \in \Omega_i$, $\tilde{x}_i \notin (A_i(\omega_i, \tilde{x}))_{a_i(\tilde{x})} \cap (P_i(\omega_i, \tilde{x}))_{p_i(\tilde{x})}$.

Proof. See in [35].

Moreover, in 1960, Fichera and Stampacchia first introduced the variational inequalities problem, this issue has been widely studied. Next, the basic concept of variational inequalities for fuzzy mappings was first introduced by Chang and Zhu [36] in 1989. In the topic of variational inequalities problem, there are many mathematicians who studied this topic (see, for example, [37, 38]). In 1993, the concept of a random variational inequality was introduced by Noor and Elsanousi [39]. Recently, Patriche [31] used the model of the Bayesian abstract fuzzy economy to prove the existence of solution for the two types of random quasi-variational inequalities with random fuzzy mappings.

5. Conclusion

The main objectives of this chapter was introduced the concept of Bayesian inference and application to some real world problems. In this chapter, we were presented about the basic concept of Bayesian inference which it can be application to the Bayesian game and a general Bayesian abstract fuzzy economy game or a fuzzy game. For application to Bayesian game, we were shown the solution of Bayesian Nash Equilibrium (BNE) for a Bayesian game with examples. Finally, we were shown the existence of Bayesian fuzzy equilibrium for a fuzzy game.

Acknowledgements

This project was supported by the Theoretical and Computation Science (TaCS) Center under Computational and Applied Science for Smart Innovation Cluster (CLASSIC), Faculty of Science, KMUTT. Moreover, Poom Kumam was supported by the Thailand Research Fund (TRF) and the King Mongkut's University of Technology Thonburi (KMUTT) under the TRF Research Scholar Award (Grant No. RSA6080047).

Author details

Wiyada Kumam[1], Plern Saipara[2] and Poom Kumam[3]*

*Address all correspondence to: poom.kum@kmutt.ac.th

1 Rajamangala University of Technology Thanyaburi (RMUTT), Thailand

2 Rajamangala University of Technology Lanna Nan (RMUTL), Thailand

3 King Mongkut's University of Technology Thonburi (KMUTT), Thailand

References

[1] Stigler S. Who discovered Bayes's theorem. The American Statistician. 1983;**37**(4):290-296

[2] Jeffreys H. Theory of Probability. 3rd ed. Oxford, England: Oxford University Press; 1961

[3] Ibrahim J, Chen M. Power prior distributions for regression models. Statistical Science. 2000;**15**:46-60

[4] Gelman A. Scaling regression inputs by dividing by two standard deviations. Statistics in Medicine. 2008;**27**:2865-2873

[5] Jaynes E. Prior probabilities. IEEE Transactions on Systems Science and Cybernetics. 1968;**4**(3):227-241

[6] Berger J, Bernardo J, Dongchu S. The formal definition of reference priors. Annals of Statistics. 2009;**37**(2):905-938

[7] Bernardo J. Reference posterior distributions for Bayesian inference (with discussion). Journal of the Royal Statistical Society, B. 1979;**41**:113-147

[8] Bernardo J. Reference analysis. In: Dey D, Rao C, editors. Handbook of Statistics. Vol. 25. Amsterdam: Elsevier; 2005. p. 17-90

[9] Irony T, Singpurwalla N. Noninformative priors do not exist: A discussion with Jose M. Bernardo. Journal of Statistical Inference and Planning. 1997;**65**:159-189

[10] Bernardo JM, Smith AFM. Bayesian Theory. New York: Wiley; 1994

[11] Carlin BP, Louis TA. Bayes and Empirical Bayes Methods for Data Analysis. London: Chapman and Hall; 1996

[12] Gelman A, Carlin JB, Stern HS, Rubin DB. Bayesian Data Analysis. London: Chapman and Hall; 1995

[13] Lin CY, Gelman A, Price PN, Krantz DH. Analysis of local decisions using hierarchical modeling, applied to home radon measurement and remediation (with discussion). Statistical Science. 1999;**14**:305-337

[14] Nash J. Equilibrium points in *n*-person games. Proceedings of the National Academy of Sciences of the United States of America. 1950;**36**(1):48-49

[15] Debreu G. A social equilibrium existence theorem. Proceedings of the National Academy of Sciences of the United States of America. 1952;**38**:886-903

[16] Arrow KJ, Debreu G. Existence of an equilibrium for a competitive economy. Econometrica. 1952;**22**:265-290

[17] Shafer W, Sonnenschein H. Equilibrium in abstract economies without ordered preferences. Journal of Mathematical Economics. 1975;**2**:345-348

[18] Borglin A, Keiding H. Existence of equilibrium actions and of equilibrium: A note on the "new" existence theorem. Journal of Mathematical Economics. 1976;**3**:313-316

[19] Huang NJ. Some new equilibrium theorems for abstract economies. Applied Mathematics Letters. 1998;**11**(1):41-45

[20] Kim WK, Tan KK. New existence theorems of equilibria and applications. Nonlinear Analysis: Theory, Methods & Applications. 2001;**47**:531-542

[21] Lin LJ, Chen LF, Ansari QH. Generalized abstract economy and systems of generalized vector quasi-equilibrium problems. Journal of Computational and Applied Mathematics. 2007;**208**:341-353

[22] Briec W, Horvath C. Nash points, Ky Fan inequality and equilibria of abstract economies in Max-Plus and B-convexity. Journal of Mathematical Analysis and Applications. 2008;**341**: 188-199

[23] Ding XP, Wang L. Fixed points, minimax inequalities and equilibria of noncompact abstract economies in FC-spaces. Nonlinear Analysis: Theory, Methods & Applications. 2008;**69**:730-746

[24] Kim WK, Kum SH, Lee KH. On general best proximity pairs and equilibrium pairs in free abstract economies. Nonlinear Analysis: Theory, Methods & Applications. 2008;**68**:2216-2227

[25] Lin LJ, Liu YH. The study of abstract economies with two constraint correspondences. Journal of Optimization Theory and Applications. 2008;**137**:41-52

[26] Ding XP, Feng HL. Fixed point theorems and existence of equilibrium points of noncompact abstract economies for L_F^*-majorized mappings in FC-spaces. Nonlinear Analysis: Theory, Methods & Applications. 2010;**72**:65-76

[27] Wang L, Cho YJ, Huang NJ. The robustness of generalized abstract fuzzy economies in generalized convex spaces. Fuzzy Sets and Systems. 2011;**176**:56-63

[28] Zadeh LA. Fuzzy sets. Information and Control. 1965;**8**:338-353

[29] Kim WK, Lee KH. Fuzzy fixed point and existence of equilibria of fuzzy games. Journal of Fuzzy Mathematics. 1998;**6**:193-202

[30] Patriche M. Bayesian abstract economy with a measure space of agents. Abstract and Applied Analysis. 2009;**2009**:1-11

[31] Patriche M. Equilibriumof Bayesian fuzzy economies and quasi-variational inequalities with random fuzzy mappings. Journal of Inequalities and Applications. 2013; 374. Article ID 58E35

[32] Patriche M. Existence of equilibrium for an abstract economy with private information and a countable space of actions. Mathematical Reports. 2013;**15**(65)(3):233-242

[33] Patriche M. Fuzzy games with a countable space of actions and applications to systems of generalized quasi-variational inequalities. Fixed Point Theory and Applications. 2014;**2014**(124)

[34] Patriche M. Equilibrium in Games and Competitive Economies. Bucharest: The Publishing House of the Romanian Academy; 2011

[35] Saipara P, Kumam P. Fuzzy games for a general Bayesian abstract fuzzy economy model of product measurable spaces. Mathematical Methods in the Applied Sciences. 2015;**39**(16): 4810-4819

[36] Chang SS, Zhu YG. On variational inequalities for fuzzy mappings. Fuzzy Sets and Systems. 1989;**32**:359-367

[37] Noor MA. Variational inequalities for fuzzy mappings III. Fuzzy Sets and Systems. 2000;**110**:101-108

[38] Park JY, Lee SY, Jeong JU. Completely generalized strongly quasivariational inequalities for fuzzy mapping. Fuzzy Sets and Systems. 2000;**110**:91-99

[39] Noor MA, Elsanousi SA. Iterative algorithms for random variational inequalities. Panamerican Mathematical Journal. 1993;**3**:39-50

Chapter 2

Node-Level Conflict Measures in Bayesian Hierarchical Models Based on Directed Acyclic Graphs

Jørund I. Gåsemyr and Bent Natvig

Additional information is available at the end of the chapter

http://dx.doi.org/10.5772/intechopen.70058

Abstract

Over the last decades, Bayesian hierarchical models defined by means of directed, acyclic graphs have become an essential and widely used methodology in the analysis of complex data. Simulation-based model criticism in such models can be based on conflict measures constructed by contrasting separate local information sources about each node in the graph. An initial suggestion of such a measure was not well calibrated. This shortcoming has, however, to a large extent been rectified by subsequently proposed alternative mutually similar tail probability-based measures, which have been proved to be uniformly distributed under the assumed model under various circumstances, and in particular, in quite general normal models with known covariance matrices. An advantage of this is that computationally costly precalibration schemes needed for some other suggested methods can be avoided. Another advantage is that noninformative prior distributions can be used when performing model criticism. In this chapter, we describe the basic framework and review the main uniformity results.

Keywords: cross-validation, data splitting, information contribution, MCMC, model criticism, pivotal quantity, preexperimental distribution, p-value

1. Introduction

Over the last decades, Bayesian hierarchical models have become an essential and widely used methodology in the analysis of complex data. Computational techniques such as Markow Chain Monte Carlo (MCMC) methods make it possible to treat very complex models and data structures. Analysis of such models gives intuitively appealing Bayesian inference based on posterior probability distributions for the parameters.

In the construction of such models, an understanding of the underlying structure of the problem can be represented by means of directed acyclic graphs (DAGs), with nodes in the graph corresponding to data or parameters, and directed edges between parameters representing conditional distributions. However, a perfect understanding of the underlying structure is usually an unachievable goal, and there is always a danger of constructing inadequate models. Box [1] suggests a pattern for the model building process where an initial candidate model is assessed for adequacy, and if necessary modified and elaborated on, leading to a new candidate that again is checked for adequacy, and so on. As a tool in this model criticism process, Ref. [1] suggests using the prior predictive distribution of some checking function or test statistic as a reference for the observed value of this checking function, resulting in a prior predictive p-value. This requires an informative and realistic prior distribution, which is not always available or even desirable. Indeed, as pointed out in Ref. [2], in an early phase of the model building process, it is often convenient to use noninformative or even improper priors and thus avoid costly and time-consuming elicitation of prior information. Noninformative priors may be used also for the inference because relevant prior information is unavailable.

There exist many other methods for checking the overall fit of the model or an aspect of the model of special interest, based on locating a test statistic or a discrepancy measure in some kind of a reference distribution. The posterior predictive p-value (ppp) of Ref. [3] uses the posterior distribution as reference and does not require informative priors. But this method uses data twice and can as a result be very conservative [2, 4–6]. Hjort et al. [5] suggest remedying this by using the ppp value as a test statistic in a prior predictive test. The computation of the resulting calibrated cppp-value is, however, very computer intensive in the general case, and again realistic, informative priors are needed. A node-level discrepancy measure suggested in Ref. [7] is subject to the same limitations. The partial posterior predictive p-value of Ref. [4] avoids double use of data and allows noninformative priors but may be difficult to compute and interpret in hierarchical models.

Comparison with other candidate models through a technique for model comparison or model choice, such as predictive methods, maximum posterior probability, Bayes factors or an information criterion, can also serve as tools for checking model adequacy indirectly when alternative candidate models exist.

In this chapter, we will, however, focus on methods for criticizing models in the absence of any particular alternatives. We will review methods for checking the modeling assumptions at each node of the DAG. The aim is to identify parts or building blocks of the model that are in discordance with reality, which may be in need of adjustment or further elaboration. O'Hagan [8] regards any node in the graph as receiving information from two disjoint subsets of the neighboring nodes. This information is represented as a conditional probability density or a likelihood or as a combination of these two kinds of information sources. Adopting the same basic perspective, our aim is to check for inconsistency between such subsets. The suggestion in Ref. [8] is to normalize these information sources to have equal height 1 and to regard the height of the curves at the point of intersection as a measure of conflict. However, as shown in Ref. [2], this measure tends to be quite conservative. Dahl et al. [9] demonstrated that it is also poorly calibrated, with false warning probabilities that vary substantially between models. Dahl et al. [9] also identified the different sources of inaccuracy and modified the measure of Ref. [8] to an approximately χ^2-distributed quantity under the assumed model by instead normalizing the information sources to probability densities. In Ref. [10], these densities were instead used to define tail probability-based conflict measures. Gåsemyr and Natvig [10] showed that these measures are uniformly distributed in quite general hierarchical normal models with fixed variances/covariances. In Ref. [11], such uniformity results were proved in various situations involving nonnormal and nonsymmetric distributions. These uniformity results indicate that the measures of Refs. [9] and [10] have comparable interpretations across different models. Therefore, they can be used without computationally costly precalibration schemes, such as the one suggested in Ref. [5]. Gåsemyr [12] focuses on some situations where the conflict measure approach can be directly compared to the calibration method of Ref. [5] and shows that the less computer-intensive conflict measure approach performs at least as well in these situations. Moreover, the conflict measure approach can be applied in models using noninformative prior distributions.

Focusing on the special problem of identifying outliers among the second-level parameters in a random-effects model, Ref. [13] defines similar conflict measures. In this setting, the group-specific means are the nodes of interest. In some models, there exist sufficient statistics for these means. Then, outlier detection at the group level can also be based on cross validation, measuring the tail probability beyond the observed value of the statistic in the posterior predictive distribution given data from the other groups. In this context, the conflict measure approach can be viewed as an extension of cross-validation to situations where sufficient statistics do not exist. Also in Ref. [13] applications to the examination of exceptionally high hospital mortality rates and to results from a vaccination program are given. In Ref. [14], this methodology is used to check for inconsistency in multiple treatment comparison of randomized clinical trials. Presanis et al. [15] apply these conflict measures in complex cases of medical evidence synthesis.

2. Directed acyclic graphs and node-specific conflict

2.1. Directed acyclic graphs and Bayesian hierarchical models

An example of a DAG discussed extensively in Ref. [8] is the random-effects model with normal random effects and normal error terms defined by

$$Y_{i,j} \sim N(\lambda_i, \sigma^2), \lambda_i \sim N(\mu, \tau^2), j = 1, \ldots, n_i, i = 1, \ldots, m. \tag{1}$$

In general, we identify the nodes or vertices of the graph with the unknown parameters $\boldsymbol{\theta}$ and the observed data $\mathbf{y}$, the latter appearing as bottom nodes and being the realizations of the random vector $\mathbf{Y}$. In the Bayesian model, the parameters, the components of $\boldsymbol{\theta}$, are also considered as random variables. In general, if there is a directed edge from node a to node b, then a is a parent of b, and b is a child of a. We denote by $\text{Ch}(a)$ the set of child nodes of a, and by $\text{Pa}(b)$ the set of parent nodes of b. More generally, b is a descendant of a if there is a directed path from a to b. The set of descendants of a is denoted by $\text{Desc}(a)$ and, for convenience, is defined to contain a itself. The directed edges encode conditional independence assumptions, indicating that, given its parents, a node is assumed to be independent of all other nondescendants. Hence, writing $\boldsymbol{\theta} = (\mathbf{v}, \boldsymbol{\mu})$, with $\boldsymbol{\mu}$ representing the vector of top-level nodes, the joint density of $(\mathbf{Y}, \boldsymbol{\theta}) = (\mathbf{Y}, \mathbf{v}, \boldsymbol{\mu})$ is

$$p(\mathbf{y}, \mathbf{v}, \boldsymbol{\mu}) = \prod_{y \in \mathbf{y}} p(y|\text{Pa}(y)) \prod_{\nu \in \nu} p(\boldsymbol{\nu}|\text{Pa}(\boldsymbol{\nu}))\pi(\boldsymbol{\mu}), \tag{2}$$

where $\pi(\boldsymbol{\mu})$ is the prior distribution of $\boldsymbol{\mu}$. The posterior distribution $\pi(\boldsymbol{\theta}\,|\,\mathbf{y})$ is the basis for the inference.

This setup can be generalized in various directions. The nodes may be allowed to represent vectors, at both the parameter and the data levels [10]. Instead of DAGs, one may consider chain graphs, as described in Ref. [16], with undirected edges representing mutual dependence as in Markov random fields. Scheel et al. [17] introduce a graphical diagnostic for model criticism in such models.

2.2. Information contributions

The representation of a Bayesian hierarchical model in terms of a DAG is often meant to reflect an understanding of the underlying structure of the problem. By looking for a conflict associated with the different nodes in the DAG, we may therefore put our understanding of this structure to test. We may also identify parts of the model that need adjustment.

The idea put forward in Ref. [8] is that for each node λ in a DAG one may in general think of each neighboring node as providing information about λ and that it is of interest to consider the possibility of conflict between different sources of information. For instance, one may want to contrast the local prior information provided by the factor $p(\lambda\,|\,\text{Pa}(\lambda))$ with the likelihood information source formed by multiplying the factors $p(\gamma\,|\,\text{Pa}(\gamma))$ for all child nodes $\gamma \in \text{Ch}(\lambda)$. The full conditional distribution of λ given all the observed and unobserved variables in the DAG, i.e.,

$$\pi(\lambda|(\mathbf{y}, \boldsymbol{\theta})_{-\lambda}) \propto p(\lambda|\text{Pa}(\lambda)) \prod_{\gamma \in \text{Ch}(\lambda)} p(\gamma|\text{Pa}(\gamma)), \tag{3}$$

is determined by these two types of factors. Here, $(\mathbf{y}, \boldsymbol{\theta})_{-\lambda}$ denotes the vector of all components of $(\mathbf{y}, \boldsymbol{\theta})$ except for λ.

Dahl et al. [9] normalize the product $\prod_{\gamma \in \text{Ch}(\lambda)} p(\gamma|\text{Pa}(\gamma))$ to a probability density function denoted by $f_c(\lambda)$, the likelihood or child node information contribution, whereas the local prior density is denoted by $f_p(\lambda)$ and called the prior or parent node information contribution. These information contributions are integrated with respect to posterior distributions for the unknown nuisance parameters to form integrated information contribution (iic) denoted by g_c and g_p. In this construction, a key to avoid the conservatism of the measure suggested in Ref. [8] is to prevent dependence between the two information sources by introducing a suitable data splitting $\mathbf{Y} = (\mathbf{Y}_p, \mathbf{Y}_c)$ and condition the parameters of f_p on $\mathbf{y}_p$ and the parameters of f_c on $\mathbf{y}_c$.

Definition 1 *For a given parameter node* λ, *denoted by* β_p *the vector whose components are* $Pa(\lambda)$, *and by* β_c *the vector whose components are*

http://dx.doi.org/10.5772/intechopen.70058

$$\cup_{\gamma \in \mathrm{Ch}(\lambda)}(\{\gamma\} \cup Pa(\gamma)) - \{\lambda\} = \mathrm{Ch}(\lambda) \cup [Pa(Ch(\lambda)) - \{\lambda\}] \tag{4}$$

Let $\mathbf{Y} = (\mathbf{Y}_p, \mathbf{Y}_c)$ *be a splitting of the data* $\mathbf{Y}$. *Define the densities* f_p, f_c, *the prior respectively likelihood information contributions, by*

$$f_p(\lambda; \boldsymbol{\beta}_p) = p(\lambda | \boldsymbol{\beta}_p), \quad f_c(\lambda; \boldsymbol{\beta}_c) \propto \prod_{\gamma \in Ch(\lambda)} p(\gamma | Pa(\gamma)) \tag{5}$$

Define the integrated information contribution densities g_p, g_c *by*

$$g_p(\lambda) = \int f_p(\lambda; \boldsymbol{\beta}_p)\pi(\boldsymbol{\beta}_p | \mathbf{y}_p) d\boldsymbol{\beta}_p, \quad g_c(\lambda) = \int f_c(\lambda; \boldsymbol{\beta}_c)\pi(\boldsymbol{\beta}_c | \mathbf{y}_c) d\boldsymbol{\beta}_c, \tag{6}$$

and denote by G_p, G_c *the corresponding cumulative distribution functions.*

Note that $\boldsymbol{\beta}_c$ may contain data nodes. The second integral in Eq. (6) is then taken only with respect to the random components of $\boldsymbol{\beta}_c$, i.e., the parameters in $\boldsymbol{\beta}_c$. If $\boldsymbol{\beta}_c$ contains no parameters, then g_c and f_c coincide. Definition 1 may also be extended to the case when λ is a vector, corresponding to a subset of parameter nodes.

Combining the set of information sources linked to a specific node in different ways leads to a modification of Definition 1 where $\boldsymbol{\beta}_c$ does not contain all child nodes of λ, the others being instead included in $\boldsymbol{\beta}_p$ together with their parent nodes. In this way, different types of conflict about the node may be revealed. This is natural, e.g., in the context of outlier detection among independent observations with a common mean. Note that $\boldsymbol{\beta}_p$ and $\boldsymbol{\beta}_c$ may then be overlapping, containing common coparents with λ. The setup is illustrated in **Figure 1** in the case when the

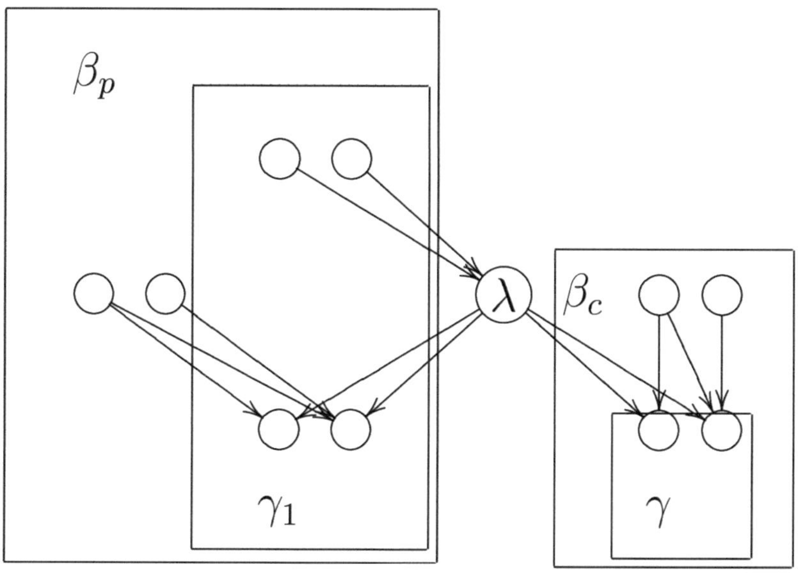

Figure 1. Part of a DAG showing information sources about λ.

set of common components, by abuse of notation denoted by $\boldsymbol{\beta}_p \cap \boldsymbol{\beta}_c$, is empty. For the general setup, Definition 1 is modified as follows.

Definition 2 *Let γ be a vector whose components are a subset of $Ch(\lambda)$, and define $\boldsymbol{\beta}_c$ as in Eq. (4). Denote by $\boldsymbol{\gamma}_1$ the rest of the child nodes of λ, and let $\boldsymbol{\beta}_p$ consist of $\boldsymbol{\gamma}_1$ together with its parent nodes in the same way as in Eq. (4), as well as $Pa(\lambda)$. The information contributions are then given by*

$$f_p(\lambda;\boldsymbol{\beta}_p) \propto p(\boldsymbol{\gamma}_1|Pa(\boldsymbol{\gamma}_1)p(\lambda|Pa(\lambda)), \tag{7}$$

$$f_c(\lambda;\boldsymbol{\beta}_c) \propto p(\boldsymbol{\gamma}|Pa(\boldsymbol{\gamma})). \tag{8}$$

In Eq. (7), $p\left(\lambda|Pa(\lambda)\right)$ is replaced by the prior density $\pi(\lambda)$ if λ is a top-level parameter. The corresponding iic densities are defined by Eq. (6) as before.

2.3. Node-specific conflict measures

The conflict measure c_λ^2 of Ref. [9] is defined as

$$c_\lambda^2 = (E^{G_p}(\lambda) - E^{G_c}(\lambda))^2/(\mathrm{var}^{G_p}(\lambda) + \mathrm{var}^{G_c}(\lambda)) \tag{9}$$

The χ_1^2-distribution is the reference distribution for this measure. For the conflict measures of Ref. [10], the uniform distribution on [0, 1] is the reference distribution. They focus on tail behavior but are based on the same iic distributions. The general distribution of information sources given in Definition 2 is also introduced in Ref. [10]. For a given pair G_p, G_c of iic distributions, let λ_p^* and λ_c^* be independent samples from G_p and G_c, respectively. Let G be the cumulative distribution function for $\delta = \lambda_p^* - \lambda_c^*$. Define

$$c_\lambda^{3+} = G(0), \quad c_\lambda^{3-} = \overline{G}(0) \stackrel{\mathrm{def}}{=} 1 - G(0) \tag{10}$$

and

$$c_\lambda^3 = 1 - 2\min(G(0), \overline{G}(0)) = 2|G(0) - 1/2|. \tag{11}$$

The c_λ^{3+}-measure and the $P_\lambda^{\mathrm{conf}}$ measure of Ref. [13] are very similar. The latter measure is aimed at detecting outlying groups or units in a three-level hierarchical model, with the second-level parameters being location parameters for group-specific data. However, the measure is interpreted as a p value, with small values indicative of conflict. Gåsemyr and Natvig [10] also defines a measure based on defining a tail area in terms of the density g of G, namely

$$c_\lambda^4 = P^G(g(\delta) > g(0)), \tag{12}$$

applicable also when λ is a vector.

Example 1. To illustrate the theory, consider the random-effects model 1, with the variance parameters σ^2, τ^2 assumed known, and with μ having the improper prior $\pi(\mu) = 1$. For simplicity, assume $n_i = n$ for all i. Suspecting the mth group of representing an outlier, let $\lambda=\lambda_m$ be the node of interest. Define the data splitting $\mathbf{Y}_p$, $\mathbf{Y}_c$ by letting $\mathbf{Y}_c = \mathbf{Y}_m = (Y_{m,1}, \dots, Y_{m,n})$, and let $\boldsymbol{\beta}_c = \mathbf{y}_c$, $\boldsymbol{\beta}_p = \mu$. Denoting the normal density function by ϕ, it is easy to see that $g_c(\lambda) = f_c(\lambda) = \phi(\lambda; \bar{y}_c, \sigma^2/n)$. Furthermore, $f_p(\lambda;\mu) = \phi(\lambda; \mu, \tau^2)$. Given $\mathbf{y}_p$, μ has the density $\pi(\mu|\mathbf{y}_p) = \phi(\mu); \sum_{i=1}^{m-1} \bar{y}i/(m-1), (1/(m-1))\tau^2 + (1/(n(m-1)))\sigma^2)$. By a standard argument

$$
\begin{aligned}
g_p(\lambda) &= \int f_p(\lambda;\mu)\pi(\mu|\mathbf{y}_p)d\mu \\
&= \phi(\lambda; \sum_{i=1}^{m-1} \bar{y}_i/(m-1), (1+1/(m-1))\tau^2 + (1/(n(m-1)))\sigma^2).
\end{aligned}
$$

It follows that $g(\delta) = \phi(\delta); \sum_{i=1}^{m=1} \bar{y}_i/(m-1) - \bar{\mathbf{y}}_c, (m/(m-1))(\tau^2 + \sigma^2/n)$. The conflict measures (Eqs. (9), (10), (11), and (12)) can hence be calculated analytically, with no simulation needed in this case.

In a simulation study of the c_λ^2-measure in Ref. [9] using a warning level equal to the 95% quantile of the χ_1^2-distribution, a false warning probability of close to 5% is obtained for a normal random-effects model with unknown variance parameters as in Eq. (1) and also in similar random-effects models with heavy-tailed t- and uniformly distributed random effects. Also with respect to detection power, this measure performs well when compared to a calibrated version of the measure given in Ref. [8], if an optimal data splitting is used. Refs. [10] and [11] prove preexperimental uniformity of the conflict measures in various situations, i.e., their distributions as functions of a $\mathbf{Y}$ which is distributed according to the assumed model are uniform, regardless of the true value of the basic parameter. Another way of stating this is that we obtain a proper p-value by subtracting these measures from 1. These results are reviewed in Section 5 of the present chapter.

2.4. Integrated information contributions as posterior distributions

In most cases, the conflict measures of Refs. [9] and [10] are based on simulated samples from G_p and G_c. Definitions 1 and 2 suggest obtaining such samples by running an MCMC algorithm to generate posterior samples of the unknown parameters in $\boldsymbol{\beta}_p$ and $\boldsymbol{\beta}_c$ and then generate samples λ_p^* and λ_c^* from the respective information contributions for each such parameter sample. If the information contributions are standard probability densities, this procedure is straightforward. If not, one may instead often use the fact that, under certain conditions on the data splitting, the distributions G_p and G_c are posterior distributions conditional on $\mathbf{y}_p$ and $\mathbf{y}_c$, respectively, the latter based on the improper prior $\pi(\lambda) = 1$, independently of the coparents.

Theorem 1 *Suppose that the data splitting satisfies*

$$\mathbf{Y}_c = \mathbf{Y} \cap [\cup_{\gamma \in Ch(\lambda) \cap \beta_c} Desc(\gamma)], \quad \mathbf{Y}_p = \mathbf{Y} - \mathbf{Y}_c, \tag{13}$$

the latter expression by abuse of notation meaning the components of $\mathbf{Y}$ *not present in* $\mathbf{Y}_c$. *Assume* λ *and the coparents* $Pa\left(Ch(\lambda) \cap \beta_p\right) - \lambda$ *are independent. We then have*

$$g_p(\lambda) = \pi(\lambda|\mathbf{y}_p)$$

and, specifying as prior density

$$\begin{aligned} \pi(\lambda|Pa(Ch(\lambda)\cap\boldsymbol{\beta}_c) - \lambda) &= 1, \\ g_c(\lambda) &= \pi(\lambda|\mathbf{y}_c). \end{aligned} \tag{14}$$

The proof is given in Appendix A in the online supporting information for Ref. [11]. Specializing to the standard setup of Definition 1, where $\mathrm{Ch}(\lambda)\subseteq\boldsymbol{\beta}_c$, we see that the requirement for Eq. (13) to hold is that $\mathbf{Y}_c$ consists of all data descendant nodes of λ. In Ref. [9], this splitting was compared with two other splittings for c_λ^2 and found to be optimal with respect to detection power. This measure was also found to be a well-calibrated measure under this splitting.

3. Noninvariance and reparametrizations

The iic distributions and the corresponding conflict measures are parametrization dependent. Based on experience so far, the conflict measures seem to be fairly robust to changes in parametrization. However, this noninvariance can be handled in a theoretically satisfactory way under certain circumstances.

Let ϕ be the parameter, in a standard parametrization, corresponding to a specific node in the DAG. Suppose for simplicity that $\mathbf{Y}_c = \mathrm{Ch}(\phi)$. Assume that there exists a sufficient statistic Y_c and an alternative parametrization λ, being a strictly monotonic function $\lambda(\phi)$, such that $Y_c - \lambda$ is a pivotal quantity, i.e., the density for Y_c given λ is of the form

$$p(y_c|\lambda) = f_{Y_c}(y_c|\lambda) = f_0(y_c - \lambda) \tag{15}$$

for some known density function f_0. Such a parametrization will be considered as a canonical or reference parametrization if it exists, as opposed to the standard parametrization involving ϕ. Accordingly, the conflict measures given in Eqs. (9)–(12) are preferably based on this reference parametrization.

By Theorem 1, samples λ_c^* from G_c may be obtained by MCMC as posterior samples from $\pi(\lambda|y_c)$ when the splitting satisfies Eq. (13) and the prior for λ satisfies Eq. (14), i.e., equals 1. According to an argument given in Section 1.3 of Ref. [18], such a prior expresses noninformativity for likelihoods of the form (Eq. (15)). Computationally, we may, however, use the standard parametrization. When generating ϕ_c^* as posterior samples from $\pi(\phi|\mathbf{Y}_c)$, the prior density $|d\lambda/d\phi|$ for ϕ must be used. Then, we may calculate $\lambda_c^* = \lambda(\phi_c^*)$. To represent the iic distribution $G_p(\lambda)$, we may calculate $\lambda_p^* = \lambda(\phi_p^*)$for samples ϕ_p^* from $\pi(\phi|\mathbf{y}_p)$ according to the given model. Now, the c_λ^4-measure can be estimated from (Eq. (12)), using a kernel density estimate of $g(\delta)$ based on corresponding samples $\delta^* = \lambda_p^* - \lambda_c^*$. However, if we limit attention to the c_λ^3-measure (Eq. (11)) and its one-sided versions (Eq. (10)), we may use the samples from $\pi(\phi|\mathbf{y}_c)$ and $\pi(\phi|\mathbf{y}_p)$ directly. To see this, note that the condition $\lambda_p^* \geq \lambda_c^*$ is equivalent to the condition ${\phi^*}_p \geq {\phi^*}_c$ (assuming that λ is increasing as a function of ϕ). Hence, the probability $G(0)$ that $\lambda_p^* - \lambda_c^* \leq 0$ can be estimated as the proportion of sample values for which ${\phi^*}_p \leq {\phi^*}_c$.

4. Extensions to deterministic nodes: Relation to cross-validation, prediction and hypothesis testing

4.1. Cross-validation and data node conflict

The model variables $\mathbf{Y}$ are represented by the bottom nodes in the DAG describing the hierarchical model. The framework can be extended to also cover conflict concerning these nodes. In this way, cross-validation can be viewed as a special case of the conflict measure approach.

Let Y_c be an element in the vector $\mathbf{Y}$ of observable random variables. We define the prior iic density $g_p(y_c)$ exactly as in Eq. (6), with λ replaced by y_c. The Dirac measure at the observed value y_c represents a degenerate iic information contribution about Y_c. This leads to the following definitions:

$$c_{y_c}^{3+} = G_p(y_c), \quad c_{y_c}^{3-} = \overline{G}_p(y_c), \tag{16}$$

$$c_{y_c}^{3} = 1 - 2\min(G_p(y_c), \overline{G}_p(y_c)), \tag{17}$$

$$c_{y_c}^{4} = P^{g_p}(g_p(Y_c) \geq g_p(y_c)). \tag{18}$$

The measures (Eqs. (16)–(18)) are called data node conflict measures. To see that these definitions are consistent with Eqs. (10)–(12), note that λ_p^* corresponds to Y_c, and λ_c^* is deterministic and corresponds to y_c. We define $X = Y_c - y_c$, corresponding to δ. We then have $g(x) = g_p(x + y_c)$. Hence,

$$G(0) = \int_{-\infty}^{0} g(x)dx = \int_{-\infty}^{y_c} g_p(y)dy = G_p(y_c),$$

and accordingly, $\overline{G}(0) = \overline{G}_p(y_c)$. It follows that Eqs. (16) and (17) are special cases of Eqs. (10) and (11). Moreover,

$$P^g(g(X) \geq g(0)) = P^{g_p}(g_p(Y_c) \geq g_p(y_c)),$$

showing that Eq. (18) is a special case of Eq. (12).

Furthermore, this correspondence between the data node conflict measures (Eqs. (16) and (17)) and the parameter node conflict measures (Eqs. (10) and (11)) can be used to motivate these latter measures. We will treat the c^{3+} measure as an example. Consider again a parameter node

λ. If λ were actually observable and known to take the value λ_c, the data node version of the c^{3+} measure could be used to measure deviations toward the right tail of G_p as

$$G_p(\lambda_c) = \int_{-\infty}^{\lambda_c} g_p(\lambda)d\lambda = \int_{-\infty}^{0} g_p(\delta + \lambda_c)d\delta.$$

Now λ is in reality not known, but we can take the expectation of this conflict with respect to the distribution G_c, which reflects the uncertainty about λ when influence from data $\mathbf{y}_p$ is removed. The result is the following theorem:

Theorem 2

$$E^{G_c}(G_p(\lambda) = c_\lambda^{3+}.$$

Proof:

$$E^{G_c}(G_p(\lambda) = \int_{-\infty}^{\infty} g_c(\lambda)\left(\int_{-\infty}^{0} g_p(\delta + \lambda)d\delta\right)d\lambda = \int_{-\infty}^{0}\left(\int_{-\infty}^{\infty} g_p(\delta + \lambda)g_c(\lambda)d\lambda\right)d\delta$$

$$= \int_{-\infty}^{0} g(\delta)d\delta = G(0) = c_\lambda^{3+}$$

by Eq. (10).

4.2. Cross-validation and sufficient statistics

Suppose the node λ of interest is the parent of the subvector $\mathbf{Y}_c$ of $\mathbf{Y}$. Suppose also that Y_c is a sufficient statistic for $\mathbf{Y}_c$. Evidently then, the measures c_λ^{3+} and $c_{Y_c}^{3+}$ address the same kind of possible conflict in the model. The following theorem, proved in Ref. [11], states that the two measures agree under certain conditions. This is a generalization of a result in Ref. [13], which also unnecessarily assumed symmetry for the conditional density of Y_c.

Theorem 3 *Suppose the conditional density for the scalar variable Y_c given the parameter λ is of the form $f_{Y_c}(y|\lambda) = f_{c,0}^2(y - \lambda)$. Then,*

$$c_{Y_c}^{3+} = c_\lambda^{3+}.$$

When a sufficient statistic exists, the cross-validatory p-value is considered by Ref. [13] as the gold standard, and the aim of their construction is to provide a measure which is generally applicable and matches cross-validation when a sufficient statistic exists.

4.3. Prediction

As mentioned in Section 2, the c^4 measure can be used to assess conflict concerning vectors of nodes. Applying this at the data node level, we may assess the quality of predictions of a subvector $\mathbf{Y}_c$ of $\mathbf{Y}$ based on a complementary subvector y_p of observations. The relevant measure is given by Eq. (18), with Y_c replaced by the vector $\mathbf{Y}_c$. This is particularly well suited to models where data accumulate as time evolves. Such a conflict measure can be used to assess the overall quality of the model. It can also be used as a tool for model comparison and model choice.

4.4. Hypothesis testing

Suppose the top-level nodes $\boldsymbol{\mu}$ appearing in Eq. (2) are assumed fixed and known according to the model, so that $\pi(\boldsymbol{\mu})$ is a Dirac measure at these fixed values of the components of $\boldsymbol{\mu}$. Hence, the DAG has deterministic nodes both at the top and at the bottom, namely the vectors $\boldsymbol{\mu}$ and $\mathbf{y}$, respectively. We may then check for a conflict concerning a component λ of $\boldsymbol{\mu}$ by introducing a random version $\tilde{\lambda}$ of λ and contrast the corresponding $g_c(\tilde{\lambda})$ with the fixed value λ. The random $\tilde{\lambda}$ has the same children and coparents as λ, and the vector $\boldsymbol{\beta}_c$, the information contribution $f_c(\tilde{\lambda};\beta_c)$ and the iic density g_c are defined as in Eqs. (4), (5) and (6). The respective conflict measures are defined as in Eqs. (16)–(18) with y_c replaced by λ and G_p and g_p replaced by G_c and g_c. If the model is rejected when the conflict exceeds a certain predefined warning level, this corresponds to a formal Bayesian test of the hypothesis $\tilde{\lambda} = \lambda$. Using the conflict measure (Eq. (18)), we may put the whole vector $\boldsymbol{\mu}$ to test in this way.

5. Preexperimental uniformity of the conflict measures

In this section, we review some results concerning the distribution of the conflict measures. If c is one of the measures (Eqs. (10), (11), (12), (16), (17) or (18)), then preexperimentally, i.e., prior to observing the data $\mathbf{y}$, c is a random variable taking a value in [0, 1]. A large value of c indicates a possible conflict in the model, and uniformity of c corresponds to $1 - c$ being a proper p-value. This does not mean that we propose a formal hypothesis testing procedure for model criticism, possibly even adjusted for multiple testing, nor that we think that a fixed significance level represents an appropriate criterion signaling the need for changing the model. A relatively large value of c may be accepted if there are convincing arguments for believing in a particular modeling aspect, while a less extreme value of c may indicate a need for adjustments in modeling aspects that are considered questionable for other reasons. But the terms "relatively large" and "less extreme" must refer to a meaningful common scale. In our view, uniformity of the conflict measure under all sources of uncertainty is the natural ideal criterion for being a well-calibrated conflict measure, the fulfillment of which ensures comparable assessment of the level of conflict across models. This means that we aim for preexperimental uniformity in cases where the prior distribution is highly noninformative, and also, as discussed in the following subsection, in cases where an informative prior represents part of the randomness in the data-generating process (aleatory uncertainty) rather than subjective (epistemic) uncertainty about the location of a fixed but unknown λ. In this chapter, we limit attention to situations where exact uniformity is achieved. The pivotality condition (Eq. (15)) turns out to be a key assumption needed to obtain such exact results. Refs. [10] and [12] provide some examples where exact uniformity is achieved in other cases.

5.1. Data-prior conflict

Consider the model

$$\mathbf{Y} \sim F_{\mathbf{Y}}(\mathbf{y}|\lambda), \lambda \sim F_{\lambda}(\lambda),$$

where F_{λ} is an arbitrary informative prior distribution. Here, we think of this prior distribution as representing aleatory rather than epistemic uncertainty. The corresponding densities are denoted by $f_{\mathbf{Y}}$ and f_{λ}. If contrasting the prior density with the likelihood $f_{\mathbf{Y}}(\mathbf{y}|\lambda)$ indicates a conflict between the prior and likelihood information contributions, we consider this a data-prior conflict. The following theorem, proved in Ref. [11], deals with this kind of conflict. Note that in this situation, the $\mathbf{Y}_p$ part of the data splitting is empty.

Theorem 4 *Suppose the conditional density for the scalar variable Y given the parameter λ is of the form $f_Y(y|\lambda) = f_0(y - \lambda)$ and that λ is generated from an arbitrary informative prior density $f_{\lambda}(\lambda)$. Then, the data-prior conflict measures about λ are preexperimentally uniformly distributed for both the c_{λ}^3- and c_{λ}^4-measures.*

The theorem obviously applies to the location parameter of normal and t-distributions with fixed variance parameters, as well as the location parameter in the skew normal distribution [19]. If the vector $\mathbf{Y}$ consists of IID normal variables, the theorem also applies to the location parameter, using as scalar variable the sufficient statistic $\overline{\mathbf{Y}}$. If the n components of $\mathbf{Y}$ are IID exponentially distributed with failure rate λ, their sum is a sufficient statistic that is gamma distributed with shape parameter n and scale parameter λ. We may then use the fact that for a variable Y which is gamma distributed with known shape parameter and unknown scale parameter λ, the quantity $\log(Y) - \log(\lambda)$ is a pivotal statistic, and uniformity is obtained by combining Theorem 4 with the approach of Section 3. In the standard parametrization, the appropriate prior distribution is $\pi(\lambda) = 1/\lambda)$. Details are given in Ref. [11], which also deals with the gamma, inverse gamma, Weibull and lognormal distributions in a similar way.

5.2. Data-data conflict

Suppose all components of $\mathbf{Y}$ have distributions determined by the same parameter λ. Suppose we want to contrast information contributions from separate parts of $\mathbf{Y}$ about λ and define the splitting $(\mathbf{Y}_p, \mathbf{Y}_c)$ accordingly. Focusing on this kind of possible conflict, we assume complete prior ignorance about λ and accordingly assume that λ has the improper prior $\pi(\lambda) = 1$. Hence, recalling Eqs. (7) and (8), we contrast the information in $f_c(\lambda; \mathbf{Y}_c)$ with that in $f_p(\lambda; \mathbf{Y}_p)$. We use the term data-data conflict in this context, since there is no prior information incorporated in f_p, and the two information contributions play symmetric roles. However, as a particular application, one may think of $\mathbf{Y}_c$ as a scalar variable representing a possible outlier.

The following theorem is proved in Ref. [11].

Theorem 5 *Suppose that the conditional densities for the scalar variables Y_p and Y_c given the parameter λ are of the form $f_{Y_p}(y|\lambda) = f_{p,0}(y - \lambda)$, $f_{Y_c}(y|\lambda) = f_{c,0}(y - \lambda)$.*

Assume λ has the improper prior $\pi(\lambda) = 1$. Then, the data-data conflict measures about λ are preexperimentally uniformly distributed for both the c^3_λ- and c^4_λ-measures.

Theorem 5 can be applied if the components of $\mathbf{Y}_c$ and $\mathbf{Y}_p$ are normally or lognormally distributed with known variance parameter, exponentially distributed, or gamma, inverse gamma or Weibull with known shape parameter, since pivotal quantities based on sufficient statistics exist for these distributions.

5.3. Normal hierarchical models with fixed covariance matrices

Allowing for each y and v appearing in Eq. (2) to be interpreted as vectors of nodes, we now assume that each conditional distribution in the decomposition (Eq. (2)) is multinormal with fixed and known covariance matrices. The random-effects model (Eq. (1)) is a simple example of this. We also assume that the top-level parameter vector $\boldsymbol{\mu}$ has the improper prior 1 and that each linear mapping $\text{Pa}(v) \to E(v|\text{Pa}(v))$ has full rank.

Now let λ be any node in the model description. It is standard to verify that, regardless of how the vector of neighboring and coparent nodes $\boldsymbol{\beta}$ is decomposed into $\boldsymbol{\beta}_p$, containing $\text{Pa}(\lambda)$, and $\boldsymbol{\beta}_c$, the densities $f_p(\lambda; \beta_p)$ and $f_c(\lambda; \beta_c)$ of Eqs. (5) and (8) are multinormal with fixed covariance matrices. Furthermore, this is true also for the iic densities g_p and g_c of Eq. (6), regardless of the data splitting. It follows that the density g of the difference δ between independent samples from g_p and g_c is multinormal with expectation $E^G(\delta) = E^{G_p}(\lambda) - E^{G_c}(\lambda)$ and covariance matrix $\text{cov}^G(\delta) = \text{cov}^{G_p}(\lambda) + E^{G_c}(\lambda)$. It follows that $\left(\delta - E^G(\delta)\right)^t \text{cov}^G(\delta)^{-1}\left(\delta - E^G(\delta)\right)$ is χ^2-distributed with $n = \dim(\lambda)$ degrees of freedom, and the probability under G that $g(\delta) < g(0)$ is easily seen to be $\Psi_n\left(E^G(\delta)^t \text{cov}^G(\delta)^{-1} E^G(\delta)\right)$, where Ψ_n is the cumulative distribution function for the χ^2_n-distribution. The preexperimental uniformity of this quantity is proved in Ref. [10].

Theorem 6 *Consider a hierarchical normal model as described above.*

i. *Let λ be an arbitrary scalar or vector parameter node. If the data splitting satisfies Eq. (13), then c^4_λ is uniformly distributed preexperimentally.*

ii. *Suppose the data splitting $(\mathbf{Y}_p, \mathbf{Y}_c)$ satisfies $Ch\left(Pa(\mathbf{Y}_c)\right) = \mathbf{Y}_c$. Then, $c^4_{\mathbf{Y}_c}$ is uniformly distributed preexperimentally.*

If λ in (i) or Y_c in (ii) are one dimensional, then G is symmetric and unimodal, and therefore, the respective c^3-measures are defined and coincide with the c^4-measures. Gåsemyr et al. [10] also show that in that case the c^{3+}- and c^{3-}-measures are uniformly distributed preexperimentally.

Example 2. Consider the following DAG model, a regression model with randomly varying regression coefficients.

$$Y_{i,j} \sim N(\mathbf{X}^t_{i,j}\xi_i, \sigma^2),\ \xi_i \sim N(\xi, \Omega),\ j = 1, \ldots, n,\ i = 1, \ldots, m,\ \pi(\xi) \propto 1. \tag{19}$$

The m units could be groups of individuals, with $y_{i,j}$ the measurement for a group member with individual covariate vector $\mathbf{X}_{i,j}$, or individuals with the successive $y_{i,j}$ representing repeated measurements over time. In this model, we could check for a possible exceptional behavior of the mth unit by means of the conflict measure $c^4_{\xi_m}$. With a data splitting for which $\mathbf{Y}_c = \mathbf{Y}_m = (Y_{m,1}, \ldots, Y_{m,n})$ the conditions for Theorem 6, part (i), are satisfied if $\dim(\xi) \leq n$, and the measure is preexperimentally uniformly distributed.

6. Concluding remarks

The assumption of fixed covariance matrices in the previous subsection is admittedly quite restrictive. In general, the presence of unknown nuisance parameters, such as parameters describing the covariance matrices in a normal model, makes the derivation of exact uniformity at least difficult and often impossible. Promising approximate results are reported in Ref. [9] for the closely related c^2_λ measure. Further empirical studies are needed in order to examine to what extent the conflict measures are approximately uniformly distributed in other situations. As an informal tool to be used in conjunction with subject matter insight, the conflict measure approach does not require exact uniformity in order to be useful.

Author details

Jørund I. Gåsemyr* and Bent Natvig

*Address all correspondence to: gaasemyr@math.uio.no

University of Oslo, Norway

References

[1] Box GEP. Sampling and Bayes' inference in scientific modelling and robustness (with discussion and rejoinder). Journal of the Royal Statistical Society. Series A. 1980;**143**:383-430

[2] Bayarri MJ, Castellanos ME. Bayesian checking of the second levels of hierarchical models. Statistical Science. 2007;**22**:322-343

[3] Gelman A, Meng X-L, Stern H. Posterior predictive assessment of model fitness via realized discrepancies (with discussion and rejoinder). Statistica Sinica. 1996;**6**:733-807

[4] Bayarri MJ, Berger JO. P values in composite null models (with discussion). The Journal of the American Statistical Association. 2000;**95**:1127-1142

[5] Hjort NL, Dahl FA, Steinbakk GH. Post-processing posterior predictive *p*-values. The Journal of the American Statistical Association. 2006;**101**:1157-1174

[6] Dahl FA. On the conservativeness of posterior predictive p-values. Statistics and Probability Letters. 2006;**76**:1170-1174

[7] Dey D, Gelfand A, Swartz T, Vlachos P. A simulation-intensive approach for checking hierarchical models. Test. 1998;**7**:325-346

[8] O'Hagan A. HSSS model criticism (with discussion). In: Green PJ, Hjort NL, Richardson S, editors. Highly Structured Stochastic Systems. Oxford: Oxford University Press; 2003. pp. 423-444

[9] Dahl FA, Gåsemyr J, Natvig B. A robust conflict measure of inconsistencies in Bayesian hierarchical models. Scandinavian Journal of Statistics. 2007;**34**:816-828

[10] Gåsemyr J, Natvig B. Extensions of a conflict measure of inconsistencies in Bayesian hierarchical models. Scandinavian Journal of Statistics. 2009;**36**:822-838

[11] Gåsemyr J. Uniformity of node level conflict measures in Bayesian hierarchical models based on directed acyclic graphs. Scandinavian Journal of Statistics. 2016;**43**:20-34

[12] Gåsemyr J. Alternatives to post-processing posterior predictive p-values. Submitted 2017

[13] Marshall EC, Spiegelhalter DJ. Identifying outliers in Bayesian hierarchical models. A simulation based approach. Bayesian Analysis. 2007;**2**:409-444

[14] Dias S, Welton NJ, Caldwell DM, Ades AE. Checking consistency in mixed treatment comparison meta-analysis. Statistics in Medicine. 2010;**29**:932-944

[15] Presanis AM, Ohlssen D, Spiegelhalter D, De Angelis D. Conflict diagnostics in directed acyclic graphs, with applications in Bayesian evidence synthesis. Statistical Science. 2013;**28**:376-397

[16] Lauritzen SL. Graphical Models. Oxford: Oxford University Press; 1996

[17] Scheel I, Green P, Rougier JC. A graphical diagnostic to identifying influential model choices in Bayesian hierarchical models. Scandinavian Journal of Statistics.2011;**38**:529-550

[18] Box GEP, Tiao GC. Bayesian Inference in Statistical Analysis. New York: Wiley; 1992

[19] Azzalini A. A class of distributions which include the normal ones. Scandinavian Journal of Statistics. 1985;**12**:171-178

Chapter 3

Classifying by Bayesian Method and Some Applications

Tai Vovan

Additional information is available at the end of the chapter

http://dx.doi.org/10.5772/intechopen.70052

Abstract

This chapter sums up and proposes some results related to classification problem by Bayesian method. We present the classification principle, Bayes error, and establish its relationship with other measures. The determination for Bayes error in reality for one and multi-dimensions is also considered. Based on training set and the object that we need to classify, an algorithm to determine the prior probability that can make to reduce Bayes error is proposed. This algorithm has been performed by the MATLAB procedure that can be applied well with real data. The proposed algorithm is applied in three domains: biology, medicine, and economics through specific problems. With different characteristics of applied data sets, the proposed algorithm always gives the best results in comparison to the existing ones. Furthermore, the examples show the feasibility and potential application in reality of the researched problem.

Keywords: Bayesian method, classification, error, prior, application

1. Introduction

Classification problem is one of the main subdomains of discriminant analysis and closely related to many fields in statistics. Classification is to assign an element to the appropriate population in a set of known populations based on certain observed variables. It is an important development direction of multivariate statistics and has applications in many different fields [25, 27]. Recently, this problem is interested by many statisticians in both theories and applied areas [14–18, 22–25]. According to Tai [22], we have four main methods to solve the classification problem: Fisher method [6, 12], logistic regression method [8], support vector machine (SVM) method [3], and Bayesian method [17]. Because Bayesian method does not require normal condition for data and can classify for two and more populations it has many advantages [22–25]. Therefore, it has been used by many scientists in their researches.

Given k populations $\{w_i\}$, with probability density functions (pdfs) and the prior probabilities respectively $\{f_i\}$ and $\{q_i\}$, $i = 1, 2, \ldots, k$, where $q_i \in (0; 1)$, $\sum_{i=1}^{k} q_i = 1$. Pham–Gia et al. [17] used the maximum function of pdfs as a tool to study about Bayesian method and obtained important results. The classification principle and Bayes error were established based on the $g_{\max}(x) = \max\{q_1 f_1(x), q_2 f_2(x), \ldots, q_k f_k(x)\}$. The relationship between the upper and lower bounds of the Bayes error and the L^1—distance of the pdfs and the overlap coefficient of the pdfs—were established. The function $g_{\max}(x)$ played a very important role in the classification problem by Bayesian method and Pham–Gia et al. [17] continued to do research on it. Using the MATLAB software, Pham–Gia et al. [18] succeeded in identifying $g_{\max}(x)$ for some cases of the bivariate normal distribution. With similar development, Tai [22] has proposed the L^1—distance of the $\{q_i f_i(x)\}$—and established its relationship with Bayes error. This distance is also used to calculate Bayes error as well as to classify new element. This research has been applied in classifying ability to repay debt of bank customers. However, we think that the survey of two Bayesian approach relevant research was not yet completed. There are some relations between Bayes error and other statistical measures.

Bayesian method has many advantages. However, to our knowledge, the field of applications of this method in practice is narrower than other methods. We can find many applications in banking and medicine using Fisher method, SVM method, logistic method [1, 3, 8, 12]. Recently, all statistics software can effectively and quickly process the classification of large data sets and multivariate statistics using either three of the methods mentioned above, whereas the Bayesian method does not have this advantage. The cause of this problem is the ambiguity in determining prior probability, in estimating pdfs, and the complexity in calculating Bayes error. Although all these issues have been discussed by many authors, the optimal methods have yet to be found [22, 25]. In this chapter, we consider to estimate the pdf and to calculate Bayes error to apply in reality. We will present the problem on how to determine the prior probability in this chapter. In case of noninformation, we normally choose prior probabilities by uniform distribution. If we have some types of past data or training set, the prior probabilities are estimated either by Laplace method: $q_i = (n_i + n/k)/(N + n)$ or by the frequencies of the sample: $q_i = n_i/N$, where n_i and N are the number of elements in the ith population and training set, respectively, n is the number of dimensions, and k is the number of groups. The above-mentioned approaches have been studied and applied by many authors [14, 15, 22, 25]. We will also propose an algorithm to determine prior probability based on the training set, classified objective, and fuzzy cluster analysis. The proposed algorithm is applied in some specific problems of biology, medicine, and economics and has advantages over existing approaches. All calculations are performed by MATLAB procedures.

The next section of this chapter is structured as follows. Section 2 presents the classification principle and Bayes error. Some results of the Bayes error are also established in this section. Section 3 resolves the related problems in real application of the Bayes method. There are estimation of pdfs and determination of Bayes error in case of one dimension and multidimension. This section also proposes an algorithm to determine prior probability. Section 4 applies the proposed algorithm in real problems and compares outcome results to those obtained using existing approaches. Section 5 concludes this chapter.

2. Classifying by Bayesian method

The classification problem by Bayesian method has been presented in many documents [15, 16, 27], where the classification principle and the Bayes error are established based on Bayes theorem. In this section, we present them via the maximum function of $q_i f_i(x)$, $i = 1, 2, \ldots, k$ that they have advantages over existing approaches in real application [17, 18, 21–25]. This section also establishes the upper and lower bounds of the Bayes error and the relationships of Bayes error with other measures in statistical pattern recognition.

2.1. Classification principle and Bayes error

Given k populations $w_1, w_2, \ldots, w_k$ with $q_i \in (0;1)$ and $f_i(x)$ are the prior probability and pdf of ith population, respectively, $i = 1, 2, \ldots, k$. According to Pham–Gia et al. [17], element x_0 will be assigned to w_i if

$$g_i(x_0) = g_{\max}(x_0),\ i = 1, 2, \ldots, k \tag{1}$$

where $g_i(x) = q_i f_i(x), g_{\max}(x) = \max\{q_1 f_1(x), q_2 f_2(x), \ldots, q_k f_k(x)\}$.

Bayes error is given by the formula:

$$Pe_{1,2,\ldots,k}^{(q)} = \sum_{i=1}^{k} \int_{R^n \setminus R_i^n} q_i f_i dx = 1 - \sum_{i=1}^{k} \int_{R_i^n} q_i f_i(x) dx, \tag{2}$$

where $R_i^n = \left\{x | q_i f_i(x) > q_j f_j(x), \forall i \neq j, i, j = 1, 2, \ldots, k\right\}, (q) = (q_1, q_2, \ldots, q_k)$.

From Eq. (2), we can prove the following result:

$$\begin{aligned}
Pe_{1,2,\ldots,k}^{(q)} &= \sum_{j=1}^{k} \int_{R^n \setminus R_j^n} q_j f_j(x) dx \\
&= \sum_{j=1}^{k} \left[\int_{R^n} q_j f_j(x) dx - \int_{R_j^n} \max_{1 \le l \le k} \{q_l f_l(x)\} dx \right] \\
&= \int_{R^n} \sum_{j=1}^{k} q_j f_j(x) dx - \sum_{j=1}^{k} \int_{R_j^n} \max_{1 \le l \le k} \{q_l f_l(x)\} dx \\
&= 1 - \int_{R^n} \max_{1 \le l \le k} \{q_l f_l(x)\} dx
\end{aligned}$$

or

$$Pe_{1,2,\ldots,k}^{(q)} = 1 - \int_{R^n} g_{\max}(x) dx. \tag{3}$$

The correct probability is determined by $Ce_{1,2,\ldots,k}^{(q)} = 1 - Pe_{1,2,\ldots,k}^{(q)}$.

For $k = 2$, we have

$$Pe_{1,2}^{(q,1-q)} = \int_{R^n} \min\{qf_1(x), (1-q)f_2(x)\}dx = \lambda_{1,2}^{(q,1-q)} = \frac{1}{2}\left[1 - \|qf_1, (1-q)f_2\|_1\right], \tag{4}$$

where

$\lambda_{1,2}^{(q,1-q)}$ is the overlap area measure of $qf_1(x)$ and $(1-q)f_2(x)$ and $\|qf_1, (1-q)f_2\|_1 = \int_{R^n} |qf_1(x) - (1-q)f_2(x)|dx$.

2.2. Some results about Bayes error

Theorem 1. Let $f_i(x)$, $i = 1, 2, \ldots, k$, $k \geq 3$ be k pdfs defined on $R^n, n \geq 1$, $q_i \in (0;1)$. We have the relationships of Bayes error with other measures as follow:

i.

$$Pe_{1,2,\ldots,k}^{(q)} \leq 1 - \frac{1}{k-1}\left(1 - \prod_{j=1}^{k} q_j^{\alpha_j} D_T(f_1, f_2, \ldots, f_k)^{\alpha}\right), \tag{5}$$

ii.

$$Pe_{1,2,\ldots,k}^{(q)} \leq \sum_{i<j} q_i^{\beta} q_j^{1-\beta} D_T\left(f_i, f_j\right)^{(\beta,1-\beta)}, \tag{6}$$

iii.

$$\left\{(k-1) - \sum_i \sum_j \|g_i, g_j\|_1\right\}/k \leq Pe_{1,2,\ldots,k}^{(q)} \leq 1 - (1/2)\max_{i<j}\left\{\|g_i, g_j\|_1\right\} - \min_i\{q_i\}, \tag{7}$$

iv.

$$0 \leq Pe_{1,2,\ldots,k}^{(q)} \leq \max_i\{q_i\}, \tag{8}$$

where

$\alpha = (\alpha_1, \alpha_2, \ldots, \alpha_k);\ \alpha_j, \beta \in (0,1), \sum_{j=1}^{k} \alpha_j = 1$, $i, j = 1, 2, \ldots, k$, and

$D_T(f_1, f_2, \ldots, f_k)^{\alpha} = \int_{R^n} \prod_{j=1}^{k} \left[f_j(x)\right]^{\alpha_j} dx$ is affinity of Toussaint [26].

***Proof*:**

i. For each $j = 1,2,\ldots,k$, we have

$$\left(\sum_{j=1}^{k} q_j f_j\right)^{\alpha_i} \geq (q_i f_i)^{\alpha_i}, i = 1, 2, \ldots, k.$$

Therefore,

$$\left(\sum_{j=1}^{k} q_j f_j\right)^{\alpha_1+\alpha_2+\ldots+\alpha_k} \geq \prod_{j=}^{k}\left(q_j f_j\right)^{\alpha_j} \Leftrightarrow \sum_{j=1}^{k} q_j f_j \geq \prod_{j=}^{k}\left(q_j f_j\right)^{\alpha_j}. \tag{9}$$

On the other hand,

$$\left(\min_{1\leq j\leq k}\left\{q_j f_j\right\}\right)^{\alpha_1} \leq (q_1 f_1)^{\alpha_1}, \ldots\ldots, \left(\min_{1\leq j\leq k}\left\{q_j f_j\right\}\right)^{\alpha_k} \leq (q_k f_k)^{\alpha_k},$$

So

$$\left(\min_{1\leq j\leq k}\left\{q_j f_j\right\}\right)^{\alpha_1+\cdots+\alpha_k} \leq \prod_{j=1}^{k}\left(q_j f_j\right)^{\alpha_j}.$$

or

$$\min_{1\leq j\leq k}\left\{q_j f_j\right\} \leq \prod_{j=1}^{k}\left(q_j f_j\right)^{\alpha_j}. \tag{10}$$

Combining Eqs. (9) and (10), we obtain

$$0 \leq \sum_{j=1}^{k} q_j f_j - \prod_{j=1}^{k}\left(q_j f_j\right)^{\alpha_j} \leq \sum_{j=1}^{k} q_j f_j - \min_{1\leq j\leq k}\left\{q_j f_j\right\}.$$

Because $\sum_{j=1}^{k} q_j f_j - \min_{1\leq j\leq k}\left\{q_j f_j\right\}$ includes $(k-1)$ terms, we have

$$\sum_{j=1}^{k} q_j f_j - \min_{1\leq j\leq k}\left\{q_j f_j\right\} \leq (k-1)\max_{1\leq j\leq k}\left\{q_j f_j\right\}.$$

Thus,

$$0 \leq \sum_{j=1}^{k} q_j f_j - \prod_{j=1}^{k}\left(q_j f_j\right)^{\alpha_j} \leq (k-1)\max_{1\leq j\leq k}\left\{q_j f_j\right\}.$$

Integrating the above relation, we obtain:

$$1-\prod_{j=1}^{k} q_j^{\alpha_j} D_T\left(f_1, f_2, \ldots, f_k\right)^{\alpha} \leq (k-1) \int_{R^n} g_{\max}(x)dx. \tag{11}$$

Using $\int_{R^n} g_{\max}(x) = 1 - Pe_{1,2,\ldots,k}^{(q)}$ for Eq. (11), we have Eq. (5).

ii. From Eq. (2), we have

$$\begin{aligned} Pe_{1,2,\ldots,k}^{(q)} &= \sum_{j=1}^{k} \int_{R^n \setminus R_j^n} q_j f_j(x)dx \\ &= \sum_{j=1}^{k} \sum_{j\neq i} \int_{R_j^n} \min\left\{q_i f_i(x),\ q_j f_j(x)\right\}dx \\ &= \sum_{i<j} \int_{R_i^n} \min\left\{q_i f_i(x), q_j f_j(x)\right\}dx. \end{aligned}$$

Since

$$\left[\min\left\{q_i f_i(x), q_j f_j(x)\right\}\right]^{\beta} \leq (q_i f_i)^{\beta} \text{ and } \left[\min\left\{q_i f_i(x), q_j f_j(x)\right\}\right]^{1-\beta} \leq (q_i f_i)^{1-\beta},$$

then

$$\min\left\{q_i f_i(x), q_j f_j(x)\right\} \leq (q_i f_i)^{\beta}\left(q_j f_j\right)^{1-\beta}.$$

Integrating the above inequality, we obtain:

$$Pe_{1,2,\ldots,k}^{(q)} \leq \sum_{i<j} \int_{R_i^n} \left[(q_i f_i(x))^{\beta}\left(q_j f_j(x)\right)^{1-\beta}\right]dx \leq \sum_{i<j} q_i^{\beta} q_j^{1-\beta} D_T\left(f_i, f_j\right)^{(\beta,1-\beta)} dx.$$

iii. We have

$$\int_{R^n} \max\{g_1(x), g_2(x), \ldots, g_k(x)\}dx \geq \max_{i<j} \int_{R^n} \max\left\{g_i(x), g_j(x)\right\}dx$$

On the other hand,

$$\begin{aligned} \max_{i<j}\left\{\int_{R^n} \max\{g_i(x), g_j(x)\}dx\right\} &= \max_{i<j}\left\{\frac{1}{2}\|g_i, g_j\|_1 + \frac{1}{2}(q_i + q_j)\right\} \\ &\geq \max_{i<j}\left\{\frac{1}{2}\|g_i, g_j\|_1\right\} + \min_{i<j}\left\{\frac{1}{2}(q_i + q_j)\right\} \\ &\geq \max_{i<j}\left\{\frac{1}{2}\|g_i, g_j\|_1\right\} + \min_{i<j}\{(q_1, q_2, \ldots, q_k)\}. \end{aligned}$$

Hence,

$$\int_{R^n} g_{\max}(x)dx \geq \frac{1}{2}\max_{i<j}\left\{\|g_i, g_j\|_1\right\} + \min_{i<j}\{(q_1, q_2, \ldots, q_k)\}. \tag{12}$$

We also have
$$\sum_{i<}\sum_{j}|g_i - g_j| \geq \sum_{j=1}^{k}\left[\max\{g_1, g_2, \cdots g_k\} - g_j\right]$$
$$= k\left[\max\{g_1, g_2, \cdots g_k\}\right] - \sum_{j=1}^{k} g_j$$

Therefore,

$$\max\{g_1, g_2, \cdots g_k\} \leq \frac{1}{k}\sum_{i<}\sum_{j}|g_i - g_j| + \frac{1}{k}\sum_{j=1}^{k} g_j. \tag{13}$$

Since

$\int_{R^n} g_i(x)dx = q_i$ and $\sum_{i=1}^{k} q_i = 1$, the inequality Eq. (13) becomes:

$$\int_{R^n} g_{\max}(x)dx \leq \frac{1}{k}\sum_{i<j}\|g_i, g_j\|_1 + \frac{1}{k}. \tag{14}$$

Replacing $\int_{R^n} g_{\max}(x) = 1 - Pe_{1,2,\ldots,k}^{(q)}$ to Eqs. (12) and (14), we have Eq. (7).

iv. We have

$q_i f_i(x) \leq \max\{q_1 f_1(x), q_2 f_2(x), \ldots, q_k f_k(x)\} \leq \sum_{i=1}^{k} q_i f_i(x)$ for all $i = 1,\ldots,k$.

Integrating the above relation, we obtain:

$$q_i \leq \int_{R^n} g_{\max}(x)dx \leq 1.$$

Above inequality is true for all $i = 1,\ldots,k$, so

$$\max\{q_i\} \leq \int_{R^n} g_{\max}(x)dx \leq 1.$$

Replacing $\int_{R^n} g_{\max}(x) = 1 - Pe_{1,2,\ldots,k}^{(q)}$ in above relation, we have Eq. (8).

From the result of Eqs. (5) and (6), with $\alpha_1 = \alpha_2 = \ldots = \alpha_k = 1/k,$, we have the relationship between Bayes error and affinity of Matusita [11]. Especially, when $k = 2$, we have the relationship between $Pe_{1,2}^{(q,1-q)}$ and Hellinger's distance.

In addition, we also have the relation between Bayes error and overlap coefficients as well as L^1–distance of $\{g_1(x), g_2(x), \ldots, g_k(x)\}$ (see Ref. [22]). For special case: $q_1 = q_2 = \ldots = q_k = 1/k$, we had established expressions about relations between Bayes error and L^1–distance of $\{f_1(x), f_2(x), \ldots, f_k(x)\}$, $Pe_{1,2,\ldots,k}^{(1/k)}$ and $Pe_{1,2,\ldots,k+1}^{(1/(k+1))}$ (see Ref. [17]).

3. Related problems in applying of Bayesian method

To apply Bayesian method in reality, we have to resolve three main problems: (i) Determine prior probability, (ii) compute Bayes error, and (iii) estimate pdfs. In this section, we propose an algorithm to solve for (i) based on fuzzy cluster analysis and classified objective that can reduces Bayes error in comparing with traditional approaches. For (ii), Bayes error is established by closed expression for general case and determine it by an algorithm to find maximum function of $g_i(x)$, $i = 1, 2, \ldots, k$ for one dimension case. The quasi-Monte Carlo method is proposed to compute Bayes error in this section. For (iii), we review the problem to estimate pdfs by kernel function method where the bandwidth parameter and kernel function are specified.

3.1. Prior probability

In the n-dimensions space, given N populations $N^{(0)} = \left\{W_1^{(0)}, W_2^{(0)}, \ldots, W_N^{(0)}\right\}$ with data set $Z = [z_{ij}]_{nxN}$. Let matrix $U = [\mu_{ik}]_{c\times n}$, where μ_{ik} is probability of the kth element belonging to w_i. We have $\mu_{ik} \in [0, 1]$ and satisfies the following conditions:

$$\sum_{i=1}^{c} \mu_{ik} = 1,\ 0 < \sum_{k=1}^{N} \mu_{ik} < N,\ 1 \le i \le c,\ 1 \le k \le N.$$

We call

$$M_{zc} = \left\{ U = [\mu_{ik}]_{cxN} | \mu_{ik} \in [0, 1], \forall i, k; \sum_{i=1}^{c} \mu_{ik} = 1, \forall k; 0 < \sum_{k=1}^{N} \mu_{ik}, \forall i \right\} \tag{15}$$

be fuzzy partitioning space of k populations,

$D_{ikA}^2 = \|z_k - v_i\|_A^2 = (z_k - v_i)^T A (z_k - v_i)$ is the matrix whose element d_{ik}^2 is the square of distance from the object z_k to the ith representative population. This representative is computed by the following formula:

$$v_i = \frac{\sum_{k=1}^{N} (\mu_{ik})^m z_k}{\sum_{k=1}^{N} (\mu_{ik})^m},\ 1 \le i \le c, \tag{16}$$

where $m \in [1,\infty)$ is the fuzziness parameter.

Given the data set Z including c known populations $w_1, w_2, \ldots, w_c$. Assume x_0 is an object that we need to classify. To identify the prior probabilities when classifying x_0, we propose the following prior probability by fuzzy clustering (PPC) algorithm:

Algorithm 1. Determining prior probability by fuzzy clustering (PPC)

Input: The data set $Z = [z_{ij}]_{n\times N}$ of c populations $\{w_1, w_2, \ldots, w_c\}$, x_0, ε, m and the initial partition matrix $U = U^{(0)} = [\mu_{ij}]_{c\times N+1}$, where $\mu_{ij} = 1$ if the jth object belongs to the w_i and $\mu_{ij} = 0$ for the opposite, $i = \overline{1,c}; j = \overline{1,N},\ \mu_{ij} = 1/c$ for $j = N + 1$.

Output: The prior probability $\mu_{i(N+1)}, i = 1, 2, \ldots c$.

Repeat:

Find the representative object of w_i: $v_i = \dfrac{\sum_{k=1}^{N}(\mu_{ik})^m z_k}{\sum_{k=1}^{N}(\mu_{ik})^m},\ 1 \le i \le c$

Compute the matrix $[D_{ik}]_{c\times N+1}$ (the pairwise distance between objects and representative objects).

Update the new partition matrix $U^{(\text{new})}$ by the following principle:

If $D_{ik} > 0$ for all $i = 1, 2, \ldots, c; k = 1, 2, \ldots, N + 1$ then

$$\mu_{ik}^{(\text{new})} = \frac{1}{\sum_{j=1}^{c}(D_{ik}/D_{jk})^{2/(m-1)}},\ i \neq j = 1, 2, \ldots, c$$

Else, $\mu_{ik}^{(\text{new})} = 0$

End;

Compute $S = \|U^{(\text{new})} - U\| = \max_{ik}\left(\left|\mu_{ik}^{(\text{new})} - \mu_{ik}\right|\right)$

$U = U^{(\text{new})}$

Until $S < \varepsilon$;

The prior probability $\mu_{i(N+1)}, i = 1, 2, \ldots c$ (the final column of the matrix U);

In the above algorithm, we have:

i. ε is a really small number and is chosen arbitrarily. The smaller ε is, the more iterations time is taken. In the examples of this chapter, we choose $\varepsilon = 0.001$.

ii. The distance matrix D_{ik} depends on the norm-inducing matrix A. When $A = I$, D_{ik} is the matrix of Euclidean distances. Besides, there are several choices of A, such as diagonal matrix or the inverse of the covariance matrix. In this chapter, we chose the Euclidean distances in the numerical examples and applications.

iii. m is the fuzziness parameter, when $m = 1$, the fuzzy clustering becomes the nonfuzzy clustering. When $m \to \infty$, the partition becomes completely fuzzy $\mu_{ik} = 1/c$. The determining of this parameter, which affects the analysis result, is difficult. Even though Yu et al. [28] proposed two rules to determine the supermom of m for clustering problems, the searching of the specific m was done by meshing method (see [2, 4, 5, 9] for more details). By this process, the best m among several of given values will be chosen. In this chapter, m is also identified by meshing method for the classification problem. The best integer m between 2 and 10 will be used.

At the end of the PPC algorithm, we obtain the prior probabilities of x_0 based on the last column of the partition matrix U $(\mu_{i(N+1)}, i = 1, 2, \ldots c)$. The PPC algorithm helps us determine the prior probabilities via the closeness degree between the classified object and the populations. Each object will receive its suitable prior probabilities.

In this chapter, Bayesian method with prior probabilities calculated by the uniform distribution approach, the ratio of samples approach, the Laplace approach, and the proposed PPC algorithm approach are respectively called BayesU, BayesR, BayesL, and BayesC.

Example 1. Given the studied marks (scale 10 grading system) of 20 students. Among them, nine students have marks that are lower than 5 (w_1: fail the exam) and 11 students have marks that are higher than 5 (w_2: pass the exam). The data are given in **Table 1**.

Assume that we need to classify the ninth object, $x_0 = 4.3$, to one in two populations. Using the PPC algorithm, we have the following final partition matrix:

$$\begin{pmatrix} 0.957 & 0.973 & 0.981 & 0.993 & 1 & 0.997 & 0.997 & 0.830 & 0.321 & 0.290 & 0.158 & 0.1 & 0.1 & 0.01 & 0.009 & 0.037 & 0.045 & 0.054 & 0.062 & 0.724 \\ 0.043 & 0.027 & 0.019 & 0.007 & 0 & 0.003 & 0.003 & 0.170 & 0.679 & 0.710 & 0.842 & 0.9 & 0.9 & 0.99 & 0.991 & 0.963 & 0.955 & 0.946 & 0.938 & 0.276 \end{pmatrix}$$

This matrix shows the prior probabilities when assigning the ninth object to w_1 and w_2 are 0.724 and 0.276, respectively. Meanwhile, the prior probabilities determined by BayesU, BayesR, and BayesL are (0.5; 0.5), (0.421; 0.579), and (0.429; 0.571), respectively.

From the data in **Table 1**, we might estimate the pdfs $f_1(x)$ and $f_2(x)$ and compute the values $q_1f_1(x)$ and $q_2f_2(x)$, where q_1 and q_2 are the calculated prior probabilities. The results of classifying x_0 by four approaches: BayesU, BayesR, BayesL, and BayesC are given in **Table 2**.

Because the actual population of x_0 is w_1, only BayesC gives the true result. The Bayes error of BayesC is also the smallest. Thus, in this example, the proposed method improves the drawback of the traditional method in determining the prior probabilities.

Objects	Marks	Groups	Objects	Marks	Groups
1	0.6	w_1	11	5.6	w_2
2	1.0	w_1	12	6.1	w_2
3	1.2	w_1	13	6.4	w_2
4	1.6	w_1	14	6.4	w_2
5	2.2	w_1	15	7.3	w_2
6	2.4	w_1	16	8.4	w_2
7	2.4	w_1	17	9.2	w_2
8	3.9	w_1	18	9.4	w_2
9	4.3	w_1	19	9.6	w_2
10	5.5	w_2	20	9.8	w_2

Table 1. The studied marks of 20 students and the actual classifications.

Methods	Priors	gmax(x0)	Populations	Bayes errors
BayesU	(0.5; 0.5)	0.0353	2	0.0538
BayesR	(0.421; 0.579)	0.0409	2	0.0558
BayesL	(0.429; 0.571)	0.0403	2	0.0557
BayesC	(0.724; 0.276)	0.0485	1	0.0241

Table 2. The results when classifying the ninth object.

3.2. Determining Bayes error

Theorem 2. Let $f_i(x)$, $i = 1, 2, \ldots, k$, $k \geq 3$ be k pdfs defined on R^n, $n \geq 1$ and let $q_i \in (0;1)$,

$$\begin{cases} R_1^n = \left\{x \in R^n : q_1 f_1(x) > q_j f_j(x), 2 \le j \le k\right\}, \\ R_k^n = \left\{x \in R^n : q_k f_k(x) > q_j f_j(x), 1 \le j \le k\right\}, \\ R_l^n = \left\{x \in R^n : q_i f_i(x) > q_l f_l(x), 1 \le i \le k, 2 \le l \le k-1, i \neq l\right\}. \end{cases} \tag{17}$$

The Bayes error is determined by

$$Pe_{1,2,\ldots,k}^{(q)} = 1 - \int_{R_1^n} q_1 f_1(x)dx - \sum_{l=2}^{k-1} \int_{R_l^n} q_l f_l(x)dx - \int_{R_k^n} q_k f_k(x)dx. \tag{18}$$

Proof:

To obtain Eq. (18), we need to prove two following results:

$$R_i^n \cap R_j^n = \phi, \ (1 \le i \neq j \le k)$$

and $\bigcup_{i=1}^{k} R_i^n = R_1^n \cup \bigcup_{i=2}^{k-1} R_i^n \cup R_k^n = R^n, \quad f_{\max}(x) = f_i(x), \quad \forall x \in R_i^n.$

Let $\overline{A} = R^n \setminus A$, we have

$$\overline{R}_{ij} = \{x \in R^n : q_i f_i(x) \le q_j f_j(x)\}, R_{ij} = \{x \in R^n : q_i f_i(x) > q_j f_j(x)\}, (1 \le i, j \le k).$$

From Eq. (17), we obtain

$$R_1^n = \bigcap_{j=2}^{k} R_{1j}, R_l^n = \bigcap_{i \neq k} \overline{R}_{il}, (2 \le l < k).$$

Therefore,

$$R_1^n \cap R_l^n = (\bigcap_{j=2}^{k} R_{ij}) \cap (\bigcap_{i \neq k} \overline{R}_{il}) \subset R_{il} \cap \overline{R}_{1l} = \phi \Rightarrow R_1^n \cap R_l^n = \phi, \ (2 \le l < k).$$

On the other hand, from antithesis style of D'Morgan, we have

$$\overline{R_1^n \cup R_l^n} = (\bigcup_{j=2}^{n} \overline{R}_{ij}) \cup (\bigcup_{i \neq k} R_{il}) \subset \overline{R}_{il} \cap R_{1l} = \phi \Rightarrow R_1^n \cup R_l^n = R^n, (2 \le l < k).$$

Similarly,

$$R_k^n \cap R_l^n = \phi, (2 \le l < k), R_1^n \cap R_k^n = \phi,$$

so

$$\bigcup_{i=1}^{k} R_i^n = R^n, \cup (\bigcup_{l=2}^{k-1} R_l^n) \cup R_k^n = R_1^n \cup (\bigcup_{l=2}^{k-1} R_l^n) \cup R_k^n$$
$$= (\bigcup_{l=2}^{k-1} R_1^n \cup R_l^n) \cup (\bigcup_{l=2}^{k-1} R_k^n \cup R_l^n) = R^n \cup R^n = R^n \Rightarrow \bigcup_{i=1}^{k} R_i^n = R^n.$$

In addition, from Eq. (17), we can directly find out

$$g_{max}(x) = g_i(x), \forall x \in R_i^n, (1 \le i \le k).$$

For $k = 2$, $q_1 = q_2 = 1/2$, we consider the two following special cases:

i. If $f_1(x)$ and $f_2(x)$ are two one-dimension normal pdfs ($N(\mu_i, \sigma_i), i = 1, 2$), without loss of generality, we suppose that $\mu_1 < \mu_2$ (for $\mu_1 \neq \mu_2$), $\sigma_1 < \sigma_2$ (for $\sigma_1 \neq \sigma_2$), then

$$Pe_{1,2}^{(1/2,1/2)} = \begin{cases} \frac{1}{2}\left[\int_{-\infty}^{x_1} f_2(x)dx + \int_{x_1}^{+\infty} f_1(x)dx\right], & if \sigma_1 = \sigma_2, \\ \frac{1}{2}\left[\int_{-\infty}^{x_2} f_1(x)dx + \int_{x_2}^{x_3} f_2(x)dx + \int_{x_3}^{+\infty} f_1(x)dx\right], & if \sigma_1 < \sigma_2, \end{cases}$$

where

$$x_1 = \frac{\mu_1 + \mu_2}{2}, \; x_2 = \frac{(\mu_1\sigma_2^2 - \mu_2\sigma_1^2) - \sigma_1\sigma_2\sqrt{(\mu_1 - \mu_2)^2 + K}}{\sigma_2^2 - \sigma_1^2},$$

$$x_3 = \frac{(\mu_1\sigma_2^2 - \mu_2\sigma_1^2) + \sigma_1\sigma_2\sqrt{(\mu_1 - \mu_2)^2 + K}}{\sigma_2^2 - \sigma_1^2}, \; K = 2(\sigma_2^2 - \sigma_1^2)\ln\left(\frac{\sigma_2}{\sigma_1}\right) \ge 0.$$

For $\mu_1 = \mu_2 = \mu$, the above result becomes:

$$Pe_{1,2}^{(1/2,1/2)} = \begin{cases} 1, & \text{if} \sigma_1 = \sigma_2, \\ \frac{1}{2}\left[\int_{-\infty}^{x_4} f_1(x)dx + \int_{x_4}^{x_5} f_2(x)dx + \int_{x_5}^{+\infty} f_1(x)dx\right] & \text{if } \sigma_1 < \sigma_2, \end{cases}$$

where $x_4 = \mu - \sigma_1\sigma_2\sqrt{E}$ and $x_5 = \mu + \sigma_1\sigma_2\sqrt{E}$ with $E = \frac{2}{\sigma_2^2-\sigma_1^2}\ln\left(\frac{\sigma_2}{\sigma_1}\right) \geq 0$.

ii. If $f_1(x)$ and $f_2(x)$ are two n-dimension normal pdfs $(N(\mu_i, \Sigma_i), n \geq 2, i = 1, 2)$ then

$$Pe_{1,2}^{(1/2,1/2)} = \frac{1}{2}\left[\int_{R_1} f_2(x)dx + \int_{R_2} f_1(x)dx\right],$$

where

$$R_1^n = \{x : d(x) \leq 0\}, R_2^n = \{x : d(x) > 0\},$$
$$d(x) = \left[\mu_1^T(\Sigma_1)^{-1} - \mu_2^T(\Sigma_2)^{-1}\right]x - \frac{1}{2}x^T\left[(\Sigma_1)^{-1} - (\Sigma_2)^{-1}\right]x - m,$$
$$m = \frac{1}{2}\left[\ln\frac{|\Sigma_1|}{|\Sigma_2|} + \mu_1^T(\Sigma_1)^{-1}\mu_1 - \mu_2^T(\Sigma_2)^{-1}\mu_2\right].$$

In case of $n = 2$, $d(x)$ can be straight lines or parabola or ellipses or hyperbola.

3.3. Maximum function in the classification problem

To classify a new element by the principle (1) and to determine Bayes error by the formula (3), we must find $g_{\max}(x)$. Some authors, such as Pham–Gia et al. [15, 17] and Tai [21, 22], have surveyed relationships between $g_{\max}(x)$ with some related quantities of classification problem. The specific expression for $g_{\max}(x)$ in some special case has been found [18]. However, the general expression for all of cases is a complex problem that has not been still found yet.

Given k pdfs $f_i(x)$ and q_i, $i = 1, 2, \ldots, k$ with $q_1 + q_2 + \ldots + q_k = 1$ and let $g_i(x) = q_i f_i(x)$, $g_{\max}(x) = \max\{g_i(x)\}$. Now, we take interest in determining $g_{\max}(x)$.

(a) For one dimension

In this case, we can find $g_{\max}(x)$ by the following algorithm:

Algorithm 2. Find the $g_{\max}(x)$ function

Input: $g_i(x) = q_i f_i(x)$, where $f_i(x)$ and q_i are the probability density function and the prior probability of w_i, $i = 1, 2, \ldots, k$, respectively.

Output: The $g_{\max}(x)$ function.

Find all roots of the equations $g_i(x) - g_j(x) = 0, i = \overline{1, k-1}, j = \overline{i+1, k}$.
Let B be the set of all roots.
For $x_{lm} \in B$ (the roof of equation $g_l(x) - g_m(x) = 0$) do
    For $p \in \{1,2,\ldots,k\} \setminus \{l,m\}$ do
        If $g_l(x_{lm}) < g_p(x_{lm})$ then $B = B \setminus \{x_{lm}\}$
        End
    End
End
Arrange the elements of B in order from smallest to largest:
$$B = \{x_1, x_2, \ldots, x_h\}, x_1 < x_2 < \ldots < x_h$$

(Determine the function $g_{\max}(x)$ in interval $(-\infty, x_1]$)

For $i = 1$ to k do

If $g_i(x_1 - \varepsilon_1) = \max\{g_1(x_1 - \varepsilon_1), g_2(x_1 - \varepsilon_1), \ldots, g_k(x_1 - \varepsilon_1)\}$ then

$g_{\max}(x) = g_i(x)$, for all $x \in (-\infty, x_1]$

End

End

(Determine the function $g_{\max}(x)$ in interval $(x_j, x_{j+1}]$, $j = \overline{1, h-1}$)

For i =1 to k do

For j =1 to h-1 do

If $g_i(x_j + \varepsilon_2) = \max\{g_1(x_j + \varepsilon_2), g_2(x_j + \varepsilon_2), \ldots, g_k(x_k + \varepsilon_2)\}$ then

$g_{\max}(x) = g_i(x)$, for all $x \in (x_j, x_{j+1}]$

End

End

End

(Determine the function $g_{\max}(x)$ in interval $(h, +\infty)$)

For $i = 1$ to k do

If $g_i(x_h + \varepsilon_3) = \max\{g_1(x_h + \varepsilon_3), g_2(x_h + \varepsilon_3), \ldots, g_k(x_h + \varepsilon_3)\}$ then

$g_{\max}(x) = g_i(x)$, for all $x \in (h, +\infty)$

End

End

In the above algorithm, ε_1, ε_2, ε_3 are the positive constants such that:

$$x_1 + \varepsilon_1 < x_2,\ x_h - \varepsilon_3 > x_{h-1},\ x_i - \varepsilon_2 < x_{i-1} \text{ and } x_i + \varepsilon_2 < x_{i+1}.$$

From this algorithm, we have written a MATLAB code to find the $g_{\max}(x)$. When $g_{\max}(x)$ is determined, we will easily calculate Bayes error by using formula (3), as well as classify a new element by principle (1).

Example 2. Given seven populations having univariate normal pdfs $\{f_1, f_2, \ldots, f_7\}$ with specific parameters as follows (**Figure 1**):

$$\mu_1 = 0.3, \mu_2 = 4.0, \mu_3 = 9.1, \mu_4 = 1.9, \mu_5 = 5.3, \mu_6 = 8, \mu_7 = 4.8,$$
$$\sigma_1 = 1.0, \sigma_2 = 1.3, \sigma_3 = 1.4, \sigma_4 = 1.6, \sigma_5 = 2, \sigma_6 = 1.9, \sigma_7 = 2.3.$$

Using codes written with $q_i = 1/7, g_i(x) = q_i f_i(x),\ i = 1, 2, .., 7,$ we have the results:

$$g_{\max}(x) = \begin{cases} g_1 & \text{if} \quad -1.28 < x \leq 0.99, \\ g_2 & \text{if} \quad 2.58 < x \leq 4.89, \\ g_3 & \text{if} \quad 8.30 < x \leq 12.52, \\ g_4 & \text{if} \quad \{-7.86 < x \leq -1.28\} \cup \{0.99 < x \leq 2.58\}, \\ g_5 & \text{if} \quad 4.89 < x \leq 6.65, \\ g_6 & \text{if} \quad \{6.65 < x \leq 8.30\} \cup \{12.52 < x \leq 23.33\}, \\ g_7 & \text{if} \quad \{x \leq -7.86\} \cup \{x > 23.33\}. \end{cases}$$

(b) For multidimension

In multidimension cases, it should be very complicated to obtain closed expression for $g_{\max}(x)$. The difficulty comes from the various forms of the intersection space curves between the pdfs surfaces. This problem has been interested by many authors in Refs. [17, 18, 21–25]. Pham–Gia

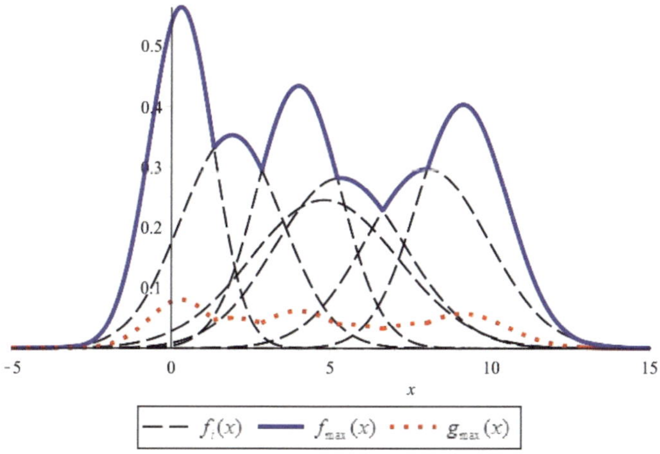

Figure 1. The graph of seven one-dimension normal pdfs, $f_{max}(x)$ and $g_{max}(x)$.

et al. [18] have attempted to find the function $g_{max}(x)$; however, it has been only established for some cases of bivariate normal distribution.

Example 3. Given the four bivariate normal pdfs $N(\mu_i, \Sigma_i)$ with the following specific parameters [16]:

$$\mu_1 = \begin{bmatrix} 40 \\ 20 \end{bmatrix}, \mu_2 = \begin{bmatrix} 48 \\ 24 \end{bmatrix}, \mu_3 = \begin{bmatrix} 43 \\ 32 \end{bmatrix}, \mu_4 = \begin{bmatrix} 38 \\ 28 \end{bmatrix},$$

$$\Sigma_1 = \begin{pmatrix} 35 & 18 \\ 18 & 20 \end{pmatrix}, \Sigma_1 = \begin{pmatrix} 28 & -20 \\ -20 & 25 \end{pmatrix}, \Sigma_1 = \begin{pmatrix} 15 & 25 \\ 25 & 65 \end{pmatrix}, \Sigma_1 = \begin{pmatrix} 5 & -10 \\ -10 & 7 \end{pmatrix}$$

With $q_1 = 0.25$, $q_2 = 0.2$, $q_3 = 0.4$, and $q_4 = 0.15$, we have the graphs of $g_i(x) = q_i f_i(x)$ and their intersection curves as shown in **Figure 2**.

Here, we do not find the expression of $g_{max}(x)$. We compute Bayes error instead by taking integration of $g_{max}(x)$ by quasi-Monte Carlo method [17]. An algorithm for doing calculations has been constructed, and a corresponding MATLAB procedure is used in Section 4.

3.4. Estimate the probability density function

There are many parameter and nonparameter methods to estimate pdfs. In the examples and applications of Section 4, we use the kernel function method, the popular one in practice nowadays. It has the following formula:

$$\widehat{f}(x) = \frac{1}{Nh_1h_2\ldots h_n} \sum_{i=1}^{N} \prod_{j=1}^{n} f_j\left(\frac{x_j - x_{ij}}{h_j}\right), \tag{19}$$

where x_j, $j = 1,2,\ldots,n$ are variables, x_j, $i = 1,2,\ldots,N$ are the ith data of the jth variable, h_j is the bandwidth parameter for the jth variable, $f_j(.)$ is the kernel function of the jth variable which is usually normal, Epanechnikov, biweight, and triweight. According to this method, the choice

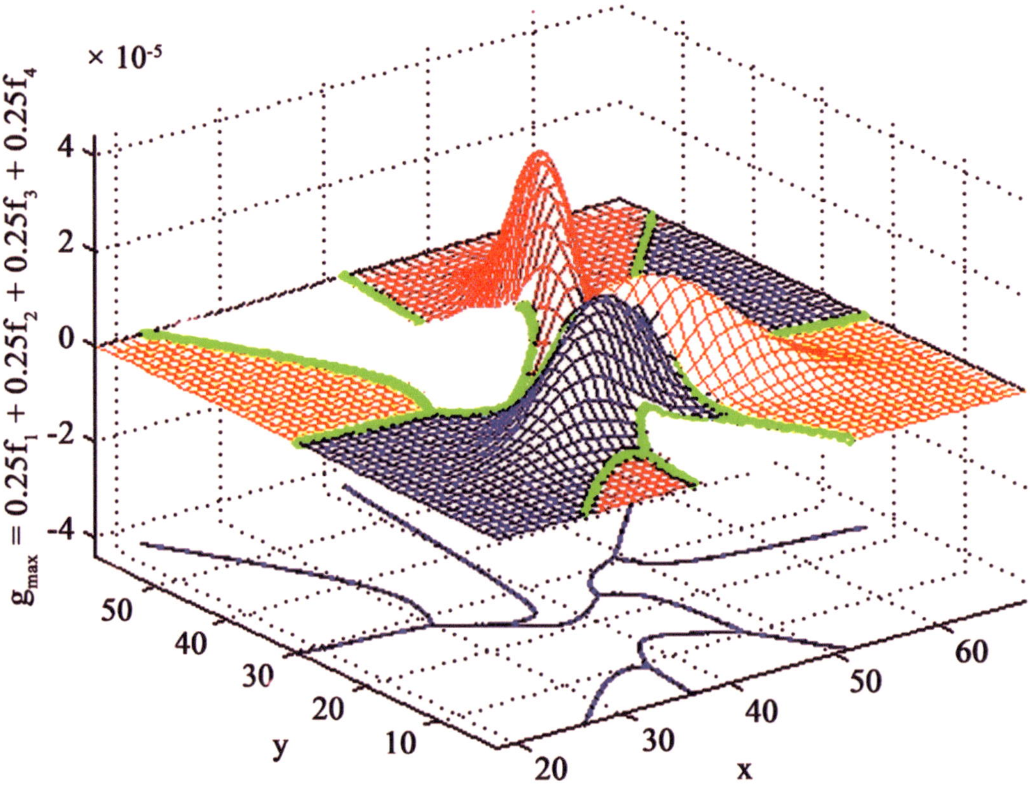

Figure 2. The graph of three bivariate normal pdfs and their $g_{\max}(x)$.

of smoothing parameter and the type of kernel function play an important role and affect the result. Although Silverman [20], Martinez and Martinez [10], and some other authors [7, 13, 27] had discussions about this problem, the optimal choice still has not been found yet. In this chapter, the smoothing parameter is from the idea of Scott [19] and the kernel function is the Gaussian one. We have also written the code by MATLAB software to estimate the pdfs in n-dimensions space using this method.

We have written the complete code for the proposed algorithm by MATLAB software. It is applied effectively for the examples of Section 4.

4. Some applications

In this section, we will consider three applications in three domains: biology, medicine, and economics to illustrate for present theories and to test established algorithms. They also show that the proposed algorithm presents more advantages than the existing ones.

Application 1. We consider classification for well-known Iris flower data, which have been presented in many documents like in Ref. [17]. These data are often used to compare the new method and existing ones in classifying. The three varieties of Iris, namely, Setosa (Se), Versicolor (Ve), and Virginica (Vi), have data in four attributes: $X1$ = sepal length, $X2$ = sepal width, $X3$ = petal length, and $X4$ = petal width.

In this application, the cases of one, two, three and four variables are respectively considered to classify for three groups (Se), (Ve), and (Vi) by Bayesian method with different prior probabilities. The purpose of this performance is to compare the results of BayesC with BayesU, BayesR, and BayesL. Because the numbers of the three groups are equal, and the results of BayesU, BayesR, and BayesL are the same. The correct probability of methods is summarized in **Table 3**.

Table 3 shows that in almost all cases, the results of proposed algorithm are better than those using other algorithms, and in the case using three variables X1, X2, and X3, it gives the best results.

Application 2. This application considers thyroid gland disease (TGD). Thyroid gland is an important and the largest gland in our body. It is responsible for the metabolism and work process of all cells. Some of the common diseases of gland thyroid are hypothyroidism, hyperthyroidism, thyroid nodules, and thyroid cancer. They are dangerous diseases. Recently, the rate of thyroid gland disease has been increasing in some poor countries. Data includes 3772 person with 3541 for ill group (I) and 231 ones for nonill group (NI). Detail for this data is given in http://www.cs.sfu.ca/wangk/ucidata/dataset/thyroid–disease, in which the surveyed variables are Age ($X1$), Query on thyroxin ($X2$), Anti-thyroid medication ($X3$), Sick ($X4$), Pregnant ($X5$), Thyroid surgery ($X6$), Thyroid Stimulating Hormone ($X7$),

Variables	B BayesU = BayesL = BayesR	BayesC
X1	0.667	0.679
X2	0.668	0.579
X3	0.903	0.916
X4	0.815	0.827
X1, X2	0.715	0.807
X1, X3	0.893	0.895
X1, X4	0.807	0.850
X2, X3	0.891	0.898
X2, X4	0.809	0.815
X3, X4	0.843	0.866
X1, X2, X3	0.892	0.919
X1, X2, X4	0.764	0.810
X1, X3, X4	0.762	0.814
X2, X3, X4	0.736	0.822
X1, X2, X3, X4	0.725	0.745

Table 3. The correct probability (%) in classifying Iris flower.

Triiodothyronine (*X*8), Total thyroxin (*X*9), T4U measured (*X*10), and Referral source (*X*11). In this application, this chapter will use random 70% of the data size (2479 elements belong to group I and 162 elements belong to group NI) as the training set to determine significant variables, to estimate pdfs, and to find suitable model. About 30% of the remaining data will be used as test set (1062 elements belong to group I and 69 elements belong to group NI). The result of Bayesian method is also compared to others.

To assess the effect of independent variables in TGD, we build the logistic regression model log ($p/1-p$) with variables Xi, $i = 1, 2, \ldots, 11$ (p is the probability of TGD). The analytical results are summarized in **Table 4**.

In **Table 4**, the three variables *X*1, *X*8, and *X*11 in bold face have statistical significance in classifying the two groups (I) and (NI) at 5% level, so we use them to classify TGD.

Applying the PPC algorithm for cases of one variable, two variables, and three variables with all prior probabilities, we obtain the results given in **Table 5**.

Table 5 shows that the correct probability is high, in which BayesC always gives the best result in all three cases of variables. BayesC gives the almost exact result with three variables. We also compare BayesC with existing methods (Fisher, SWM, and logistic) for all the above three cases. All cases show that BayesC is more advantageous than others in reducing Bayes error.

Variable	Sig.	Variable	Sig.
X1	**0.000**	X7	0.304
X2	0.279	**X8**	**0.000**
X3	0.998	X9	0.995
X4	0.057	X10	0.999
X5	0.997	**X11**	**0.000**
X6	0.997	Const	0.992

Table 4. Value Sigs of logistic regression model.

Cases	Variables	BayesU	BayesR	BayesL	BayesC
One variable	X1	91.13	97.47	97.46	97.97
	X8	90.72	98.51	98.50	98.65
	X11	90.53	97.48	97.47	98.19
Two variables	X1, X8	98.73	98.77	98.77	99.78
	X1, X11	98.11	98.65	97.65	99.44
	X8, X11	98.71	98.77	98.77	99.82
Three variables	X1, X8, X11	98.35	98.89	98.89	99.96

Table 5. The correct probability (%) in classifying TGD by Bayesian method from training set.

Using the best results for each case of methods from **Table 6**, classifying for test set (1131 elements), we have the results given in **Table 7**.

From **Table 7**, we see that with the test set, BayesC also gives the best result.

Application 3. This application considers the problem of repaying bank debt (RBD) by customers. In bank credit operations, determining the repayment ability of customers is really important. If the lending is too easy, the bank may have bad debt problems. In contrast, the bank will miss a good business. Therefore, in the current years, the classification of credit application on assessing the ability to repay bank debt has been specially studied and has been a difficult problem in Vietnam. In this section, we appraise this ability of companies in Can Tho city (CTC), Vietnam by using the proposed approach. We collect a data on 214 enterprises operating in key sectors as agriculture, industry, and commerce, including 143 cases of good debt (G) and 71 cases of bad debt (B). Data are provided by responsible organizations of CTC. Each company is evaluated by 13 independent variables in the expert opinion. The specific variables are given in **Table 8**.

Because of sensitive problem, author has to conceal real data and use training data set. The steps to perform in this application are similar as in Application 2. Training set has 100 elements belonging to group G and 50 elements belonging to group B, and the test set has 43 elements belonging to group G and 21 elements belonging to group B. With training set, the logistic regression model shows only three variables $X1$, $X4$, and $X7$ have statistical significance at 5% level, so we use these three variables to perform BayesU, BayesR, BayesL, and BayesC. Their results are given in **Table 9**.

From **Table 9**, we see that BayesC gives the highest probability in all the cases. We also use logistic method, Fisher, and SVM with training set to find the best results. We have the correct probability given in **Table 10**.

Methods	One variable	Two variables	Three variables
Logistic	93.90	93.90	93.90
Fisher	72.30	73.60	71.70
SVM	93.87	93.87	93.87
BayesC	98.65	99.82	99.96

Table 6. The correct probability (%) for optimal models of methods in classifying TGD.

Methods	Correct numbers	False numbers	Correct probability
Logistic	835	296	73.8
Fisher	835	296	73.8
SVM	1062	69	90.9
BayesC	1062	69	93.9

Table 7. Compare the correct probability (%) in classifying TGD from test set.

Using the best model for each case of methods from **Table 10** to classify the test set (67 elements), we obtain the results given in **Table 11**.

Once again from **Table 11**, we see that with test data, BayesC also gives the best result.

Xi	**Independent variables**	**Detail**
X1	Financial leverage	Total debt/total equity
X2	Reinvestment	Total debt/total equity
X3	Roe	Net profit/equity
X4	Interest	(Net income + depreciation)/total assets
X5	Floating capital	(Current assets − current liabilities)/total assets
X6	Liquidity	(Cash + Short-term investments)/current liabilities
X7	Profits	Net profit/total assets
X8	Ability	Net sales/Total assets
X9	Size	Logarithm of total assets
X10	Experience	Years in business activity
X11	Agriculture	Agricultural and forestry sector
X12	Industry	Industry and construction
X13	Commerce	Trade and services

Table 8. The surveyed independent variables.

Cases variables		**BayesU**	**BayesR**	**BayesL**	**BayesC**
One variable	X1	86.21	86.14	84.13	87.13
	X4	81.12	82.91	86.16	88.19
	X7	83.21	84.63	83.14	84.52
Two variables	X1, X4	87.25	88.72	87.19	89.06
	X1, X7	88.16	88.34	83.26	89.56
	X4, X7	89.25	89.04	89.02	91.34
Three variables	X1, X5, X7	91.15	91.53	90.17	93.18

Table 9. The correct probability (%) in classifying RBD by Bayesian method from training set.

Methods	**One variable**	**Two variables**	**Three variables**
Logistic	84.04	88.29	88.69
Fisher	84.73	80.73	79.32
SWM	82.34	82.03	83.07
BayesC	88.19	91.34	93.18

Table 10. The correct probability (%) for optimal models of methods in classifying RBD.

Methods	Correct numbers	False numbers	Correct probability
Logistic	53	11	82.81
Fisher	52	12	81.25
SVM	53	11	82.81
BayesC	57	7	89.06

Table 11. Compare the correct probability (%) in classifying RBD from test set.

5. Conclusion

This chapter presents the classification algorithm by Bayesian method in both theory and application aspect. We establish the relations of Bayes error with other measures and consider the problem to compute it in real application for one and multidimensions. An algorithm to determine the prior probabilities which may decrease Bayes error is proposed. The researched problems are applied in three different domains: biology, medicine, and economics. They show that the proposed approach has more advantages than existing ones. In addition, a complete procedure on MATLAB software is completed and is effectively used in some real applications. These examples show that our works present potential applications for research on real problems.

Author details

Tai Vovan

Address all correspondence to: vvtai@ctu.edu.vn

College of Natural Sciences, Can Tho University, Can Tho City, Vietnam

References

[1] Altman DG. Statistics in medical journals: Development in 1980s. Statistical Medicine. 1991;**10**:546-551. DOI: 10.1002/sim.4780101206

[2] Bora DJ, Gupta AK. Impact of exponent parameter value for the partition matrix on the performance of fuzzy C means Algorithm. International Journal of Scientific Research in Computer Science Applications and Management Studies. 2014;**3**:1-6. DOI: arXiv:1406.4007

[3] Cristiani S, Shawe TJ. An introduction to support vector machines and other kernel-based learning method. 2nd ed. London: Cambridge University; 2000. p. 204. DOI: 10.1108/k.2001.30.1.103.6

[4] Cannon RL, Dave JV, Bezdek JC. Efficient implementation of the fuzzy c-means clustering algorithms. IEEE Transactions on Pattern Analysis and Machine Intelligence. 1986;**2**: 248-255. DOI: 10.1109/TPAMI.1986.4767778

[5] Fadili MJ, et al. On the number of clusters and the fuzziness index for unsupervised FCA application to BOLD fMRI time series. Medical Image Analysis. 2001;**5**(1):55-67. DOI: 10.1016/S1361-8415(00)00035-9

[6] Fisher RA. The statistical utilization of multiple measurements. Annals of Eugenic. 1936;**7**: 376-386. DOI: 10.1111/j.1469-1809.1938.tb02189

[7] Ghosh AK. Classification using kernel density estimates. Technometrics. 2006;**48**:120-132. DOI: 10.1198/004017005000000391

[8] Jan YK, Cheng CW, Shih YH. Application of logistic regression analysis of home mortgage loan prepayment and default. ICIC Express Letters. 2010;**2**:325-331. DOI: 325-331. 10.12783/ijss.2015.03.014

[9] Hall LO, et al. A comparison of neural network and fuzzy clustering techniques in segmenting magnetic resonance images of the brain. IEEE Transactions on Neural Networks. 1992;**3**(5):672-682. DOI: 10.1109/72.159057

[10] Martinez WL, Martinez AR. Computational statistics handbook with Matlab. 1st ed. Boca Raton: CRC Press; 2007. DOI: 1198/tech.2002.s89

[11] Matusita K. On the notion of affinity of several distributions and some of its applications. Annals of the institute of Statistical Mathematics. 1967;**19**:181-192. DOI: 10.1007/BF02481018

[12] Marta E. Application of Fisher's method to materials that only release water at high temperatures. Portugaliae Etecfochlmlca Acta. 2001;**15**:301-311. DOI: 10.1016/S0167-7152(02)00310-3

[13] McLachlan GJ, Basford KE. Mixture Models: Inference and Applications to Clustering. 1st ed. New York: Marcel Dekker; 1988. DOI: 10.2307/2348072

[14] Miller G, Inkret WC, Little TT. Bayesian prior probability distributions for internal dosimetry. Radiation Protection Dosimetry. 2001;**94**:347-352. DOI: 10.1093/oxfordjournals.rpd.a006509

[15] Pham-Gia T, Turkkan T. Bounds for the Bayes error in classification: A Bayesian approach using discriminant analysis. Statistical Methods and Applications. 2006;**16**:7-26. DOI: 10.1007/s10260-006-0012-x

[16] Pham–Gia T, Turkkan N, Bekker A. Bayesian analysis in the L1–norm of the mixing proportion using discriminant analysis. Metrika. 2008;**64**:1-22. DOI: 10.1007/s00184-006-0027-1

[17] Pham–Gia T, Turkkan N, Tai VV. Statistical discrimination analysis using the maximum function. Communications in Statistics-Simulation and Computation. 2008;**37**:320-336. DOI: 10.1080/03610910701790475

[18] Pham–Gia T, Nhat ND, Phong, NV. Statistical classification using the maximum function. Open Journal of Statistics. 2015;**15**:665-679. DOI: 10.4236/ojs.2015.57068

[19] Scott DW. Multivariate density estimation: Theory, practice, and visualization. 1st ed. New York: Wiley; 1992. DOI: 10.1002/9780470316849

[20] Silverman BW. Density Estimation for Statistics and Data Analysis. 1st ed. Boca Raton: CRC Press; 1986. DOI: 10.1007/978-1-4899-3324-9

[21] Tai VV, Pham–Gia T. Clustering probability distributions. Journal of Applied Statistics. 2010;**37**:1891-1910. DOI: 10.1080/02664760903186049

[22] Tai VV. L^1–distance and classification problem by Bayesian method. Journal of Applied Statistics. 2017; **44**(3):385-401. DOI: 10.1080/02664763.2016.1174194

[23] Tai VV, Thao NT, Ha CN. The prior probability in classifying two populations by Bayesian method. Applied Mathematics Engineering and Reliability. 2016;**6**:35-40. DOI: 10.1201/b21348-7

[24] Thao NT, Tai VV. Fuzzy clustering of probability density functions. Journal of Applied Statistics. 2017;**44**(4):583-601. DOI: 0.1080/02664763.2016.117750

[25] Thao NT, Tai VV. A new approach for determining the prior probabilities in the classification problem by Bayesian method. Advances in data analysis and classification. Forthcoming. DOI: 10.1007/s11634-016-0253

[26] Toussaint GT. Some inequalities between distance: Measures for feature evaluation. IEEE Transactions on Computers. 1972;**21**:405-410. DOI: 10.1109/TC1972.5008990

[27] Webb AR. Statistical Pattern Recognition. 1st ed. New York: Wiley; 2003. DOI: 10.1109/34.824819

[28] Yu J, Cheng Q, Huang H. Analysis of the weighting exponent in the FCM. IEEE Transactions on Systems, Man, and Cybernetics, Part B. 2004;**34**(1):634-639

Chapter 4

Hypothesis Testing for High-Dimensional Problems

Naveen K. Bansal

Additional information is available at the end of the chapter

http://dx.doi.org/10.5772/intechopen.70210

Abstract

For high-dimensional hypothesis problems, new approaches have emerged since the publication. The most promising of them uses Bayesian approach. In this chapter, we review some of the past approaches applicable to only law-dimensional hypotheses testing and contrast it with the modern approaches of high-dimensional hypotheses testing. We review some of the new results based on Bayesian decision theory and show how Bayesian approach can be used to accommodate directional hypotheses testing and skewness in the alternatives. A real example of gene expression data is used to demonstrate a Bayesian decision theoretic approach to directional hypotheses testing with skewed alternatives.

Keywords: multiple directional hypotheses, false discovery rate, familywise error rate, gene expression, skew-normal distribution

1. Introduction

In today's world, most of the statistical inference problems involve high-dimensional multiple hypothesis testing. Whenever we collect data, we collect data on multiple features, involving very high-dimensional variables in some cases. For example, gene expression data consist of gene expressions on thousands of genes; image data consist of image expressions on multiple voxels. The statistical analysis for these types of data involves multiple hypotheses testing (MHT). It is well known that univariate methods cannot be applied to simultaneously test hypotheses on the multiple features. The reason for this is that the error rates for the univariate analysis get multiplied under MHT, and as a result the actual error rate can be very high. To understand the main issue of multiplicity, consider the following example. Suppose there are, say, 100 misspelled words in a book, and each of these words occurs in 5% of the pages. You pick a page at random. For each misspelled word, the probability is certainly 0.05 of finding that word in the page. However, the probability is much higher that you will find at least one of the 100 misspelled words. If these words were independently distributed, then the probability of finding at least one misspelled word is $1-(0.95)^{100}\approx 0.995$. If the placements of the misspelled words were positively dependent, then the probability will be lower than 0.995. For example, if we take an extreme case of dependence that they all occur together, then the probability will be 0.05. The same phenomenon occurs in the MHT. The statistical inference, based on the error rate of individual hypothesis testing, can lead to very high error rate for the combined hypotheses. Thus, for the MHT, adjustment in the error rate needs to be made. Note that the adjustment rate may depend on the dependent structure, but due to the complexity of the dependent structure in high dimension, dependency is usually ignored in the literature [1].

The statistical inference depends on how we define the error rate for the combined hypotheses testing. Let us suppose that there are m hypotheses testing $H_0^i\, vs.\, H_a^i$, $i=1,2,\ldots,m$. If we do not want to make even one false discovery, then we should control the familywise error rate (*FWER*), which is defined as

$$FWER = \Pr\left(\text{Falsely Reject } H_0^i \text{ for at least one } i,\ i=1,2,\ldots,m\right) \tag{1}$$

There are many methods for controlling $FWER \le \alpha_F$ (=0.05, e.g.). A simplest method is the Bonferroni's procedure. Let T_i be the test statistics for testing $H_0^i\, vs.\, H_a^i$ with the corresponding p-values p_i. Then, Bonferroni's procedure rejects H_0^i if $p_i < \alpha_F/m$. To see the proof of this, suppose I_0 be the set of all i for which H_0^i is true, and suppose $p_j < \alpha_F/m$ for at least one $\in I_0$. Then using Boole's inequality, we have, from Eq. (1),

$$FWER = Pr\left\{\bigcup_{i\in I_0} \left(p_i < \alpha_F/m\right)\right\} \le \sum_{i\in I_0} Pr\{p_i < \alpha_F/m\} \tag{2}$$

Now, since, under $H_0^i, p_i \sim U(0,1)$, $\Pr\{p_i < \alpha_F/m\} = \alpha_F/m$. Then, assuming that there are m_0 number of elements in I_0, we have, from Eq. (2),

$$FWER \le \frac{m_0\alpha_F}{m} \le \alpha_F$$

Holm [2] gave a modified version of Bonferroni's procedure which also controls the familywise error rate. Holm's Bonferroni Procedure is the following: First rank all the p-values, $p_{(1)} \le p_{(2)} \le \ldots \le p_{(m)}$, and let $H_0^{(1)}, H_0^{(2)}, \ldots, H_0^{(m)}$ be their associated null hypotheses. Let l be the smallest index such that $p_{(l)} > \alpha_F/(m-l+1)$. Then, reject only those null hypotheses that are associated with $H_0^{(1)}, H_0^{(2)}, \ldots, H_0^{(l-1)}$. Note that the selected hypotheses have p-values with $p_{(1)} < \alpha_F/m, p_{(2)} < \alpha_F/(m-1), \ldots, p_{(l-1)} < \alpha_F/(m-l+2)$, and thus more powerful than Bonferroni's procedure, since hypotheses that are selected under Bonferroni's procedure will also be selected under Holm's procedure.

The above Bonferroni type procedures are not very satisfactory when m is very high. Let us suppose $m = 10,000$ (this is actually not very high for most of the high-dimensional problems), and suppose we want to control FWER by $\alpha_F = 0.05$. Then, for Holm's procedure, the smallest

http://dx.doi.org/10.5772/intechopen.70210

p-value has to be lower than 0.000005 in order to reject at least one hypothesis, which may be very hard to achieve. The problem is not really with Holm's procedure; the problem is with the use of FWER as an error rate. For a high-dimensional problem, it is unrealistic to seek for a procedure which will not make at least one false discovery. Benjamini and Hochberg [1] proposed a new approach called false discovery rate (FDR) and proposed a procedure that works much better for high-dimensional MHT.

In Section 2, we review the FDR procedure and Bayesian procedures for two-sided alternatives. An extension of directional hypotheses is presented in Section 3. In Section 3, we also discuss Bayesian procedures under skewed alternatives. In Section 4, the problem of directional hypotheses is considered by converting p-values to normally distributed test statistics. We also discuss, in Section 4, a Bayes procedure under skew-normal alternatives. An application using real data of gene expressions is also discussed in Section 4. Some concluding remarks are made in Section 5.

2. False discovery rate (FDR), Benjamini and Hochberg's (BH) procedure, and Bayesian procedures

For each of the hypothesis testing $H_0^i\, vs.\, H_a^i$, suppose a statistical procedure either rejects the null hypothesis H_0^i or fails to reject H_0^i. For the sake of simplicity, we equate fail to reject H_0^i as accepting the null H_0^i . However, for small sample size case, it will be unwise to make a conclusion of accepting H_0^i. From now on, rejections of the null will be called discoveries. **Table 1** shows the possible outcomes by a procedure, where, for example, V is the total number of discoveries, among them V_0 is the number of false discoveries.

Thus, the proportion of the false discoveries is $V_0/\max(V,1)$. The FDR is defined as the expected proportion of false discoveries, that is,

$$FDR = E\left[\frac{V_0}{\max(V, 1)}\right]. \tag{3}$$

If, for example, $FDR = 0.05$, then we can expect on the average 5% of all discoveries to be false. In other words, under repeated experiments on the average, we make 5% of the false discoveries (in a frequentist's sense). Note that $FDR \le FWER = P(V_0 \ge 1)$ as the following inequality shows:

	Accept H_0	**Reject H_0**	**Total**
H_0 is true	U_0	V_0	m_0
H_a is true	U_a	V_a	$m-m_0$
	U	V	m

Table 1. Total number of decisions made.

$$FDR = E\left[\frac{V_0}{\max(V,1)}\right] = E\left[\frac{V_0}{\max(V,1)} I(V_0 \geq 1)\right] \leq E[I(V_0 \geq 1)] = P(V_0 \geq 1).$$

Thus, we are likely to make a higher number of discoveries under FDR approach than under FWER, since if a procedure controls FWER ($\leq \alpha$), then it also controls FDR (($\leq \alpha$), but not vice versa.

2.1. Benjamini and Hochberg's procedure

Benjamini and Hochberg [1] proposed the following BH procedure which controls the FDR.

Let p_i be the p-value for the ith hypothesis under a test statistic T_i. Suppose $T_1, T_2, \ldots, T_m$ are independently distributed. Let $p_{[1]} < p_{[2]} < \ldots < p_{[m]}$ be the ordered p-values with the corresponding null hypotheses be denoted by $H_0^{(1)}, H_0^{(2)}, \ldots, H_0^{(m)}$. Let

$$i_0 = \max\left\{i : p_{[i]} \leq \frac{i}{m}\alpha\right\}$$

Then, reject $H_0^{(i)}$ for all $i \leq i_0$.

This procedure controls $FDR \leq \frac{m_0}{m}\alpha \leq \alpha$. Since m_0 is unknown, having the upper bound of $\frac{m_0}{m}\alpha$ is not very useful. If m_0 can be estimated reliably, a better bound is possible.

The above result was proven in [1], under the independence of the test statistics. Hochberg and Yekitieli [3] extended the result to positively correlated test statistics, and they also sharpened the BH procedure with new i_0 defined as

$$i_0 = \max\left\{i : p_{[i]} \leq \frac{1}{mc(m)}\alpha\right\},$$

where $c(m) = \sum_{i=1}^{m} \frac{1}{i}$.

2.2. Bayesian procedures

Under Bayesian setting, we assume that H_0^i and H_a^i, $i = 1, 2, \ldots, m$ are generated probabilistically with

$$P(H_0^i) = p \text{ and } P(H_a^i) = 1 - p$$

Under this setting, [4] developed a concept of local false discovery rate (fdr). If $T_i, i = 1, 2, \ldots, m$ are test statistics with pdf $T_i | H_0 \sim f_0(t)$ and $T_i | H_a \sim f_a(t)$. Then, marginally, $T_i \sim f(t) = pf_0(t) + (1 - p)f_a(t)$, and

$$fdr(t) = P(H_0^i | T_i = t) = \frac{pf_0(t)}{f(t)} \tag{4}$$

The idea is that if $T_i \in [t, t+\delta t]$, where $\delta t \rightarrow 0$, then $fdr(t)$ represents that the proportion of the times H_0^i will be true. If t is very high, then $fdr(t)$ will be very small indicating the probability of H_0^i to be very small (i.e., the false discovery rate will be very small). In Eq. (3), p and $f(t)$ are unknown, which can be estimated (see [4]).

Storey [5] proposed a positive false discovery rate

$$pFDR = E\left[\frac{V_0}{V} | V > 0\right], \tag{5}$$

where expectation is with respect to the distribution of $(T_i, \theta_i), i = 1, 2, \ldots, m$. Under the assumption that $T_1, T_2, \ldots T_m$ are identically and independently distributed, [6] proved that

$$pFDR(\Gamma) = P(H_0 | T \in \Gamma),$$

for a procedure that rejects H_0^i when $T_i \in \Gamma$. Based on this, $q - value$ for the multiple hypothesis (analogous to p-value for a single hypothesis) is defined as the smallest value of $pFDR(\Gamma)$ such that the observed $T_i = t_i \in \Gamma$, see [6]. Under most cases, $q - value(t_i) = P(H_0 | T_i > t_i)$. This gives a procedure under multiple hypothesis that rejects H_0^i if $q - value(t_i) < \alpha$.

3. Directional hypotheses testing

As described earlier, the null hypothesis H_0^i is either accepted or rejected. In most cases, however, rejection of null hypotheses is not sufficient. After rejecting H_0^i, finding the direction of the alternatives may also be important. A detailed discussion of the directional hypotheses can be found in [7].

Direction hypotheses testing involves testing H_0^i against directional hypotheses H_-^i and H_+^i, and the objective is to obtain selection region $\{T_i \in \Gamma_-\}$ for selecting H_-^i and selection region $\{T_i \in \Gamma_+\}$ for selecting H_+^i. In other words, H_0^i will be rejected if $T_i \in \Gamma_-$ or $T_i \in \Gamma_+$, and the direction H_-^i or H_+^i is determined according to $T_i \in \Gamma_-$ or $T_i \in \Gamma_+$, respectively. Analogous to **Table 1**, we now have

Table 2 illustrates the number of cases possible when accepting H_0 or selecting H_- or selecting H_+. For example, out of V times when selecting H_-, V_0 times errors are made when in fact H_0 is

	Accept H_0	Select H_-	Select H_+	Total
H_0 is true	U_0	V_0	W_0	m_0
H_- is true	U_-	V_-	W_-	m_-
H_+ is true	U_+	V_+	W_+	m_+
Total	U	V	W	m

Table 2. Number of decisions under directional hypotheses.

true, and V_+ times errors are made when in fact H_+ is true. In other words, when selecting H_-, not only H_0 is falsely rejected V_0 times but the direction is also falsely selected V_+ times. This leads to a concept of directional false discovery rate *DFDR* defined as

$$DFDR = E\left[\frac{V_0 + V_+ + W_0 + W_-}{\max(V + W,\, 1)}\right]. \tag{6}$$

This is analogous to *FDR* for two-sided alternatives. For most cases, [8] showed that DFDR-controlling procedures for directional hypotheses can be treated as FDR-controlling procedure for two-sided multiple hypotheses with direction determined by the sign of the test statistics.

Bansal and Miescke [9] considered a decision theoretic formulation to multiple hypotheses problems. The approach assumes parametric modeling. Suppose the model for the observed data $\boldsymbol{x}$ be represented by $P(\boldsymbol{x};\, \boldsymbol{\theta}, \eta)$, where $\boldsymbol{\theta} = (\theta_1, \theta_2, \ldots, \theta_m)'$ is a parameter vector of interest, and η is a nuisance parameter. The problem of interest is to test

$$H_0^i : \theta_i = 0\, vs. H_-^i : \theta_i < 0 \; or \; H_+^i : \theta_i > 0 \tag{7}$$

Let the loss function of a decision rule $d(\boldsymbol{x}) = (d_1(\boldsymbol{x}), d_2(\boldsymbol{x}), \ldots, d_m(\boldsymbol{x}))$ is given by

$$L(\boldsymbol{\theta},\, d(\boldsymbol{x})) = \sum_{i=1}^{m} l_i(\boldsymbol{\theta},\, d_i(\boldsymbol{x})), \tag{8}$$

where $l_i(\boldsymbol{\theta}, d_i(\boldsymbol{x}))$ is an individual loss of d_i. Here, $d_i \in \{-1, 0, 1\}$ with $d_i = 0$, $d_i = -1$, and $d_i = 1$ means accepting H_0^i, selecting H_-^i and selecting H_+^i, respectively. Note that for the "0-1" loss, that is, when $l_i = 0$ for correct decision, and $l_i = 1$ for the incorrect decision, L is the total number of incorrect decisions. Thus, minimizing the $E[L(\boldsymbol{\theta}, d(\boldsymbol{X}))]$ for the "0-1" loss amounts to minimizing the expected number of incorrect decisions.

Now, suppose under the Bayesian setting, $\theta_i, i = 1, 2, \ldots, m$ are generated from

$$\pi(\theta) = p_-\pi_-(\theta) + p_0 I(\theta = 0) + p_+\pi_+(\theta), \tag{9}$$

where π_- is the prior density over $(-\infty, 0)$ and π_+ is the prior density over $(0, \infty)$. A special case of prior (9) is that $\pi_-(\theta) = \pi_+(-\theta)$. In this case, p_- and p_+ reflect the skewness in the alternative hypotheses. For example, if $p_- = p_+$, then we have a symmetric case. In this case, the selection of H_- or H_+, after rejecting H_0, based on the sign of the test statistics makes sense. On the other hand, if $p_- < p_+$, then it reflects that more of the θ_is are positives than negatives. For many gene expressions data analyses, this presents a useful case when over-expressed genes may occur more frequently than under-expressed genes as a result of gene mutation (naturally or as a result of external factors). For specific examples, see [9, 10].

From now on, we focus on the "0-1" loss. The results can be easily extended to other loss functions. The "0-1" loss can be written as

$$L(\boldsymbol{\theta}, \boldsymbol{d}) = \sum_{i=1}^{m} \left[1 - \sum_{j=-1}^{1} I(d_i = j) I\left(v_i^{\theta} = j\right) \right],$$

where $v_i^{\theta} \in \{-1, 0, 1\}$ is an indicator variable indicating $\theta_i < 0$ when $v_i^{\theta} = -1$, $\theta_i = 0$ when $v_i^{\theta} = 0$, and $\theta_i > 0$ when $v_i^{\theta} = 1$. It is easy to see that minimizing the posterior expected loss yields the selection rule that selects H_{-}^{i}, H_{0}^{i}, or H_{+}^{i} according to $\max\left\{v_i^{(-)}, v_i^{(0)}, v_i^{(+)}\right\}$, where $v_i^{(-)} = P\left(H_i^{(-)} | \boldsymbol{x}\right)$, $v_i^{(0)} = P\left(H_i^{(0)} | \boldsymbol{x}\right)$, and $v_i^{(+)} = P\left(H_i^{(+)} | \boldsymbol{x}\right)$.

3.1. The constrained Bayes rule

The Bayes procedure described earlier accommodates skewness in the prior, but no type of false discovery rates is controlled. In order to control a false discovery rate, we need to obtain a constrained Bayes rule that minimizes the posterior expected loss subject to a constraint on the false discovery rate.

The directional false discovery rate (6) is defined in a frequentist's manner, in which expectation is with respect to $\boldsymbol{X} | \theta$. Let us define Eq. (6) as *BDFDR* when expectation is taken with respect to $\boldsymbol{X} | \theta$ and then further expectation is taken with respect to $\boldsymbol{\theta}$. We define posterior version of Eq. (6) as *PDFDR* when the expectation is taken with respect to the posterior distribution of $\boldsymbol{\theta} | X = x$. It can be shown that

$$PDFDR = 1 - \frac{\sum_{i=1}^{m} \left\{ I(d_i = -1) v_i^{(-)} + I(d_i = +1) v_i^{(+)} \right\}}{(|D_{-}| + |D_{+}|) \vee 1} \tag{10}$$

Here, $|D_{-}| = \sum_{i=1}^{m} I(d_i = -1)$ and $|D_{+}| = \sum_{i=1}^{m} I(d_i = 1)$.

A constrained Bayes rule can be obtained by minimizing the posterior expected loss subject to the constraint that $PDFDR \leq \alpha$. There can be many approaches to obtain the constraint minimization. We present, here, an approach given in [9], which is as follows:

Consider the sets D_{-}^{B} and D_{+}^{B} of indices that selects $H_i^{(-)}$ and $H_i^{(+)}$, respectively, according to the unconstraint Bayes rule, that is, when $v_i^{(-)} = \max\left\{v_i^{(0)}, v_i^{(+)}\right\}$ and $v_i^{(+)} = \max\left\{v_i^{(0)}, v_i^{(-)}\right\}$, respectively. Define $\xi_i = v_i^{(-)}$ for $i \in D_{-}^{B}$, and $\xi_i = v_i^{(+)}$ for $i \in D_{+}^{B}$, and then rank all ξ_i, $i \in D_{-}^{B} \cup D_{+}^{B}$ from the lowest to the highest. Let the ranked values be denoted by $\xi_{[1]} \leq \xi_{[2]} \leq \ldots \leq \xi_{[\widehat{k}]}$, where $\widehat{k} = \left|D_{-}^{B} \bigcup D_{+}^{B}\right|$. Denote

$$\widehat{i}_0 = \max\left\{ j \leq \widehat{k} : \frac{1}{j} \sum_{i=1}^{j} \xi_{[\hat{k}-i+1]} \geq 1 - \alpha \right\}.$$

Let $\boldsymbol{D}_{\xi}$ denotes the set of indices corresponding to $\boldsymbol{\xi}_{[\widehat{k}]} \geq \boldsymbol{\xi}_{[\widehat{k}-1]} \geq \ldots \geq \boldsymbol{\xi}_{[\widehat{k}-\widehat{i}_0+1]}$. Now, select $\boldsymbol{H}_{-}^{i}$ for $\boldsymbol{i} \in \boldsymbol{D}_{-}^{B} \cap \boldsymbol{D}_{\xi}$, and $\boldsymbol{H}_{+}^{i}$ for $\boldsymbol{i} \in \boldsymbol{D}_{+}^{B} \bigcap \boldsymbol{D}_{\xi}$.

3.2. Estimating mixture parameters

The above procedure requires estimation of the parameters (p_{-}, p_0, p_{+}) and estimation of the nuisance parameter η. Note that marginally,

$$X_i \sim p_{-}f_{-}(x_i|\eta) + p_0 f_0(x_i|\eta) + p_{+}f_{+}(x_i|\eta),$$

where $f_0(x_i|\eta) = f(x_i|0, \eta)$, and

$$f_{-}(x_i|\eta) = \int_{-\infty}^{0} f(x_i|\theta,\eta)\pi_{-}(\theta)d\theta, f_{+}(x_i|\eta) = \int_{0}^{\infty} f(x_i|\theta,\eta)\pi_{+}(\theta)d\theta$$

and $X_1, X_2, \ldots, X_m$ are independently distributed. Estimates of the parameters of the mixed density can be obtained by using EM algorithm. It is easy to see that the EM estimators of (p_{-}, p_0, p_{+}) follows the following iterative scheme:

$$p_{-}^{(j+1)} = \frac{1}{m}\sum_{i=1}^{m} \frac{p_{-}^{(j)}f_{-}(x_i|\eta)}{p_{-}^{(j)}f_{-}(x_i|\eta) + p_0^{(j)}f_0(x_i|\eta) + p_{+}^{(j)}f_{+}(x_i|\eta)},$$

$$p_0^{(j+1)} = \frac{1}{m}\sum_{i=1}^{m} \frac{p_0^{(j)}f_{-}(x_i|\eta)}{p_{-}^{(j)}f_{-}(x_i|\eta) + p_0^{(j)}f_0(x_i|\eta) + p_{+}^{(j)}f_{+}(x_i|\eta)},$$

$$p_{+}^{(j+1)} = \frac{1}{m}\sum_{i=1}^{m} \frac{p_{+}^{(j)}f_{-}(x_i|\eta)}{p_{-}^{(j)}f_{-}(x_i|\eta) + p_0^{(j)}f_0(x_i|\eta) + p_{+}^{(j)}f_{+}(x_i|\eta)}$$

Estimation of η can also be estimated iteratively by using EM algorithm or by different means. See [9] for more details.

4. Bayes rules by converting *p*-values to normally distributed test statistics

Let $T_i, i = 1, 2, \ldots, m$ be independently and identically distributed test statistics. Let $P_i = P(T_i \leq t_i|H_0^i)$ be the corresponding p-values. Note that under H_0^i, $P_i \sim U(0, 1)$. Let $X_i = \Phi^{-1}(P_i)$ be the corresponding z-score. Then, under H_0^i, $X_i \sim N(0,1)$. Efron [11] suggested using $X_i \sim N(0,\sigma^2)$ under H_0^i with σ^2 appropriately estimated. Efron pointed out that, in practice, σ^2 may not be equal to 1 due to possible correlation among multiple components. Under the alternative, we assume that $X_i \sim N(\theta_i,\sigma^2)$, where θ_is are generated with distribution described in Eq. (9). It is true that this is a big leap in making this assumption. In practice, this assumption can be tested, however, and if true, it can lead to very powerful results. [9] assumed that $\pi_{+}(\theta)$ is a truncated normal distribution $N(0, \sigma^2/\omega)$, and $\pi_{-}(\theta) = \pi_{+}(-\theta)$, where ω is some positive constant depending upon how inflated we believe the alternative θ_is are. It can be seen that

http://dx.doi.org/10.5772/intechopen.70210

$$v_i^{(-)} \propto p\_T_-(x_i),\ v_i^{(+)} \propto p_+T_+(x_i),\ \text{and}\ v_i^{(0)} \propto p_0 \tag{11}$$

with the proportionality constant $[p\_T\_(x_i) + p_+T_+(x_i) + p_0\}^{-1}$. Also, $T\_(x_i) = T_+(-x_i)$, and

$$T_+(x_i) = \exp\left\{\frac{x_i^2}{2(1+\omega)\sigma^2}\right\}\Phi\left(\frac{x_i}{\sigma\sqrt{1+\omega}}\right) \tag{12}$$

In order to apply the Bayes procedure as discussed in Section 3, all we need are Eqs. (11) and (12). For computation details, see [9].

4.1. Skew-normal alternatives

In the above discussions, we assumed that θ_is are generated from distribution with pdf (9). [12] considered the case when θ_is are generated from a skew-normal distribution under the alternative hypotheses. The skew-normal distribution was first introduced in [13]. It has an important property that if $(\xi_1,\xi_2) \sim$ Bivariate Norma with mean 0, then the distribution of $\xi_1|\xi_2 > 0 \sim Skew - normal$. Its pdf is given by

$$g_+(\xi_1) = 2\frac{1}{\sigma_1}\phi\left(\frac{\xi_1}{\sigma_1}\right)\Phi\left(\lambda\frac{\xi_1}{\sigma_1}\right),$$

and is denoted by $SN(0,\sigma_1,\lambda)$. Here, λ is a skew parameter. If $\lambda = 0$, then this distribution is $N(0,\sigma_1)$. The implication of this result is the following: suppose within a normal system an outcome follows a normal distribution, but if a correlated factor starts exerting a positive effect, then the outcome variable will start following a skew-normal distribution. For example, consider RNAs experiments and assume that genes are in a normal state. Suppose a gene mutation occurs at a later state and it starts exerting positive effect on the affected genes. In this case, based on the above property of skew-normal distribution, we can assume that the expressions of the affected genes will follow a skew-normal distribution.

Under this formulation, we assume that $\theta_i = 1, 2, \ldots, m$ are generated from

$$\pi_\lambda(\theta_i) = pI(\theta_i = 0) + (1-p)\frac{2}{\sigma_1}\phi\left(\frac{\theta_i}{\sigma_1}\right)\Phi\left(\lambda\frac{\theta_i}{\sigma_1}\right)$$

Now, similar to Eq. (11), it can be seen that

$$v_i^{(-)} \propto (1-p)T_-(x_i),\ v_i^{(+)} \propto (1-p)T_+(x_i),\ v_i^{(0)} \propto p$$

with proportionality constant $[(1-p)(T_+(x_i) + T\_(x_i) + p]^{-1}$, where

$$T_+(x_i) = \frac{2}{\sigma_1}\int_0^\infty \exp\left(\frac{x_i\theta}{\sigma^2}\right)\phi\left(\sqrt{\frac{1}{\sigma_1^2}+\frac{1}{\sigma^2}}\theta\right)\Phi\left(\frac{\lambda\theta}{\sigma_1}\right)d\theta,$$

and

$$T_{-}(x_i) = \frac{2}{\sigma_1}\int_{-\infty}^{0} \exp\left(\frac{x_i\theta}{\sigma^2}\right)\phi\left(\sqrt{\frac{1}{\sigma_1^2}+\frac{1}{\sigma^2}}\theta\right)\Phi\left(\frac{\lambda\theta}{\sigma_1}\right)d\theta.$$

The sets D_{-}^{B} and D_{+}^{B} can be written as

$$D_{-}^{B} = \{i : x_i < -c_1\} \text{ and } D_{+}^{B} = \{i : x_i > c_2\}$$

where $c_1 > 0$ and $c_2 > 0$ are determined as shown in **Figure 1** by considering the point of intersections of $y = p/(1 - p)$ and $y = T_{-}(x)$, and $y = p/(1 - p)$ and $y = T_{+}(x)$, respectively. Note that when $\lambda > 0$, the intersection point Q (as shown in the figure) will be to the left of $x = 0$, and when $\lambda < 0$, Q will be to the right of $x = 0$. Thus, when $\lambda > 0, c_1 > c_2$ and the opposite is true when $\lambda < 0$. When $\lambda = 0, T_{-}(x) = T_{+}(-x)$ and thus $c_1 = c_2$. If $\lambda \to \infty$, $T_{-}(x) \to 0$ and thus D_{B}^{-} is an empty set which is equivalent to a one-tailed test. As discussed in Section 3, the procedure based on Eq. (13) by itself does not control *BDFDR*. However, c_1 and c_2 can be further shrunk so that the resulting procedure achieves $BDFDR \leq \alpha$; see [12] for details.

To illustrate the above procedure, and to compare it with the standard FDR procedure (BY) of [8], which selects the direction based on the sign of the test statistics, we consider a HIV data described in [14]. For detailed analysis, see [12]. Here, we describe the analysis very briefly. The data consist of eight microarrays, four from cells of HIV-infected subjects and four from uninfected subjects, each with expression levels of 7680 genes. For each gene, we obtained a two-sample t-statistic, comparing the infected versus the uninfected subjects, which is then transformed to a z-value, where $z_i = \Phi^{-1}\{F_6(t_i)\}$. Here, $F_6(\cdot)$ denotes the cumulative distribution

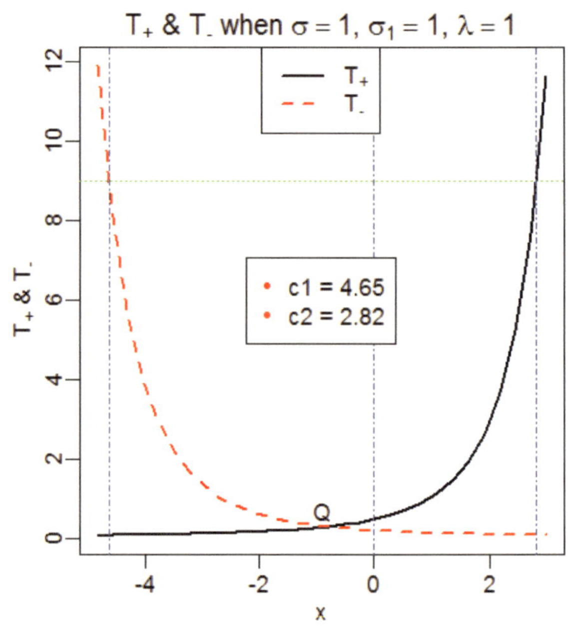

Figure 1. Graph of $T_{+}(x)$ and $T_{-}(x)$ with cutoff values $-c_1$ and c_2 such that $T_{+}(x) \geq \frac{p}{1-p}$ and $T_{-}(x) \geq \frac{p}{1-p}$.

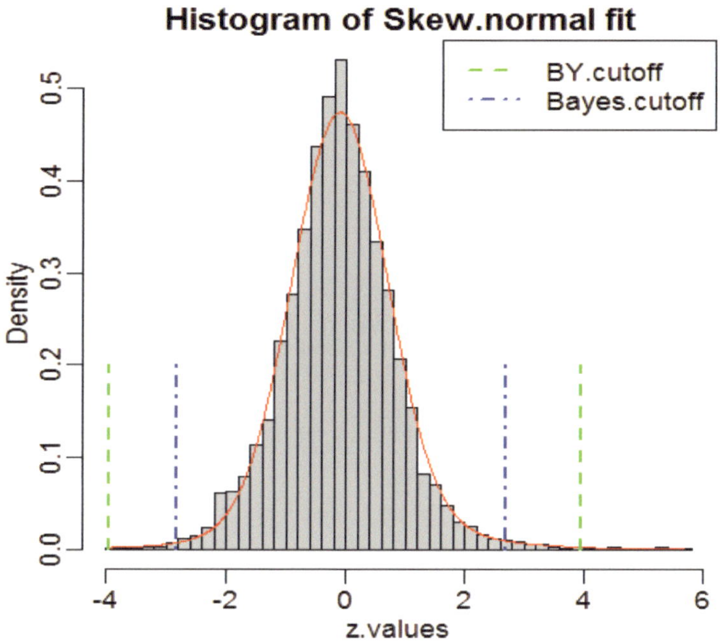

Figure 2. Histogram of the HIV data with cutoff points by BY and the Bayes method under skew-normal prior.

function (cdf) of t -distribution with six degrees of freedom. **Figure 2** shows the histogram of the z-values with a skew-normal fit. Although the null distribution of Z_i should be $N(0,1)$. However, as suggested in [11], we use the null distribution as $N(-0.11, 0.75^2)$. Thus, we formulate our problem as testing hypotheses (7) with test statistics $Z_i \sim N(-0.11 + \theta_i, 0.75^2)$.

BY procedure resulted in cutoffs $(-3.94, 3.94)$, which resulted in 18 total discoveries with two genes declared as under-expressed and 16 as over-expressed. For the constrained Bayes rule, we first used the EM algorithm to obtain the parameter estimates as $\widehat{p} = 0.9$, $\widehat{\sigma} = 0.79$, $\widehat{\sigma_1} = 1.54$, and $\widehat{\lambda} = 0.22$. The Bayes procedure ended up with cut-off points $(-2.82, 2.70)$ with a total of 86 discoveries (under-expressed genes: 23 and over-expressed genes: 63). Note that the number of discoveries by the Bayes rule is much higher than by the BY procedure.

5. Concluding remarks

There are many different methods of testing multiple hypotheses. Methodologies, however, depend on the criteria we choose. When the dimension of multiple hypotheses is not very high, the familywise error rate (FWER) is an appropriate criterion which safeguards against making even one false discovery. However, when the dimension of multiple hypotheses is very high, the FWER is not very useful; instead, a false discover rate (FDR) criterion is a good approach. Although FDR was originally defined as a frequentist's concept, it can be re-interpreted in a Bayesian framework. The Bayesian framework brings many advantages. For example, a decision-theoretic formulation is easy to implement, directional hypotheses are easy to handle, and the skewness in the alternatives is easy to implement. Drawback is that we need to make an assumption about the prior distributions under the alternatives. Some work has been done based on nonparametric priors; however, much more work is needed.

Author details

Naveen K. Bansal

Address all correspondence to: naveen.bansal@mu.edu

Department of Mathematics, Statistics, and Computer Science, Marquette University, Milwaukee, WI, USA

References

[1] Benjamini Y, Hochberg Y. Controlling the false discovery rate: A practice and powerful approach to multiple testing. Journal of the Royal Statistical Society B. 1995;**57**(1):289-300

[2] Holm S. A simple sequentially rejective multiple test procedure. Scandinavian Journal of Statistics. 1979;**6**(2):65-70

[3] Hochberg B, Yekitieli D. The control of the false discovery rate in multiple testing under dependency. Annals of Statistics. 2001;**29**(4):1165-1188

[4] Efron B, Tibshirani R, Storey JD, Tusher V. Empirical Bayes analysis of a microarray experiment. Journal of the American Statistical Association. 2001;**96**(456):1151-1160

[5] Storey JD. A direct approach to false discovery rates. Journal of the Royal Statistical Society B. 2002;**64**(3):479-498

[6] Storey JD. The positive false discovery rate: A Bayesian interpretation and the q value. The Annals of Statistics. 2003;**31**(6):2013-2035

[7] Shaffer JP. Multiplicity, directional (Type III) errors, and the null hypothesis. Psychological Methods. 2002;**7**(3):356-369

[8] Benjamini Y, Yekutieli D. False discovery rate controlling confidence intervals for selected parameters. Journal of American Statistical Association. 2005:71-80

[9] Bansal NK, Miescke KJ. A Bayesian decision theoretic approach to directional multiple hypotheses problems. Journal of Multivariate Analysis. 2013:205-215

[10] Bansal NK, Jiang H, Pradeep P. A Bayesian methodology for detecting targeted genes under two related experiments. Statistics in Medicine. 2015;**34**(25):3362-3375

[11] Efron B. Correlation and large-scale simultaneous significance testing. Journal of the American Statistical Association. 2007:93-103

[12] Bansal NK, Hamedani GG, Maadooliat M. Testing multiple hypotheses with skewed alternatives. Biometrics. 2016;**72**(2):494-502

[13] Azzalini A. A class of distributions which includes the normal ones. Scandinavian Journal of Statistics. 1985;**12**(2):171-178

[14] van't Wout AB, Lehrman GK, Mikheeva SA, O'Keeffe GC, Katze MG, Bumgarner RE, Mullins JI. Cellular gene expression upon human immunodeficiency virus type 1 infection of CD4+-T-cell lines. Journal of Virology. 2003;**77**(2):1392-1402

Chapter 5

Bayesian vs Frequentist Power Functions to Determine the Optimal Sample Size: Testing One Sample Binomial Proportion Using Exact Methods

Valeria Sambucini

Additional information is available at the end of the chapter

http://dx.doi.org/10.5772/intechopen.70168

Abstract

In order to avoid the drawbacks of sample size determination procedures based on classical power analysis, it is possible to define analogous criteria based on 'hybrid classical-Bayesian' or 'fully Bayesian' approaches. We review these conditional and predictive procedures and provide an application, when the focus is on a binomial model and the analysis is performed through exact methods. The distinction between analysis and design prior distributions is essential for the practical implementation of the criteria: some guidelines for choosing these priors are discussed, and their impact on the required sample size is examined.

Keywords: analysis and design prior distributions, binomial proportion, Bayesian power functions, conditional and predictive approach, sample size determination, saw-toothed behaviour of power

1. Introduction

The calculation of an adequate sample size is a crucial aspect in the design of experiments. Researchers need to select the appropriate number of participants required to ensure ethically and scientifically valid results. If samples are too large, time and resources are wasted, often for minimal gain. On the other hand, too small samples may lead to inaccurate results. Therefore, sample size determination (SSD) plays a very important role in the design aspect of studies in many fields, especially in the context of clinical trials where, in addition to economical problems, investigators have to deal with important ethical implications.

Sample size determination (SSD) methods, when the focus is on hypothesis testing, are typically related to the concept of *power function*. Let us denote the parameter of interest by θ and let us assume that we are interested in testing $H_0 : \theta \in \Theta_0$ versus $H_1 : \theta \in \Theta_1$, where Θ_0 and Θ_1 form a partition of the parameter space Θ. The most widely used frequentist SSD criterion consists in choosing the minimal sample size that guarantees a given power, for a fixed type I error rate, under the assumption that θ is equal to a suitable *design value*, $\theta^D \in \Theta_1$. In practice, the idea is to ensure a sufficiently large probability of obtaining a statistically significant result (i.e. of rejecting the null hypothesis), when the true value of θ belongs to the alternative hypothesis and is equal to θ^D. In many textbooks (see [1–3], among others) sample size formulas, derived using this procedure, are provided in many occurring situations, under different hypothesis testing and based on both categorical and quatitative data.

In the frequentist criterion described above, a crucial role is played by the design value that the trial is designed to detect with high probability, whose uncertainty is not accounted for. In fact, the local optimality is one of the most criticized aspects of the method. Moreover, this frequentist procedure does not allow to take into account pre-experimental information about θ, for instance available from previous studies. By adopting a 'hybrid classical-Bayesian approach' or a 'fully Bayesian approach', it is possible to define analogous criteria for sample size selection that allow the researcher to avoid the problem of the local optimality or/and to introduce possible prior information in the SSD process.

In this chapter, we illustrate how to construct frequentist and Bayesian power functions, based on both conditional and predictive approaches, and how to use them to determine the optimal sample size. An essential element of the method is the use of two different prior distributions for the parameter of interest, which play two distinct roles in the criteria. The importance of this distinction in sample size determination problems has been stressed by several authors (see, for instance, [4–9] among others). The rest of the chapter is organized as follows: in Section 2, we review both the frequentist conditional and predictive procedures based on power analysis to determine the optimal sample size. Section 3 provides a description of analogous methods based on Bayesian power functions. Then, in Section 4, we formalize different SSD criteria that depend on the shape of the power curves as a function of the sample size and, as a consequence, on the nature of the data distributions. Furthermore, in Section 5, we illustrate an application of the frequentist and Bayesian SSD procedures, when the parameter of interest is a single binomial proportion. Finally, Section 6 contains a brief final discussion.

2. Frequentist power functions and SSD methods

Let us consider a parameter of interest θ and assume that we are interested in testing $H_0 : \theta \in \Theta_0$ versus $H_1 : \theta \in \Theta_1$, where Θ_0 and Θ_1 form a partition of the parameter space Θ. Moreover, let Y_n be the random result of the experiment that is typically a suitable statistic used to summarize the data relevant to the parameter θ. In the notation, we have highlighted that Y_n depends on the sample size n. Finally, we denote by $f_n(y_n | \theta)$ the sampling distribution of Y_n.

The power function is defined as the probability of obtaining a statistically significant result that leads to reject the null hypothesis H_0, when the actual value of the parameter is θ. In a frequentist approach, the investigator is firstly required to specify a fixed level α for the type

http://dx.doi.org/10.5772/intechopen.70168

I error probability that one is willing to tolerate. This significance level is typically set equal to 0.05 and is used to obtain the rejection region of H_0, denoted by R_{H_0}, that represents an appropriate subset of outcomes that—if observed—lead to the rejection of H_0. Therefore, given a frequentist test of size α, Y_n is considered a statistically significant result if it belongs to R_{H_0}. Consequently, in general terms, the power function is defined as

$$\eta(n,\theta) = \mathbb{P}_\theta(Y_n \in R_{H_0}), \tag{1}$$

where $\mathbb{P}_\theta$ is the probability measure associated with a suitable distribution of Y_n.

In order to exploit the frequentist power function in Eq. (1) for sample size determination purposes, investigators can adopt two different approaches: the conditional and the predictive one. The conditional approach is certainly the most widely known and used, when performing sample size calculations based on pre-study power analysis. It requires the specification of a suitable *design value* for θ, denoted by θ^D, that belongs to the alternative hypothesis and is considered a relevant value important to detect. By assuming that the true value of the parameter is equal to θ^D, we obtain the *frequentist conditional power* given by

$$\eta_F^C(n,\theta^D) = \mathbb{P}_{f_n(\cdot|\theta^D)}(Y_n \in R_{H_0}), \tag{2}$$

where $\mathbb{P}_{f_n(\cdot|\theta^D)}$ is the probability measure associated with the sampling distribution of Y_n when $\theta = \theta^D$. Since θ^D has to be selected within the subspace Θ_1, the conditional frequentist power can be interpreted as the probability of correctly rejecting H_0, when the true value of the parameter belongs to the alternative hypothesis and is exactly equal to θ^D. Then, the sample size determination criterion consists in choosing the minimal sample size that guarantees a desired level for $\eta_F^C(n,\theta^D)$. In practice, the idea is to ensure a sufficiently large probability of rejecting H_0, when the true θ belongs to the alternative hypothesis and, more specifically, it is equal to $\theta^D \in \Theta_1$.

The SSD procedure based on the power function in Eq. (2) is strongly affected by the choice of θ^D. In order to account for uncertainty in the specification of the design value and to avoid local optimality, it is natural to incorporate Bayesian concepts into the sample size determination process. By adopting a 'hybrid classical-Bayesian approach', it is possible to model uncertainty on the appropriate design value for θ through the elicitation of a prior distribution, denoted by $\pi^D(\theta)$ and called *design prior*. This prior is used to compute the marginal or prior predictive distribution of the data by averaging the sampling distribution as follows:

$$m_n^D(y_n) = \int_\Theta f_n(y_n|\theta)\pi^D(\theta)d\theta. \tag{3}$$

Therefore, the design prior cannot be a non-informative improper distribution in order to have $m_n^D(y_n)$ well defined. In any case, the elicitation of a non-informative $\pi^D(\theta)$ would not be reasonable choice. In fact, the design prior is used to introduce uncertainty on the suitable design value for θ that we need to specify when using the SSD procedure previously described and the possible guessed values have to belong to the subspace Θ_1. Thus, $\pi^D(\theta)$ serves to describe a design scenario of interest that supports values of θ under the alternative hypothesis:

it has to be an informative distribution that assigns a negligible probability to values of θ under the null hypothesis.

Once the design prior has been elicited, the idea is to average the conditional frequentist power with respect to it by computing

$$\begin{aligned}\int_{\Theta}\eta_F^C(n,\theta)\pi^D(\theta)d\theta &= \int_{\Theta}\left[\int_{R_{H_0}} f_n(y_n|\theta)dy_n\right]\pi^D(\theta)d\theta \\ &= \int_{R_{H_0}} m_n^D(y_n)dy_n.\end{aligned} \tag{4}$$

This leads to the *frequentist predictive power* that is given by

$$\eta_F^P(n,\pi^D) = \mathbb{P}_{m_n^D(\cdot)}(Y_n \in R_{H_0}), \tag{5}$$

where $\mathbb{P}_{m_n^D(\cdot)}$ is the probability measure associated with the marginal distribution of Y_n obtained using $\pi^D(\theta)$. The power function in Eq. (5) expresses the probability of making a correct decision by rejecting H_0, when θ actually belongs to the subspace defined under the alternative hypothesis, where we can assume that it is distributed according to the design prior. Therefore, the corresponding SSD criterion requires to select the minimum n to achieve a desired level for $\eta_F^P(n,\pi^D)$.

Note that if $\pi^D(\theta)$ is chosen as a point mass distribution centred on θ^D, no uncertainty on the relevant design values is taken into account and the marginal distribution coincides with the sampling one. In this case, there is no difference between the frequentist power functions obtained under the conditional and the predictive approach.

3. Bayesian power functions and SSD methods

In the previous section, we have described how to select the sample size through power functions by assuming that a frequentist analysis will be performed at the end of the study. In both the frequentist conditional and predictive powers, the decision about the two hypotheses is based on the construction of the rejection region of H_0 of a classical test of fixed size α. A major limitation to the fully classical and the hybrid classical-Bayesian approaches previously introduced is the inability to incorporate past experience and information about the unknown parameter, as well as expert prior opinions. The use of a 'fully Bayesian approach' allows to take into account important knowledge and belief about θ when planning the study.

It is well known that the information available before starting the study can be expressed by introducing a prior distribution for θ, $\pi^A(\theta)$, which in this context is typically called *analysis prior* to distinguish it from the design prior. It is worth pointing out that $\pi^A(\theta)$ is the usual prior distribution employed in a Bayesian analysis: it formalizes pre-experimental knowledge, often represented by historical data, and subjective opinions of experts and is used to compute the posterior distribution of the parameter, $\pi_n^A(\theta|y_n) \propto f_n(y_n|\theta)\pi^A(\theta)$. Moreover, it is often chosen as a non-informative distribution to avoid the inclusion of external evidence in the posterior inference.

Let us recall that, in general terms, a power function is defined as the probability of obtaining a significant result, i.e. a result that leads to the rejection of the null hypothesis. Then, to exploit this function as a useful tool to determine the optimal sample size, we need to compute it under the assumption that the alternative hypothesis is true. In practice, we have to consider a design scenario where the true θ belongs to Θ_1, so that the power function represents the probability of making a correct decision. Therefore, to define power functions from a Bayesian point of view, first of all we need to decide when we reject the null hypothesis in a Bayesian setting, that is we have to establish the condition for the 'Bayesian significance'. Following Spiegelhalter et al. [10], we define the result Y_n as 'significant from a Bayesian perspective' if the corresponding posterior probability that θ belongs to the alternative hypothesis is sufficiently large, that is if

$$\mathbb{P}_{\pi_n^A(\cdot|Y_n)}(\theta \in \Theta_1) > \lambda, \tag{6}$$

where $\mathbb{P}_{\pi_n^A(\cdot|Y_n)}$ denotes the probability measure associated with the posterior distribution of θ computed using the analysis prior and $\lambda \in (0, 1)$ represents a suitably specified threshold. Let us stress that, since we are dealing with a pre-experimental problem, the posterior probability in Eq. (6) is a random variable, depending on a random result that has not yet been observed. In order to construct Bayesian power functions, we need to compute the probability of obtaining a Bayesian significant result. Similar to what we have seen in the frequentist case, we can use two alternative distributions of the data, according to the approach we decide to adopt.

The *conditional approach* realizes the pre-experimental assumption that the alternative hypothesis is true, by fixing a design value $\theta^D \in \Theta_1$, which is considered relevant and important to detect. Then the sampling distribution of Y_n conditional on θ^D, $f_n(\cdot|\theta^D)$, is used to compute the probability of getting Bayesian significance. In this way, we obtain the *Bayesian conditional power*

$$\eta_B^C(n, \theta^D) = \mathbb{P}_{f_n(\cdot|\theta^D)}\left(\mathbb{P}_{\pi_n^A(\cdot|Y_n)}(\theta \in \Theta_1) > \lambda\right). \tag{7}$$

The *predictive approach*, instead, aims at avoiding the problem of local optimality in the SSD procedure by introducing a design prior for θ, $\pi^D(\theta)$, that accounts for additional uncertainty involved in the choice of the design values θ^D. Then, the prior predictive distribution of Y_n, $m_n^D(\cdot)$, is computed and used in place of the sampling distribution conditional on θ^D. This leads to the *Bayesian predictive power*

$$\eta_B^P(n, \pi^D) = \mathbb{P}_{m_n^D(\cdot)}\left(\mathbb{P}_{\pi_n^A(\cdot|Y_n)}(\theta \in \Theta_1) > \lambda\right). \tag{8}$$

Both the power functions in Eqs. (7) and (8) express the probability of rejecting H_0 under a Bayesian framework, assuming that the true θ actually belongs to H_1. In fact, we assume that θ is equal to a specific value under the alternative hypothesis (conditional approach) or that θ is in the specific subspace defined under the alternative hypothesis, where we can assume that it is distributed according to the design prior (predictive approach). The sample size determination criteria, therefore, require to select the minimal sample size to ensure a sufficiently large level for $\eta_B^C(n,\theta^D)$ or $\eta_B^P(n,\pi^D)$. Moreover, note that, when the specified design prior distribution assigns the whole mass probability to θ^D, the two Bayesian power functions coincide, leading to the same optimal sample size.

4. SSD criteria according to the nature of the distribution of Y_n

In this section, we explicitly formalize the SSD criteria based on frequentist and Bayesian power functions, according to the nature of the random result Y_n. When Y_n has a continuous distribution, each of the power functions previously introduced shows a monotonically increasing behaviour as a function of n. In this case, the SSD criteria sensibly select the minimum sample size to guarantee the desired level of power, that is

$$n_F^C = \min\{n \in \mathbb{N}: \eta_F^C(n,\theta^D) > \gamma\}, \tag{9}$$

$$n_F^P = \min\{n \in \mathbb{N}: \eta_F^P(n,\pi^D) > \gamma\}, \tag{10}$$

$$n_B^C = \min\{n \in \mathbb{N}: \eta_B^C(n,\theta^D) > \gamma\}, \tag{11}$$

$$n_B^P = \min\{n \in \mathbb{N}: \eta_B^P(n,\pi^D) > \gamma\}, \tag{12}$$

for a conveniently chosen threshold $\gamma \in (0, 1]$. Let us remark that in the notation for the optimal sample sizes, as well as in the notations for the power functions, the subscripts are used to specify the approach (frequentist or Bayesian) adopted at the analysis stage. The superscripts, instead, indicate the appoach (conditional or predictive) used to represent the design expectations. An application of the criteria formalized above is provided by Gubbiotti and De Santis [11], where it is assumed that the statistic Y_n follows a normal distribution with mean equal to θ and known variance.

However, it may happen that $\eta_F^C(n,\theta^D)$, $\eta_F^P(n,\pi^D)$, $\eta_B^C(n,\theta^D)$ and $\eta_B^P(n,\pi^D)$ are not monotonically increasing functions of the sample size: this occurs when dealing with discrete distributions of Y_n. In these cases, the power functions show a basically increasing behaviour as a function of n, but with some small fluctuations. A suitable SSD criterion has to take into account this kind of behaviour. For instance, instead of selecting the smallest sample size that attains the condition of interest, it can be considered more appropriate to select the smallest sample size in such a way that the condition is fulfilled also for all the sample size values greater than it. Given a threshold $\gamma \in (0, 1)$, the corresponding SSD criteria are

$$n_F^C = \min\{n^* \in \mathbb{N}: \eta_F^C(n,\theta^D) > \gamma,\ \forall n \geq n^*\}, \tag{13}$$

$$n_F^P = \min\{n^* \in \mathbb{N}: \eta_F^P(n,\pi^D) > \gamma,\ \forall n \geq n^*\}, \tag{14}$$

$$n_B^C = \min\{n^* \in \mathbb{N}: \eta_B^C(n,\theta^D) > \gamma,\ \forall n \geq n^*\}, \tag{15}$$

$$n_B^P = \min\{n^* \in \mathbb{N}: \eta_B^P(n,\pi^D) > \gamma,\ \forall n \geq n^*\}. \tag{16}$$

http://dx.doi.org/10.5772/intechopen.70168

In this way, it is possible to avoid the paradox of having the condition of interest fulfilled for the selected sample size, but not satisfied for some larger values of n any longer.

5. Single binomial proportion using exact methods

In this section, we focus on exact procedures for one-sample testing problem with binary response. For instance, in a clinical context, we could be interested in evaluating the efficacy of a new experimental treatment or drug that is received at the same dose by all the n patients enrolled in the trial. No comparisons with other therapies are involved. A binary response variable, which assumes value 1 if clinicians classify the patient as a responder to the therapy and 0 otherwise, is considered and, therefore, the parameter of interest θ is the true response rate (i.e. an unknown proportion). In these one-arm studies, θ is compared with a fixed target value, say θ_0, that should ideally represent the response rate for the current 'gold standard' therapy and that is typically obtained through historical data. Values of θ greater than θ_0 suggest that the experimental drug can be considered sufficiently effective and, therefore, the following hypotheses are considered

$$H_0 : \theta = \theta_0 \quad \text{and} \quad H_1 : \theta > \theta_0. \tag{17}$$

This kind of single-arm studies is typically conducted in phase II of clinical trials, whose primary goal is not to definitively assess the efficacy of new drugs, but to screen out those that are ineffective. In practice, in the clinical development process of a new drug, phase II aims at avoiding that not sufficiently promising treatments reach phase III, where randomized controlled trials, based on large patients groups, are generally conducted.

It is important to point out that the power functions based on exact procedures usually do not have explicit forms. Hence, exact formulas for sample size calculations cannot be obtained. However, it is possible to proceed numerically by evaluating the conditions of interest for different increasing or decreasing values of the sample size, until reaching the optimal one. In the following sections, we provide the expressions of the frequentist and Bayesian power functions for non-comparative studies with binary responses. The saw-toothed shape of the power curves as a function of n is shown and, hence, the conservative criteria illustrated in the previous section are adopted. All the graphical and numerical results have been obtained by using the R programming language [12].

5.1. Frequentist conditional power

In the statistical context described above, the number of responders out of the n patients treated with the new drug (i.e. the number of successes in n trials) is the natural statistic Y_n we have to consider and its sampling distribution is

$$f_n(y_n|\theta) = \text{bin}(y_n; n, \theta), \quad \text{for } y_n = 0, \dots, n, \tag{18}$$

where bin($\cdot$; n, θ) denotes the probability mass function of a binomial distribution of parameters n and θ.

Let us consider the two hypotheses in Eq. (17). For a fixed significance level α and assuming that H_0 is true, there exists a non-negative integer r between 0 and n such that

$$\sum_{i=r}^{n} \text{bin}(i; n, \theta_0) \leq \alpha \quad \text{and} \quad \sum_{i=r-1}^{n} \text{bin}(i; n, \theta_0) > \alpha. \tag{19}$$

Then, the rejection region at α level is $R_{H_0} = \{y_n \in \{0, 1, ..., n\} : y_n \geq r\}$, where the critical value r can be expressed in symbols by

$$r = \min\left\{ k \in \{0, 1, ..., n\} : \sum_{i=k}^{n} \text{bin}(i; n, \theta_0) \leq \alpha \right\}. \tag{20}$$

For a given design value θ^D, that has to be specified under the alternative hypothesis, the frequentist conditional power is provided by

$$\begin{aligned} \eta_F^C(n, \theta^D) &= \mathbb{P}_{f_n(\cdot|\theta^D)}(Y_n \in R_{H_0}) \\ &= \sum_{y_n=r}^{n} \text{bin}(y_n; n, \theta^D). \end{aligned} \tag{21}$$

In practice, $\eta_F^C(n, \theta^D)$ is obtained by the sum of the probabilities of the all the outcomes that belong to R_{H_0}, when we assume that the true θ is equal to the design value.

Figure 1 shows the behaviour of the frequentist conditional power as a function of n, when θ_0 = 0.2, θ^D = 0.4 and α = 0.05. It is evident that $\eta_F^C(n, \theta^D)$ is not a monotonically increasing function of the sample size, because of the discrete nature of the sampling distribution of Y_n.

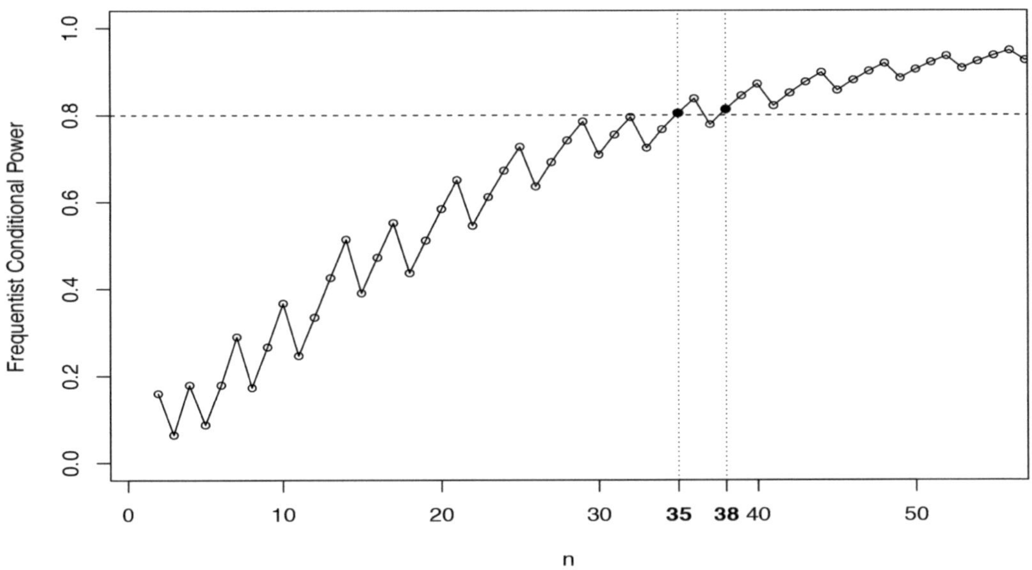

Figure 1. Behaviour of $\eta_F^C(n, \theta^D)$ as a function of n, when θ_0 = 0.20, θ^D = 0.4 and α = 0.05.

http://dx.doi.org/10.5772/intechopen.70168

The reasons for this saw-toothed behaviour can be clarified by the numerical results presented in **Table 1**. Here, for all the possible values of the sample size between 3 and 50, we provide not only the level of the frequentist conditional power used to obtain **Figure 1**, but also the corresponding critical value *r* and the actual value for the type I error probability. Obviously, this latter value is always below the fixed threshold 0.05. Note that whenever the sample size is increased by one unit, the corresponding critical value *r* may also increase or it may remain constant. In the second case, both the actual type I error rate and the conditional frequentist power grow up; otherwise, if also the critical value changes by one unit, they both get smaller. To help in reading the table, the colours white and grey are used alternately to highlight blocks

n	r	$\eta_F^C(n,\theta^D)$	Actual type I error rate	n	r	$\eta_F^C(n,\theta^D)$	Actual type I error rate
3	3	0.0640	0.0080	27	10	0.6913	0.0304
4	3	0.1792	0.0272	28	10	0.7412	0.0391
5	4	0.0870	0.0067	29	10	0.7853	0.0493
6	4	0.1792	0.0170	30	11	0.7085	0.0256
7	4	0.2898	0.0333	31	11	0.7546	0.0327
8	5	0.1737	0.0104	32	11	0.7954	0.0411
9	5	0.2666	0.0196	33	12	0.7242	0.0216
10	5	0.3669	0.0328	34	12	0.7669	0.0274
11	6	0.2465	0.0117	35	12	0.8048	0.0344
12	6	0.3348	0.0194	36	12	0.8380	0.0424
13	6	0.4256	0.0300	37	13	0.7783	0.0231
14	6	0.5141	0.0439	38	13	0.8136	0.0288
15	7	0.3902	0.0181	39	13	0.8446	0.0355
16	7	0.4728	0.0267	40	13	0.8715	0.0432
17	7	0.5522	0.0377	41	14	0.8219	0.0242
18	8	0.4366	0.0163	42	14	0.8509	0.0298
19	8	0.5122	0.0233	43	14	0.8762	0.0362
20	8	0.5841	0.0321	44	14	0.8979	0.0436
21	8	0.6505	0.0431	45	15	0.8570	0.0250
22	9	0.5460	0.0201	46	15	0.8807	0.0304
23	9	0.6116	0.0273	47	15	0.9012	0.0366
24	9	0.6721	0.0362	48	15	0.9187	0.0437
25	9	0.7265	0.0468	49	16	0.8851	0.0256
26	10	0.6358	0.0232	50	16	0.9045	0.0308

Table 1. Numerical calculations related to **Figure 1**: sample sizes, corresponding critical values, frequentist conditional power and actual values for the type I error rate, when $\theta_0 = 0.20$, $\theta^D = 0.4$ and $\alpha = 0.05$.

of sample sizes with the same critical value: within each block both the power and the actual type I rate monotonically raise as n increases. But, in correspondence with the first sample size of the subsequent block, they both decrease. This determines the basically increasing behaviour of the power as a function of n, with some small fluctuations, which is represented in **Figure 1**. For additional discussion about the saw-toothed shape of the frequentist power function, the reader is referred to Chernick and Liu [13].

Now, the problem of which sample size we should select arises because of the non-monotonic behaviour of $\eta_F^C(n, \theta^D)$. If we set the desired threshold γ for the power equal to 0.8, we have that the smallest sample size that meets the power requirement is $n = 35$. At that sample size, the critical value is 12 and the power level is 0.8048. Then for $n = 36$, the critical value is still 12 and the power increases to 0.8380. However, the power drops below 0.8 to 0.7783, when $n = 37$, at which $r = 13$, and rises again over 0.8 when $n = 38$. Then $\eta_F^C(n, \theta^D)$ never decreases below 0.8 for sample sizes greater than 38. Therefore, instead of selecting the smallest n that attains the power condition, it can be more appropriate to consider the more conservative sample size criterion formalized in Section 4, according to which the optimal sample size is selected as

$$n_F^C = \min\{n^* \in \mathbb{N}: \eta_F^C(n, \theta^D) > \gamma, \ \forall n \geq n^*\}. \tag{22}$$

The criterion ensures that the power will not decrease below the desired threshold for any larger sample size: in our specific case, it consists in selecting $n = 38$, instead of $n = 35$.

5.2. Frequentist predictive power

In order to model uncertainty in the specification of the design value, we need to adopt the hybrid classical-Bayesian approach described previously. We introduce a beta design prior density for θ, $\pi^D(\theta) = \text{beta}(\theta; \alpha^D, \beta^D)$, that is used to obtain the prior predictive distribution of the data. It is well known that by averaging the binomial sampling $f_n(y_n | \theta)$ with respect to the beta design prior, we obtain the following marginal distribution

$$m_n^D(y_n) = \text{beta-bin}(y_n; \alpha^D, \beta^D, n), \quad \text{for } y_n = 0, \dots, n, \tag{23}$$

where beta-bin($\cdot$; α^D, β^D, n) denotes the probability mass function of a beta-binomial distribution with parameters (α^D, β^D, n).

The design prior $\pi^D(\theta)$ can be elicited in many different ways. One useful possibility consists in (i) setting the prior mode equal to the fixed design value θ^D, which investigators would choose within the subset under H_1 when using the conditional approach, and (ii) regulating the concentration of the distribution around its mode according to the degree of uncertainty one wishes to express. This can be done by using for the hyperparameters of $\pi^D(\theta)$ the following expressions:

$$\alpha^D = n^D\theta^D + 1 \quad \text{and} \quad \beta^D = n^D(1 - \theta^D) + 1, \tag{24}$$

where θ^D is the prior mode and n^D is a design parameter that can be interpreted as *prior sample size*. The larger the n^D, the smaller the variance of the beta design prior. Therefore, we need to increase n^D if we want to reduce uncertainty on the guessed values of θ. More specifically, if we set $n^D = \infty$, the design prior of θ assigns all the probability mass to θ^D: in this case, no uncertainty is involved and the marginal distribution of the data coincides with the sampling distribution conditional on θ^D. We thus must set $n^D < \infty$ to distinguish between conditional and predictive approaches. In particular, once a prior mode θ^D has been selected, the researcher can choose n^D by assuring a large level (say very close to 1) for $\mathbb{P}_{\pi^D(\cdot)}(\theta > \theta_0)$, that is the probability assigned by $\pi^D(\theta)$ to the event $\theta > \theta_0$. Let us assume, for instance, that $\theta_0 = 0.2$ and consider three possible choices for θ^D (i.e. 0.3, 0.4 and 0.5). For each of them, we compute the smallest n^D such that $\mathbb{P}_{\pi^D(\cdot)}(\theta > \theta_0)$ is about equal to 0.999, and the behaviour of the corresponding design priors is shown in **Figure 2(a)**. Clearly, if the prior mode approaches θ_0, we need to increase n^D to guarantee that $\mathbb{P}_{\pi^D(\cdot)}(\theta > \theta_0) \simeq 0.999$. Moreover, for a fixed prior mode θ^D, if we decided to decrease the value of n^D with respect to the one used in the graph, $\mathbb{P}_{\pi^D(\cdot)}(\theta > \theta_0)$ would decrease. In fact, n^D has been specified in order to express the minimum degree of prior enthusiasm about the efficacy of the treatment necessary to have the prior probability that θ exceeds the target θ_0 at least equal to the chosen level 0.999. An alternative way of proceeding consists in choosing n^D by ensuring a fixed level for the prior probability assigned to a symmetrical interval around the prior mode. For instance, if we set $\theta^D = 0.4$, we can find that 255, 111 and 60 are the values of n^D such that it is about equal to 0.999 the probability that $\pi^D(\theta)$ assigns to the intervals (0.3, 0.5), (0.25, 0.55) and (0.2, 0.6), respectively. The corresponding design prior distributions are shown in **Figure 2(b)**. It is important to point out that all the design densities, represented in both the graphs of **Figure 2**, express uncertainty in the suitable design value that it is worthwhile to consider when applying the SSD criteria based on power analysis. Thus, all the distributions assign a negligible probability to values of θ smaller than θ_0, which are those values specified under H_0.

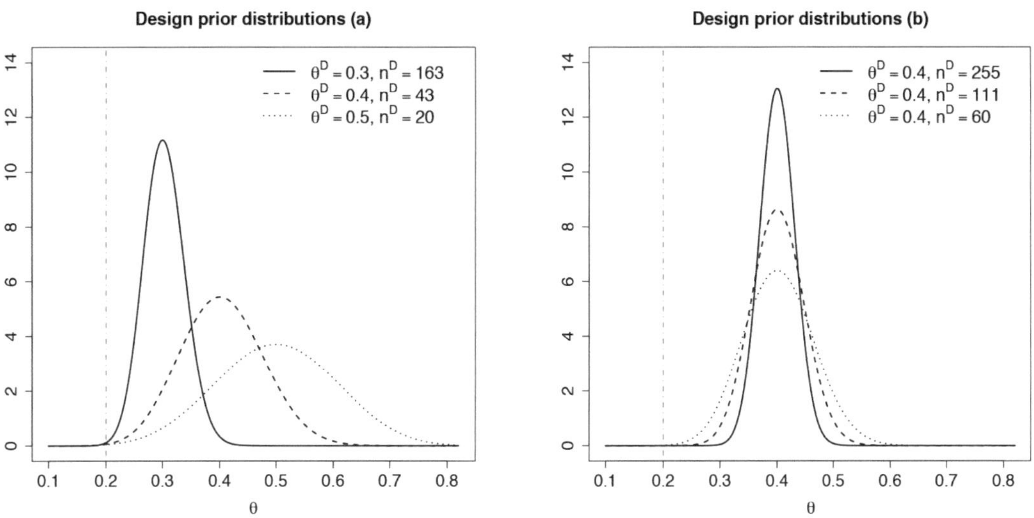

Figure 2. Possible choices of the design prior distribution, when $\theta_0 = 0.2$.

Once $\pi^D(\theta)$ has been specified, the frequentist predictive power can be obtained by computing the probability of rejecting the null hypothesis at α level with respect to $m_n^D(y_n)$. Hence, we have

$$\eta_F^P(n,\pi^D) = \mathbb{P}_{m_n^D(\cdot)}(Y_n \in R_{H_0}) = \sum_{y_n=r}^{n} \text{beta-bin}(y_n;\alpha^D,\beta^D,n), \tag{25}$$

where r is the critical value provided in Eq. (20). In practice $\eta_F^P(n,\pi^D)$ is given by the sum of the probabilities of the all the outcomes inside R_{H_0}, computed under a design scenario according to which the true θ belongs to the interval $(\theta_0, 1)$, where it is distributed according to the design prior density. Let us remark again that if the design prior is a point mass distribution on θ^D (i.e. $n^D = \infty$), we have that the frequentist power functions, conditional and predictive coincide.

Similarly to the frequentist conditional power, also the predictive one presents a saw-toothed shape as a function of n, since $m_n^D(y_n)$ is a discrete distribution. Therefore, we suggest to adopt the conservative approach previously described and to select

$$n_F^P = \min\{n^* \in \mathbb{N} : \eta_F^P(n,\pi^D) > \gamma, \ \forall n \geq n^*\}, \tag{26}$$

for a fixed desired threshold γ. **Figure 3** shows the behaviour of the frequentist predictive power as a function of n for different choices of the design prior, when $\theta_0 = 0.2$ and $\alpha = 0.05$. More specifically, we consider the three $\pi^D(\theta)$ plotted in **Figure 2(b)** that are all centred on $\theta^D = 0.4$, but with different degrees of concentrations regulated by the n^D value. In each graph, we highlight which is the optimal sample size obtained according to the criterion in Eq. (26) when $\gamma = 0.8$. Note that the larger the n^D, the smaller the degree of uncertainty we introduce through the design prior and, as a consequence, the smaller the optimal sample size. In fact, we obtain the optimal values 46, 42 and 39, for n^D equal to 60, 111 and 255, respectively. If we set $n^D = \infty$, we would retrieve the conditional criterion in Eq. (22), where no uncertainty is considered in specifying the design value, and the optimal n would be equal to 38 (see

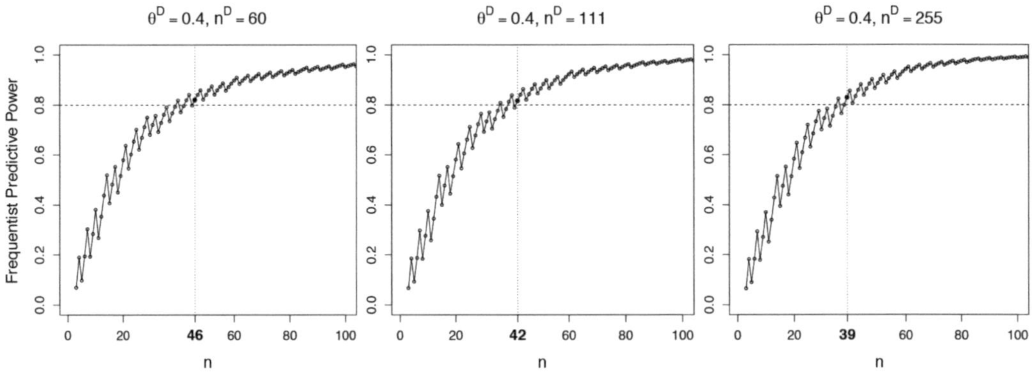

Figure 3. Behaviour of $\eta_F^P(n,\pi^D)$ as a function of n for different choices of the design prior distribution, when $\theta_0 = 0.2$ and $\alpha = 0.05$.

Figure 1). Moreover, let us fix again $\theta_0 = 0.2$, $\alpha = 0.05$ and $\gamma = 0.8$ and consider the three design prior distributions in **Figure 2(a)**, which are characterized by different prior modes. The evident difference between the prior scenarios represented by these design priors clearly affects the optimal sample size: we obtain the optimal values 157, 46 and 23, for (θ^D, n^D) = (0.3, 163), (θ^D, n^D) = (0.4, 43) and (θ^D, n^D) = (0.5, 20), respectively.

5.3. Bayesian conditional power

When we decide to adopt a Bayesian approach to establish the statistical significance of the result, we need to introduce an analysis prior distribution for θ. In our specific case, it is computationally convenient to specify a beta analysis prior, $\pi^A(\theta)$ = beta$(\theta; \alpha^A, \beta^A)$: in this way, from conjugate analysis we obtain that the corresponding posterior distribution is still a beta density with updated parameters,

$$\pi_n^A(\theta|y_n) = \text{beta}(\theta; \alpha^A + y_n, \beta^A + n - y_n). \tag{27}$$

Through $\pi^A(\theta)$, the researcher can incorporate in the SSD procedure pre-experimental knowledge, as well as sceptical or enthusiastic expert prior opinions about the efficacy of the experimental treatment. However, one of the most common ways of proceeding is to choose a non-informative—or based on very weak information–density, to let the posterior distribution be based almost entirely on the evidence in the data. We could, therefore, specify $\pi^A(\theta)$ = beta $(\theta; 1, 1)$ or consider the non-informative Jeffreys prior. Alternatively, if we want to use informative analysis prior distributions, we can express the hyperparameters in terms of the prior mode θ^A and the prior sample size n^A, that is

$$\alpha^A = n^A\theta^A + 1 \quad \text{and} \quad \beta^A = n^A(1 - \theta^A) + 1. \tag{28}$$

In this way, for instance, it is possible to express scepticism or optimism about large treatment effects by setting θ^A less or higher than the target θ_0, respectively. Obviously, when $\theta^A < \theta_0$, the larger the n^A, the larger the degree of scepticism we wish to express; while, when $\theta^A > \theta_0$ larger values of n^A are used to increase the degree of enthusiasm we desire to take into account. However, the value $n^A = 1$ is often used to have a weakly informative prior distribution. The upper panel of **Figure 4** shows three possible choices for the analysis prior when $\theta_0 = 0.2$. These distributions are obtained by fixing the prior mode θ^A and, then, selecting n^A so that $\mathbb{P}_{\pi^A(\cdot)}(\theta > \theta_0)$ (i.e. the probability assigned by $\pi^A(\theta)$ to the event $\theta > \theta_0$) is about equal to a desired level. More specifically, we have considered (i) a sceptical prior mode $\theta^A = 0.1$ and $\mathbb{P}_{\pi^A(\cdot)}(\theta > \theta_0) \simeq 0.4$, (ii) a neutral prior mode $\theta^A = 0.2$ and $\mathbb{P}_{\pi^A(\cdot)}(\theta > \theta_0) \simeq 0.6$ and finally (iii) an enthusiastic prior mode $\theta^A = 0.3$ and $\mathbb{P}_{\pi^A(\cdot)}(\theta > \theta_0) \simeq 0.8$. The corresponding values of n^A are 7, 14 and 4, respectively. These densities will be used to illustrate how the optimal sample sizes based on Bayesian powers are affected by the information formalized through the analysis priors.

The random result Y_n is defined as 'significant' from a Bayesian perspective, if the corresponding posterior probability that $\theta > \theta_0$ is sufficiently large. In symbols, we decide to reject the null hypothesis, on the basis of the result Y_n, if the following condition is satisfied.

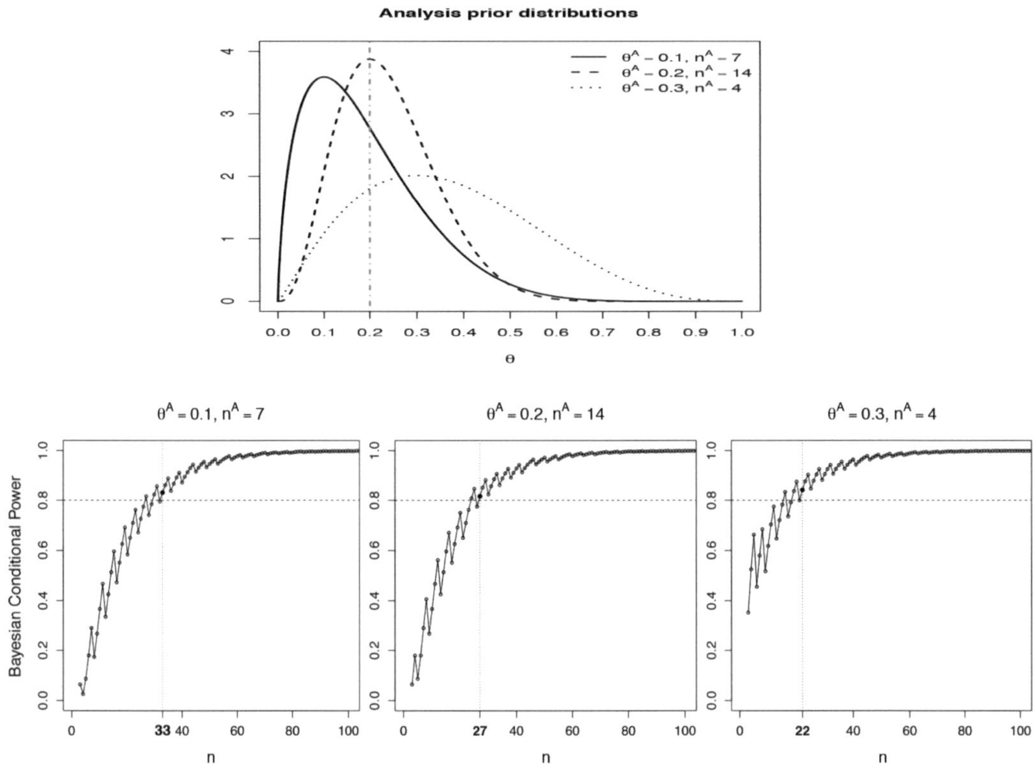

Figure 4. Upper panel: possible choices of the analysis prior distribution, when $\theta_0 = 0.2$. Lower panel: behaviour of $\eta_C^B(n, \theta^D)$ as a function of n for each of the analysis prior distributions represented in the upper panel, when $\theta_0 = 0.2$, $\theta^D = 0.4$ and $\lambda = 0.9$.

$$\mathbb{P}_{\pi_n^A(\cdot|Y_n)}(\theta > \theta_0) > \lambda, \tag{29}$$

where $\mathbb{P}_{\pi^A(\cdot|Y_n)}$ is the probability measure associated with the posterior distribution in Eq. (27) and $\lambda \in (0, 1)$ is a pre-specified threshold. It is worth noting that, for a given value of n, the posterior quantity $\mathbb{P}_{\pi_n^A(\cdot|Y_n)}(\theta > \theta_0)$ is an increasing function of Y_n. As a consequence, we can find a non-negative integer $\tilde{r}$ between 0 and n, such that

$$\mathbb{P}_{\pi_n^A(\cdot|\tilde{r})}(\theta > \theta_0) > \lambda \quad \text{and} \quad \mathbb{P}_{\pi_n^A(\cdot|\tilde{r}-1)}(\theta > \theta_0) \leq \lambda, \tag{30}$$

and we can claim that H_0 is rejected if the observed number of responders y_n is equal to or greater than $\tilde{r}$. In practice, $\tilde{r}$ represents the smallest number of successes such that the condition for the Bayesian significance is satisfied, and in symbols it can be expressed by

$$\tilde{r} = \min\left\{k \in \{0, 1, \ldots, n\} : \mathbb{P}_{\pi_n^A(\cdot|k)}(\theta > \theta_0) > \lambda\right\}. \tag{31}$$

By considering a fixed design value θ^D greater than θ_0, the Bayesian conditional power is therefore obtained as

$$\eta_B^C\left(n,\theta^D\right) = \mathbb{P}_{f_n(\cdot|\theta^D)}\left(\mathbb{P}_{\pi_n^A(\cdot|Y_n)}(\theta > \theta_0) > \lambda\right)$$
$$= \sum_{y_n=\tilde{r}}^{n} \text{bin}(y_n; n, \theta^D). \tag{32}$$

Essentially, it is given by the sum of the probabilities of all the Bayesian significant results, computed assuming that the true θ is equal to θ^D.

Since we are dealing with discrete data, also this power function is not monotonically increasing as a function of n. Let us assume that $\theta_0 = 0.20$, $\theta^D = 0.4$ and $\lambda = 0.9$. The detailed calculations shown in **Table 2** can help to understand why $\eta_B^C\left(n,\theta^D\right)$ has the typical saw-toothed behaviour. For each sample size between 3 and 50, the table provides the corresponding value of $\tilde{r}$, the level of the Bayesian conditional power and the posterior probability that θ exceeds θ_0 conditional on the result $\tilde{r}$. Clearly, these latter values are always larger than the threshold λ that is 0.9. The white and grey colours are used alternately to highlight blocks of sample sizes with the same value of $\tilde{r}$ associated. When the sample size grows, but $\tilde{r}$ remains constant, $\mathbb{P}_{\pi_n^A(\cdot|\tilde{r})}(\theta > \theta_0)$ decreases, while $\eta_B^C\left(n,\theta^D\right)$ increases. However, when both n and $\tilde{r}$ are simultaneously increased by one unit, $\mathbb{P}_{\pi_n^A(\cdot|\tilde{r})}(\theta > \theta_0)$ jumps up, while the Bayesian power drops.

Because of the saw-toothed nature of the power curve, for a fixed threshold γ, the optimal sample size is selected using the conservative criterion, that is

$$n_B^C = \min\left\{n^* \in \mathbb{N}: \eta_B^C\left(n,\theta^D\right) > \gamma, \ \forall n \geq n^*\right\}. \tag{33}$$

The lower panel of **Figure 4** shows the behaviour of the Bayesian conditional power as a function of n for each of the three analysis prior density plotted in the upper panel, when $\theta_0 = 0.2$, $\theta^D = 0.4$ and $\lambda = 0.9$. In each graph, it is indicated the optimal sample size according to the criterion in Eq. (33) for $\gamma = 0.8$. As expected, as we move from sceptical prior opinions towards more enthusiastic beliefs about the efficacy of the experimental treatment, the required sample size decreases.

5.4. Bayesian predictive power

Besides introducing pre-experimental information, if we also wish to model uncertainty on the design value, we have to consider the Bayesian predictive power. Therefore, as described in Section 5.3, we elicit an analysis prior distribution to obtain the beta posterior density $\pi_n^A\left(\theta|y_n\right)$. Moreover, following the indications provided in Section 5.2, we introduce a design prior distribution to construct the marginal distribution $m_n^D\left(y_n\right)$.

The Bayesian predictive power is computed by adding the probabilities of all the Bayesian significant results, computed under the design scenario expressed through the design prior. Thus, we have

$$\eta_B^P\left(n,\pi^D\right) = \mathbb{P}_{m_n^D(\cdot)}\left(\mathbb{P}_{\pi_n^A(\cdot|Y_n)}(\theta > \theta_0) > \lambda\right)$$
$$= \sum_{y_n=\tilde{r}}^{n} \text{beta-bin}\left(y_n; \alpha^D, \beta^D, n\right), \tag{34}$$

n	$\tilde{r}$	$\eta_B^C(n,\theta^D)$	$\mathbb{P}_{\pi_n^A(\cdot\mid\tilde{r})}(\theta > \theta_0)$	n	$\tilde{r}$	$\eta_B^C(n,\theta^D)$	$\mathbb{P}_{\pi_n^A(\cdot\mid\tilde{r})}(\theta > \theta_0)$
3	3	0.0640	0.9263	27	9	0.8161	0.9077
4	4	0.0256	0.9703	28	10	0.7412	0.9464
5	4	0.0870	0.9558	29	10	0.7853	0.9354
6	4	0.1792	0.9377	30	10	0.8237	0.9230
7	4	0.2898	0.9159	31	10	0.8566	0.9092
8	5	0.1737	0.9618	32	11	0.7954	0.9460
9	5	0.2666	0.9476	33	11	0.8310	0.9356
10	5	0.3669	0.9304	34	11	0.8617	0.9239
11	5	0.4672	0.9102	35	11	0.8877	0.9110
12	6	0.3348	0.9559	36	12	0.8380	0.9460
13	6	0.4256	0.9422	37	12	0.8667	0.9362
14	6	0.5141	0.9260	38	12	0.8911	0.9252
15	6	0.5968	0.9075	39	12	0.9118	0.9131
16	7	0.4728	0.9518	40	13	0.8715	0.9464
17	7	0.5522	0.9388	41	13	0.8945	0.9371
18	7	0.6257	0.9237	42	13	0.9140	0.9267
19	7	0.6919	0.9065	43	13	0.9305	0.9153
20	8	0.5841	0.9491	44	13	0.9441	0.9028
21	8	0.6505	0.9367	45	14	0.9164	0.9381
22	8	0.7102	0.9226	46	14	0.9320	0.9284
23	8	0.7627	0.9067	47	14	0.9450	0.9176
24	9	0.6721	0.9474	48	14	0.9558	0.9059
25	9	0.7265	0.9357	49	15	0.9336	0.9394
26	9	0.7745	0.9225	50	15	0.9460	0.9301

Table 2. Numerical calculations to explain the saw-toothed behaviour of $\eta_B^C(n,\theta^D)$ as a function of n: sample sizes, the corresponding value of $\tilde{r}$, the Bayesian conditional power and the posterior probability that $\theta > \theta_0$ when the observed result is equal to $\tilde{r}$ successes, for $\theta_0 = 0.20$, $\theta^D = 0.4$ and $\lambda = 0.9$.

where $\tilde{r}$ is given in Eq. (31). Obviously, also $\eta_B^P(n,\pi^D)$ shows the typical saw-toothed behaviour as a function of n, because of the discrete nature of the beta-binomial marginal distribution of y_n. Therefore, given a desired threshold γ and according to the suitable conservative approach previously used, we select the optimal sample size as

$$n_B^P = \min\{n^* \in \mathbb{N}: \eta_B^P(n,\pi^D) > \gamma, \ \ \forall n \geq n^*\}. \tag{35}$$

http://dx.doi.org/10.5772/intechopen.70168

		$\theta^A = 0.1$	$\theta^A = 0.2$	$\theta^A = 0.3$
θ^D	n^D	$n^A = 7$	$n^A = 14$	$n^A = 4$
(a) Design prior distributions in **Figure 2(a)**				
0.3	163	120	109	94
0.4	43	37	31	22
0.5	20	21	18	11
(b) Design prior distributions in **Figure 2(b)**				
0.4	60	37	31	22
0.4	111	33	31	22
0.4	255	33	27	22

Table 3. n_B^P for different choices of the analysis and the design priors, when $\theta_0 = 0.2$ and $\lambda = 0.9$.

In **Table 3** we provide the values of n_B^P, for different choices of the analysis and the design prior densities. More specifically, we consider the three analysis priors plotted in the upper panel of **Figure 4** and the design prior distributions represented in both the panels of **Figure 2**, when $\theta_0 = 0.2$ and $\lambda = 0.9$. Similarly to what we have seen for the Bayesian conditional power, the sample sizes obtained under the sceptical analysis prior are uniformly larger than those obtained under the more enthusiastic distributions. As regard the impact of the design priors, it is straightforward to see that the stronger the degree of uncertainty on the appropriate design value expressed by $\pi^D(\theta)$, the larger the required sample size. For instance, for a fixed prior mode of the design prior, n_B^P increases as n^D get smaller (see **Table 3(b)**, where $\theta^D = 0.4$). However, let us note that more evident changes in the sample size can be appreciated when we compare the effects of design priors based on different prior modes (see the results in **Table 3(a)**, where the design priors represent very distant design scenarios).

These Bayesian predictive SSD procedures, which include the conditional ones as a special case, have been exploited in Ref. [8] to construct single-arm two-stage design for phase II of clinical trials based on binary data. In Ref. [14], instead, an extension to the randomized case has been presented, while in Ref. [15] the same procedures have been implemented by adding the possibility of taking into account uncertainty in the historical response rate.

6. Conclusions

Especially in clinical research, the pre-experimental power analysis is one of the most commonly used methods for sample size calculations. It is tacitly implied that the power function is constructed under a frequentist framework. However, it is possible to introduce Bayesian concepts in the power analysis to provide more flexibility to the sample size determination process.

When the power function is used as a tool to obtain the appropriate sample size, the general idea is to ensure a large probability of correctly rejecting the null hypothesis H_0, when it is actually false because the true θ belongs to H_1. Therefore, the conjecture that the alternative hypothesis is true represents an essential element of the method. It can be realized by assuming that the true θ is equal to a fixed design value θ^D, suitably selected inside H_1 (conditional approach); alternatively, we can introduce uncertainty on the guessed design value by introducing a design prior distribution that assigns negligible probability to values of θ under H_0 (predictive approach). Moreover, the decision about the rejection of H_0 can be made under a frequentist framework or by performing a Bayesian analysis. In the latter case, it is possible to incorporate in the methodology pre-experimental information possibly available through the specification of an analysis prior distribution. By combining frequentist and Bayesian procedures of analysis, with both the conditional and predictive approaches, we obtain the four power functions described in this chapter. Let us remark that the Bayesian predictive power is the one that allows to add more flexibility to the sample size calculations. At the same time, it let the researcher take into account prior knowledge, as well uncertainty on the design value. However, no design uncertainty can be involved by considering a point-mass design distribution. On the other hand, if no information is available, it is possible to elicit a non-informative analysis prior and let the analysis be based entirely on the data.

Author details

Valeria Sambucini

Address all correspondence to: valeria.sambucini@uniroma1.it

Department of Statistical Sciences, Sapienza Università di Roma, Sapienza, Italy

References

[1] Ryan TP. Sample Size Determination and Power. Haboken: Wiley; 2013

[2] Chow SC, Wang H, Shao J. Sample Size Calculations in Clinical Research. 2nd ed. Boca Raton: Chapman and Hall/CRC; 2008

[3] Julious SA. Sample Sizes for Clinical Trials. Boca Raton: Chapman and Hall/CRC; 2010.

[4] Wang F, Gelfand AE. A simulation-based approach to Bayesian sample size determination for performance under a given model and for separating models. Statistical Science. 2002;**17**(2):193-208. DOI: 10.1214/ss/1030550861

[5] De Santis F. Sample size determination for robust Bayesian analysis. Journal of the American Statistical Association. 2006;**101**(473):278-291. DOI: 10.1198/016214505000000510

[6] Sahu SK, Smith TMF. A Bayesian method of sample size determination with practical applications. Journal of the Royal Statistical Society: Series A. 2006;**169**:235-253. DOI: 10.1111/j.1467-985X.2006.00408.x

[7] Brutti P, De Santis F, Gubbiotti S. Robust Bayesian sample size determination in clinical trials. Statistics in Medicine. 2008;**27**(13):2290-2306. DOI: 10.1002/sim.3175

[8] Sambucini V. A Bayesian predictive two-stage design for phase II clinical trials. Statistics in Medicine. 2008;**27**(8):1199-1224. DOI: 10.1002/sim.3021

[9] Sambucini V. A Bayesian predictive strategy for an adaptive two-stage design in phase II clinical trials. Statistics in Medicine. 2010;**29**(13):1430-1442. DOI: 10.1002/sim.3800

[10] Spiegelhalter DJ, Abrams KR, Myles JP. Bayesian Approaches to Clinical Trials and Health-Care Evaluation. New York: Wiley; 2004

[11] Gubbiotti S, De Santis F. Classical and Bayesian power functions: Their use in clinical trials. Biomedical Statistics and Clinical Epidemiology. 2008;**2**(3):201-211. DOI: 10.1198/016214505000000510

[12] R Core Team. R: A Language and Environment for Statistical Computing, R Foundation for Statistical Computing, Vienna, Austria. 2016. Available from: http://www.R-project.org

[13] Chernick MR, Liu CY. The saw-toothed behavior of power versus sample size and software solutions: Single binomial proportion using exact methods. The American Statistician. 2002;**56**(2):149-155. DOI: 10.1198/000313002317572835

[14] Cellamare M, Sambucini V. A randomized two-stage design for phase II clinical trials based on a Bayesian predictive approach. Statistics in Medicine. 2015;**34**(6):1059-1078. DOI: 10.1002/sim.6396

[15] Matano F, Sambucini V. Accounting for uncertainty in the historical response rate of the standard treatment in single-arm two-stage designs based on Bayesian power functions. Pharmaceutical Statistics. 2016;**15**(6):517-530. DOI: 10.1002/pst.1788

Chapter 6

Converting Graphic Relationships into Conditional Probabilities in Bayesian Network

Loc Nguyen

Additional information is available at the end of the chapter

http://dx.doi.org/10.5772/intechopen.70057

Abstract

Bayesian network (BN) is a powerful mathematical tool for prediction and diagnosis applications. A large Bayesian network can constitute many simple networks, which in turn are constructed from simple graphs. A simple graph consists of one child node and many parent nodes. The strength of each relationship between a child node and a parent node is quantified by a weight and all relationships share the same semantics such as prerequisite, diagnostic, and aggregation. The research focuses on converting graphic relationships into conditional probabilities in order to construct a simple Bayesian network from a graph. Diagnostic relationship is the main research object, in which sufficient diagnostic proposition is proposed for validating diagnostic relationship. Relationship conversion is adhered to logic gates such as AND, OR, and XOR, which are essential features of the research.

Keywords: diagnostic relationship, Bayesian network, transformation coefficient

1. Introduction

Bayesian network (BN) is a directed acyclic graph (DAG) consists of a set of nodes and a set of arcs. Each node is a random variable. Each arc represents a relationship between two nodes. The strength of a relationship in a graph can be quantified by a number called *weight*. There are some important relationships such as prerequisite, diagnostic, and aggregation. The difference between BN and normal graph is that the strength of every relationship in BN is represented by a conditional probability table (CPT) whose entries are conditional probabilities of a child node given parent nodes. There are two main approaches to construct a BN, which are as follows

- The first approach aims to learn BN from training data by learning machine algorithms.
- The second approach is that experts define some graph patterns according to specific relationships and then, BN is constructed based on such patterns along with determined CPTs.

This research focuses on the second approach in which relationships are converted into CPTs. Essentially, relationship conversion aims to determine conditional probabilities based on weights and meanings of relationships. We will have different ways to convert graphic weights into CPTs for different relationships. It is impossible to convert all relationships but some of them such as diagnostic, aggregation, and prerequisite are mandatory ones that we must specify as computable CPTs of BN. Especially, these relationships are adhered to logic X-gates [1] such as AND-gate, OR-gate, and SIGMA-gate. The X-gate inference in this research is derived and inspired from noisy OR-gate described in the book "Learning Bayesian Networks" Neapolitan ([2], pp. 157–159). Díez and Druzdzel [3] also researched OR/MAX, AND/MIN, and noisy XOR inferences but they focused on canonical models, deterministic models, and ICI models whereas I focused on logic gate and graphic relationships. So, their research is different from mine but we share the same result that is AND-gate model. In general, my research focuses on applied probability adhered to Bayesian network, logic gates, and Bayesian user modeling [4]. The scientific results are shared with Millán and Pérez-de-la-Cruz [4].

Factor graph [5] represents factorization of a global function into many partial functions. If joint distribution of BN is considered as the global function and CPTs are considered as partial functions, the sumproduct algorithm [6] of factor graph is applied into calculating posterior probabilities of variables in BN. Pearl's propagation algorithm [7] is very successful in BN inference. The application of factor graph into BN is only realized if all CPT (s) of BN are already determined whereas this research focuses on defining such CPTs firstly. I did not use factor graph for constructing BN. The concept "X-gate inference" only implies how to convert simple graph into BN. However, the arrange sum with a fixed variable mentioned in this research is the "*not-sum*" ([6], p. 499) of factor graph. Essentially, X-gate probability shown in Eq. (10) is as same as λ message in the Pearl's algorithm ([6], p. 518) but I use the most basic way to prove the X-gate probability.

As default, the research is applied in learning context in which BN is used to assess students' knowledge. Evidences are tests, exams, exercises, etc. and hypotheses are learning concepts, knowledge items, etc. Note that diagnostic relationship is very important to Bayesian evaluation in learning context because it is used to evaluate student's mastery of concepts (knowledge items) over entire BN. Now, we start relationship conversion with a research on diagnostic relationship in the next section.

2. Diagnostic relationship

In some opinions like mine, the diagnostic relationship should be from hypothesis to evidence. For example, disease is hypothesis and symptom is evidence. The symptom must be conditionally dependent on disease. Given a symptom, calculating the posterior probability of disease is essentially to diagnose likelihood of such disease ([8], p. 1666). Inversely, the arc from evidence to hypothesis implies prediction where evidence and hypothesis represent observation and event, respectively. Given an observation, calculating the posterior probability of the event is essentially to predict/assert such event ([8], p. 1666). **Figure 1** shows diagnosis and prediction.

The weight w of the relationship between X and D is 1. **Figure 1** depicts simplest graph with two random variables. We need to convert diagnostic relationship into conditional probabilities in order to construct a simplest BN from the simplest graph. Note that hypothesis is binary but evidence can be numerical. In learning context, evidence D can be test, exam, exercise, etc. The conditional probability of D given X (likelihood function) is $P(D|X)$. The posterior probability of X is $P(X|D)$, which is used to evaluate student's mastery over concept (hypothesis) X given evidence D. Eq. (1) specifies CPT of D when D is binary (0 and 1)

$$P(D|X) = \begin{cases} D \text{ if } X = 1 \\ 1 - D \text{ if } X = 0 \end{cases} \tag{1}$$

Eq (1) is our first relationship conversion. It implies

$$P(D|X = 0) + P(D|X = 1) = D + 1 - D = 1$$

Evidence D can be used to diagnose hypothesis X if the so-called *sufficient diagnostic proposition* is satisfied, as seen in **Table 1**.

The concept of sufficient evidence is borrowed from the concept of sufficient statistics and it is inspired from equivalence of variables T and T' in the research ([4], pp. 292-295). The proposition can be restated that evidence D is only used to assess hypotheses if it is sufficient evidence. As a convention, the proposition is called *diagnostic condition* and hypotheses have uniform distribution. The assumption of hypothetic uniform distribution ($P(X = 1) = P(X = 0)$) implies that we cannot assert whether or not given hypothesis is true before we observe its evidence.

In learning context, D can be totally used to assess student's mastery of X if diagnostic condition is satisfied. Derived from such condition, Eq. (2) specifies transformation coefficient k given uniform distribution of X.

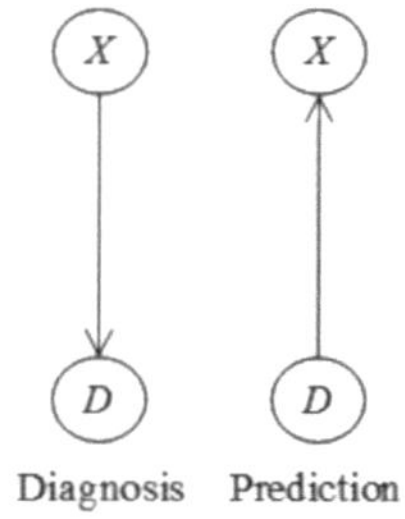

Figure 1. Diagnosis and prediction with hypothesis X and evidence D.

D is equivalent to X in diagnostic relationship if $P(X \mid D) = kP(D \mid X)$ given uniform distribution of X and the *transformation coefficient* k is independent from D. In other words, k is constant with regards to D and so D is called *sufficient evidence*.

Table 1. Sufficient diagnostic proposition.

$$k = \frac{P(X|D)}{P(D|X)} \tag{2}$$

We need to prove that Eq. (1) satisfies diagnostic condition. Suppose the prior probability of X is uniform.

$$P(X = 0) = P(X = 1)$$

we have

$$P(X|D) = \frac{P(D|X)P(X)}{P(D)} = \frac{P(D|X)P(X)}{P(D|X = 0)P(X = 0) + P(D|X = 1)P(X = 1)}$$

$$\text{(due to Bayes'rule)}$$

$$= \frac{P(D|X)P(X)}{P(X)\left(P(D|X = 0) + P(D|X = 1)\right)}$$

$$\left(\text{due to } P(X = 0) = P(X = 1)\right)$$

$$= \frac{P(D|X)}{P(D|X = 0) + P(D|X = 1)} = 1 * P(D|X)$$

$$\left(\text{due to } P(D|X = 0) + P(D|X = 1) = 1\right) \blacksquare$$

It is easy to infer that the transformation coefficient k is 1, if D is binary. In practice, evidence D is often a test whose grade ranges within an interval $\{0, 1, 2, \ldots, \eta\}$. Eq. (3) specifies CPT of D in this case

$$P(D|X) = \begin{cases} \frac{D}{S} \text{if } X = 1 \\ \frac{\eta}{S} - \frac{D}{S} \text{if } X = 0 \end{cases} \tag{3}$$

Where

$$D \in \{0, 1, 2, \ldots, \eta\}$$

$$S = \sum_{D=0}^{n} D = \frac{\eta(\eta + 1)}{2}$$

As a convention, $P(D|X) = 0, \forall D \notin \{0, 1, 2, \ldots, \eta\}$. Eq. (3) implies that if student has mastered concept ($X = 1$), the probability that she/he completes the exercise/test D is proportional to her/his mark on D $\left(P(D|X) = \frac{D}{S}\right)$. We also have

$$P(D|X=0) + P(D|X=1) = \frac{D}{S} + \frac{\eta - D}{S} = \frac{\eta}{S} = \frac{2}{(\eta+1)}$$

$$\sum_{D=0}^{\eta} P(D|X=1) = \sum_{D=0}^{\eta} \frac{D}{S} = \frac{\sum_{D=0}^{\eta} D}{S} = \frac{S}{S} = 1$$

$$\sum_{D=0}^{\eta} P(D|X=0) = \sum_{D=0}^{\eta} \frac{\eta - D}{S} = \frac{\sum_{D=0}^{\eta} (\eta - D)}{S} = \frac{\sum_{D=0}^{\eta} \eta - \sum_{D=0}^{\eta} D}{S} = \frac{\eta(\eta+1) - S}{S} = \frac{2S - S}{S} = 1$$

We need to prove that Eq. (3) satisfies diagnostic condition. Suppose the prior probability of X is uniform.

$$P(X=0) = P(X=1)$$

The assumption of prior uniform distribution of X implies that we do not determine if student has mastered X yet. Similarly, we have

$$P(X|D) = \frac{P(D|X)P(X)}{P(D)} = \frac{P(D|X)}{P(D|X=0) + P(D|X=1)} = \frac{\eta+1}{2} P(D|X) \blacksquare$$

So, the transformation coefficient k is $\frac{\eta+1}{2}$ if D ranges in $\{0, 1, 2, \ldots, \eta\}$.

In the most general case, discrete evidence D ranges within an arbitrary integer interval $\{a, a+1, a+2, \ldots, b\}$. In other words, D is bounded integer variable whose lower bound and upper bound are a and b, respectively. Eq. (4) specifies CPT of D, where $D \in \{a, a+1, a+2, \ldots, b\}$.

$$P(D|X) = \begin{cases} \frac{D}{S} \text{ if } X = 1 \\ \frac{b+a}{S} - \frac{D}{S} \text{ if } X = 0 \end{cases} \tag{4}$$

Where

$$D \in \{a, a+1, a+2, \ldots, b\}$$
$$S = a + (a+1) + (a+2) + \ldots + b = \frac{(b+a)(b-a+1)}{2}$$

Note, $P(D|X) = 0, \forall D \notin \{a, a+1, a+2, \ldots, b\}$. According to the diagnostic condition, we need to prove the equality $P(X|D) = kP(D|X)$, where

$$k = \frac{b-a+1}{2}$$

Similarly, we have

$$P(X|D) = \frac{P(D|X)P(X)}{P(D)} = \frac{P(D|X)}{P(D|X=0)+P(D|X=1)} = \frac{b-a+1}{2}P(D|X) \blacksquare$$

If evidence D is continuous in the real interval $[a, b]$ with note that a and b are real numbers, Eq. (5) specifies probability density function (PDF) of continuous evidence $D \in [a, b]$. The PDF $p(D|X)$ replaces CPT in case of continuous random variable.

$$p(D|X) = \begin{cases} \dfrac{2D}{b^2-a^2} \text{ if } X=1 \\ \dfrac{2}{b-a} - \dfrac{2D}{b^2-a^2} \text{ if } X=0 \end{cases}$$

where

$$D \in [a, b] \text{ where } a \text{ and } b \text{ are real numbers}$$

$$S = \int_a^b D\mathrm{d}D = \frac{b^2-a^2}{2} \tag{5}$$

As a convention, $[a, b]$ is called domain of continuous evidence, which can be replaced by open or half-open intervals such as (a, b), $(a, b]$, and $[a, b)$. Of course we have $p(D|X) = 0$, $\forall D \notin [a, b]$. In learning context, evidence D is often a test whose grade ranges within real interval $[a, b]$.

Functions $p(D|X=1)$ and $p(D|X=0)$ are valid PDFs due to

$$\int_D p(D|X=1)\mathrm{d}D = \int_a^b \frac{2D}{b^2-a^2}\mathrm{d}D = \frac{1}{b^2-a^2}\int_a^b 2D\mathrm{d}D = 1$$

$$\int_D p(D|X=0)\mathrm{d}D = \frac{2}{b-a}\int_a^b \mathrm{d}D - \frac{1}{b^2-a^2}\int_a^b 2D\mathrm{d}D = 1.$$

According to the diagnostic condition, we need to prove the equality

$$P(X|D) = kp(D|X)$$

where,

$$k = \frac{b-a}{2}$$

When D is continuous, its probability is calculated in ε-vicinity where ε is very small number. As usual, ε is bias if D is measure values produced from equipment. The probability of D given X, where $D + \varepsilon \in [a, b]$ and $D - \varepsilon \in [a, b]$ is

$$P(D|X) = \int_{D-\varepsilon}^{D+\varepsilon} p(D|X)\mathrm{d}D = \begin{cases} \displaystyle\int_{D-\varepsilon}^{D+\varepsilon} \frac{2D}{b^2 - a^2}\mathrm{d}D \text{ if } X = 1 \\ \displaystyle\int_{D-\varepsilon}^{D+\varepsilon} \left(\frac{2}{b-a} - \frac{2D}{b^2 - a^2}\right)\mathrm{d}D \text{ if } X = 0 \end{cases}$$

$$= \begin{cases} \dfrac{4\varepsilon D}{b^2 - a^2} \text{ if } X = 1 \\ \dfrac{4\varepsilon}{b-a} - \dfrac{4\varepsilon D}{b^2 - a^2} \text{ if } X = 0 \end{cases} = 2\varepsilon p(D|X)$$

In fact, we have

$$P(X|D) = \frac{P(D|X)P(X)}{P(D|X=0)P(X=0) + P(D|X=1)P(X=1)} = \frac{P(D|X)}{P(D|X=0) + P(D|X=1)}$$

$$\left(\text{due to Bayes}' \text{rule and the assumption } P(X=0) = P(X=1)\right)$$

$$= \frac{b-a}{4\varepsilon} P(D|X) = kp(D|X) \ \blacksquare$$

In general, Eq. (6) summarizes CPT of evidence of single diagnostic relationship.

$$P(D|X) = \begin{cases} \dfrac{D}{S} \text{ if } X = 1 \\ \dfrac{M}{S} - \dfrac{D}{S} \text{ if } X = 0 \end{cases}$$

$$k = \frac{N}{2}$$

Where,

$$N = \begin{cases} 2 \text{ if } D \in \{0, 1\} \\ \eta + 1 \text{ if } D \in \{0, 1, 2, \ldots, \eta\} \\ b - a + 1 \text{ if } D \in \{a, a+1, a+2, \ldots, b\} \\ b - a \text{ if } D \text{ continuous and } D \in [a, b] \end{cases}$$

$$M = \begin{cases} 1 \text{ if } D \in \{0, 1\} \\ \eta \text{ if } D \in \{0, 1, 2, \ldots, \eta\} \\ b + a \text{ if } D \in \{a, a+1, a+2, \ldots, b\} \\ b + a \text{ if } D \text{ continuous and } D \in [a, b] \end{cases}$$

$$S = \sum_D D = \frac{NM}{2} = \begin{cases} 1 \text{ if } D \in \{0, 1\} \\ \dfrac{\eta(\eta+1)}{2} \text{ if } D \in \{0, 1, 2, \ldots, \eta\} \\ \dfrac{(b+a)(b-a+1)}{2} \text{ if } D \in \{a, a+1, a+2, \ldots, b\} \\ \dfrac{b^2 - a^2}{2} \text{ if } D \text{ continuous and } D \in [a, b] \end{cases} \quad (6)$$

In general, if the conditional probability $P(D|X)$ is specified by Eq. (6), the diagnostic condition will be satisfied. Note that the CPT $P(D|X)$ is the PDF $p(D|X)$ in case of continuous evidence. The diagnostic relationship will be extended with more than one hypothesis. The next section will mention how to determine CPTs of a simple graph with one child node and many parent nodes based on X-gate inferences.

3. X-gate inferences

Given a simple graph consisting of one child variable Y and n parent variables X_i, as shown in **Figure 2**, each relationship from X_i to Y is quantified by normalized weight w_i where $0 \leq w_i \leq 1$. A large graph is an integration of many simple graphs. **Figure 2** shows the DAG of a simple BN. As aforementioned, the essence of constructing simple BN is to convert graphic relationships of simple graph into CPTs of simple BN.

Child variable Y is called target and parent variables X_is are called sources. Especially, these relationships are adhered to X-gates such as AND-gate, OR-gate, and SIGMA-gate. These gates are originated from logic gate [1]. For instance, AND-gate and OR-gate represent prerequisite relationship. SIGMA-gate represents aggregation relationship. Therefore, relationship conversion is to determined X-gate inference. The simple graph shown in **Figure 2** is also called X-gate graph or X-gate network. Please distinguish the letter "X" in the term "X-gate inference" which implies logic operators (AND, OR, XOR, etc.) from the "variable X".

All variables are binary and they represent events. The probability $P(X)$ indicates event X occurs. Thus, $P(X)$ implicates $P(X = 1)$ and $P(\text{not}(X))$ implicates $P(X = 0)$. Eq. (7) specifies the simple NOT-gate inference.

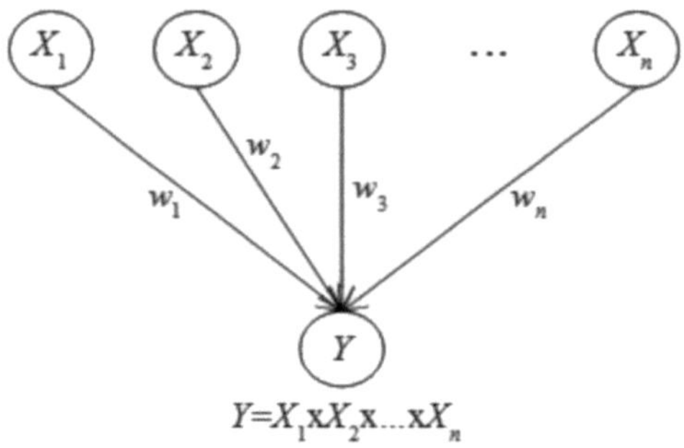

Figure 2. Simple graph or simple network.

$$P\Big(\text{not}(X)\Big) = P(\overline{X}) = P(X = 0) = 1 - P(X = 1) = 1 - P(X)$$
$$P\Big(\text{not}\Big(\text{not}(X)\Big)\Big) = P(X) \tag{7}$$

X-gate inference is based on three assumptions mentioned in Ref. ([2], p. 157), which are as follows

- *X-gate inhibition*: Given a relationship from source X_i to target Y, there is a factor I_i that inhibits X_i from being integrated into Y. Factor I_i is called inhibition of X_i. That the inhibition I_i is turned off is prerequisite of X_i integrated into Y.
- *Inhibition independence*: Inhibitions are mutually independent. For example, inhibition I_1 of X_1 is independent from inhibition I_2 of X_2.
- *Accountability*: X-gate network is established by accountable variables A_i for X_i and I_i. Each X-gate inference owns particular combination of A_is.

Figure 3 shows the extended X-gate network with accountable variables A_is ([2], p. 158).

The strength of each relationship from source X_i to target Y is quantified by a weight $0 \le w_i \le 1$. According to the assumption of inhibition, probability of $I_i = OFF$ is p_i, which is set to be the weight w_i.

$$p_i = w_i$$

If notation w_i is used, we focus on the strength of relationship. If notation p_i is used, we focus on probability of *OFF* inhibition. In probabilistic inference, p_i is also prior probability of $X_i = 1$. However, we will assume each X_i has uniform distribution later on. Eq. (8) specifies probabilities of inhibitions I_is and accountable variables A_is.

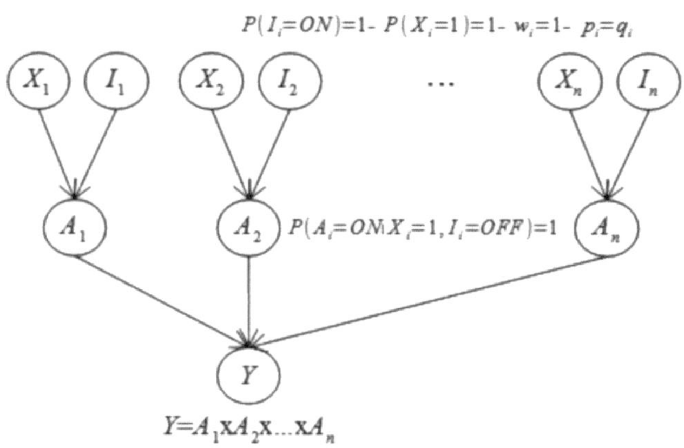

Figure 3. Extended X-gate network with accountable variables A_is.

$$\begin{aligned}
P(I_i = OFF) &= p_i = w_i \\
P(I_i = ON) &= 1 - p_i = 1 - w_i \\
P(A_i = ON|X_i = 1, I_i = OFF) &= 1 \\
P(A_i = ON|X_i = 1, I_i = ON) &= 0 \\
P(A_i = ON|X_i = 0, I_i = OFF) &= 0 \\
P(A_i = ON|X_i = 0, I_i = ON) &= 0 \\
P(A_i = OFF|X_i = 1, I_i = OFF) &= 0 \\
P(A_i = OFF|X_i = 1, I_i = ON) &= 1 \\
P(A_i = OFF|X_i = 0, I_i = OFF) &= 1 \\
P(A_i = OFF|X_i = 0, I_i = ON) &= 1
\end{aligned} \tag{8}$$

According to Eq. (8), given probability $P(A_i{=}ON \mid X_i{=}1, I_i{=}OFF)$, it is assured 100% confident that accountable variables A_i is turned on if source X_i is 1 and inhibition I_i is turned off. Eq. (9) specifies conditional probability of accountable variables A_i (s) given X_i (s), which is corollary of Eq. (8).

$$\begin{aligned}
P(A_i = ON|X_i = 1) &= p_i = w_i \\
P(A_i = ON|X_i = 0) &= 0 \\
P(A_i = OFF|X_i = 1) &= 1 - p_i = 1 - w_i \\
P(A_i = OFF|X_i = 0) &= 1
\end{aligned} \tag{9}$$

Appendix A1 is the proof of Eq. (9). As a definition, the set of all X_is is complete if and only if

$$P(X_1 \cup X_2 \cup \cdots \cup X_n) = P(\Omega) = \sum_{i=1}^{n} w_i = 1$$

The set of all X_is is mutually exclusive if and only if

$$X_i \cap X_j = \varnothing, \forall i \neq j$$

For each X_i, there is only one A_i and vice versa, which establishes a bijection between X_is and A_is. Obviously, the fact that the set of all X_is is complete is equivalent to the fact that the set of all A_i (s) is complete. We will prove by contradiction that "the fact that the set of all X_i (s) is mutually exclusive is equivalent to the fact that the set of all A_i (s) is mutually exclusive." Suppose $X_i \cap X_j = \varnothing, \forall i \neq j$ but $\exists i \neq j$: $A_i \cap A_j = B \neq \varnothing$. Let $B^{-1} \neq \varnothing$ be preimage of B. Due to $B \subseteq A_i$ and $B \subseteq A_j$, we have $B^{-1} \subseteq X_i$ and $B^{-1} \subseteq X_j$, which causes that $X_i \cap X_j = B^{-1} \neq \varnothing$. There is a contradiction and so we have

$$X_i \cap X_j = \varnothing, \forall i \neq j \Rightarrow A_i \cap A_j = \varnothing, \forall i \neq j$$

By similar proof, we have

$$A_i \cap A_j = \varnothing, \forall i \neq j \Rightarrow X_i \cap X_j = \varnothing, \forall i \neq j \ \blacksquare$$

The extended X-gate network shown in **Figure 3** is interpretation of simple network shown in **Figure 2**. Specifying CPT of the simple network is to determine the conditional probability $P(Y = 1 \mid X_1, X_2, \ldots, X_n)$ based on extended X-gate network. The X-gate inference is represented by such probability $P(Y = 1 \mid X_1, X_2, \ldots, X_n)$ specified by Eq. (10) ([2], p. 159).

$$P(Y|X_1, X_2, \ldots, X_n) = \sum\nolimits_{A_1, A_2, \ldots, A_n} P(Y|A_1, A_2, \ldots, A_n) \prod\nolimits_{i=1}^{n} P(A_i|X_i) \tag{10}$$

Appendix A2 is the proof of Eq. (10). It is necessary to make some mathematical notations because Eq. (10) is complicated, which is relevant to arrangements of X_i (s). Given the set $\Omega = \{X_1, X_2, \ldots, X_n\}$ where all variables are binary, **Table 2** specifies binary arrangements of Ω.

Given $\Omega = \{X_1, X_2, \ldots, X_n\}$ where $|\Omega| = n$ is cardinality of Ω.
Let $a(\Omega)$ be an arrangement of Ω which is a set of n instances $\{X_1=x_1, X_2=x_2, \ldots, X_n=x_n\}$ where x_i is 1 or 0. The number of all $a(\Omega)$ is $2^{|\Omega|}$. For instance, given $\Omega = \{X_1, X_2\}$, there are $2^2=4$ arrangements as follows:

$$a(\Omega) = \{X_1 = 1, X_2 = 1\}, a(\Omega) = \{X_1 = 1, X_2 = 0\}, a(\Omega) = \{X_1 = 0, X_2 = 1\}, a(\Omega) = \{X_1 = 0, X_2 = 0\}.$$

Let $a(\Omega:\{X_i\})$ be the arrangement of Ω with fixed X_i. The number of all $a(\Omega:\{X_i\})$ is $2^{|\Omega|-1}$. Similarly, for instance, $a(\Omega:\{X_1, X_2, X_3\})$ is an arrangement of Ω with fixed X_1, X_2, X_3. The number of all $a(\Omega:\{X_1, X_2, X_3\})$ is $2^{|\Omega|-3}$.
Let $c(\Omega)$ and $c(\Omega:\{X_i\})$ be the number of arrangements $a(\Omega)$ and $a(\Omega:\{X_i\})$, respectively. Such $c(\Omega)$ and $c(\Omega:\{X_i\})$ are called arrangement counters. As usual, counters $c(\Omega)$ and $c(\Omega:\{X_i\})$ are equal to $2^{|\Omega|}$ and $2^{|\Omega|-1}$, respectively but they will vary according to specific cases.
Let $\sum_a F\left(a(\Omega)\right)$ and $\prod_a F\left(a(\Omega)\right)$ denote sum and product of values generated from function F acting on every $a(\Omega)$. The number of arrangements on which F acts is $c(\Omega)$.

Let x denote the X-gate operator, for instance, x = ⊙ for AND-gate, x = ⊕ for OR-gate, x = not ⊙ for NAND-gate, x = not ⊕ for NOR-gate, x = ⊗ for XOR-gate, x = not ⊗ for XNOR-gate, x = ⊎ for U-gate, x = + for SIGMA-gate. Given an x-operator, let $s(\Omega:\{X_i\})$ and $s(\Omega)$ be sum of all $P(X_1 x X_2 x \ldots x X_n)$ through every arrangement of Ω with and without fixed X_i, respectively.

$$s(\Omega) = \sum_a P\left(X_1 x X_2 x \ldots x X_n | a(\Omega)\right) = \sum_a P\left(Y = 1 | a(\Omega)\right)$$
$$s(\Omega : \{X_i\}) = \sum_a P\left(X_1 x X_2 x \ldots x X_n | a(\Omega : \{X_i\})\right) = \sum_a P\left(Y = 1 | a(\Omega : \{X_i\})\right)$$

For example, $s(\Omega)$ and $s(\Omega:\{X_i\})$ for OR-gate are:

$$s(\Omega) = \sum_a P\left(X_1 \oplus X_2 \oplus \ldots \oplus X_n | a(\Omega)\right)$$
$$s(\Omega : \{X_i\}) = \sum_a P\left(X_1 \oplus X_2 \oplus \ldots \oplus X_n | a(\Omega : \{X_i\})\right)$$

Such $s(\Omega)$ and $s(\Omega:\{X_i\})$ are called arrangement sum. They are acting function F.
Note that Ω can be any set of binary variables.

Table 2. Binary arrangements.

It is not easy to produce all binary arrangements of Ω. **Table 3** shows a code snippet written by Java programming language for producing such all arrangements.

Each element of the list "*arrangements*" is a binary arrangement $a(\Omega)$ presented by an array of bits (0 and 1). The method "*create(int*[] *a, int i*)" which is recursive method, is the main one that generates arrangements. The method call "*ArrangementGenerator.parse*(2, *n*)" will list all possible binary arrangements.

Eq. (11) specifies the connection between $s(\Omega{:}\{X_i = 1\})$ and $s(\Omega{:}\{X_i = 0\})$, between $c(\Omega{:}\{X_i = 1\})$ and $c(\Omega{:}\{X_i = 0\})$.

$$\begin{aligned} s(\Omega : \{X_i = 1\}) + s(\Omega : \{X_i = 0\}) &= s(\Omega) \\ c(\Omega : \{X_i = 1\}) + c(\Omega : \{X_i = 0\}) &= c(\Omega) \end{aligned} \tag{11}$$

It is easy to draw Eq. (11) when the set of all arrangements $a(\Omega{:}\{X_i = 1)$ is complement of the set of all arrangements $a(\Omega{:}\{X_i = 0)$.

Let K be a set of X_is whose values are 1 and let L be a set of X_is whose values are 0. K and L are mutually complementary. Eq. (12) determines sets K and L.

$$\begin{cases} K = \{i : X_i = 1\} \\ L = \{i : X_i = 0\} \\ K \cap L = \varnothing \\ K \cup L = \{1, 2, \ldots, n\} \end{cases} \tag{12}$$

The AND-gate inference represents prerequisite relationship satisfying AND-gate condition specified by Eq. (13).

$$P(Y = 1 | A_i = OFF \text{ for some } i) = 0 \tag{13}$$

From Eq. (10), we have

$$\begin{aligned} P(Y = 1 | X_1, X_2, \ldots, X_n) &= \sum_{A_1, A_2, \ldots, A_n} P(Y = 1 | A_1, A_2, \ldots, A_n) \prod_{i=1}^{n} P(A_i | X_i) \\ &= \prod_{i=1}^{n} P(A_i = ON | X_i) \\ &\left(\text{Due to } P(Y = 1 | A_i = OFF \text{ for some } i) = 0\right) \\ &= \left(\prod_{i \in K} P(A_i = ON | X_i = 1)\right)\left(\prod_{i \notin K} P(A_i = ON | X_i = 0)\right) \\ &= \left(\prod_{i \in K} p_i\right)\left(\prod_{i \notin K} 0\right) = \begin{cases} \prod_{i=1}^{n} p_i \text{ if all } X_i(s) \text{ are } 1 \\ 0 \text{ if there exists at least one } X_i = 0 \end{cases} \end{aligned}$$

(Due to Eq. (9))

```
public class ArrangementGenerator {
private ArrayList<int[]> arrangements;
private int n;
private int r;

private ArrangementGenerator(int n, int r) {
        this.n = n;
        this.r = r;
        this.arrangements = new ArrayList();
}
private void create(int[] a, int i) {
      for(int j = 0; j < n; j++) {
            a[i] = j;
          if(i < r - 1)
                create(a, i + 1);
          else if(i == r -1) {
                int[] b = new int[a.length];
                for(int k = 0; k < a.length; k++) b[k] = a[k];
                arrangements.add(b);
          }
      }
}
public int[] get(int i) {
      return arrangements.get(i);
}
public long size() {
      return arrangements.size();
}
public static ArrangementGenerator parse(int n, int r) {
      ArrangementGenerator arr =
            new ArrangementGenerator(n, r);
      int[] a = new int[r];
      for(int i=0; i<r; i++) a[i] = -1;
      arr.create(a, 0);
      return arr;
}
}
```

Table 3. Code snippet generating all binary arrangements.

In general, Eq. (14) specifies AND-gate inference.

$$P(X_1 \odot X_2 \odot \ldots \odot X_n) = P(Y = 1|X_1, X_2, \ldots, X_n) = \begin{cases} \prod_{i=1}^{n} p_i \text{ if all } X_i(s) \text{ are } 1 \\ 0 \text{ if there exists at least one } X_i = 0 \end{cases}$$
$$P(Y = 0|X_1, X_2, \ldots, X_n) = \begin{cases} 1 - \prod_{i=1}^{n} p_i \text{ if all } X_i(s) \text{ are } 1 \\ 1 \text{ if there exists at least one } X_i = 0 \end{cases} \tag{14}$$

The AND-gate inference was also described in ([3], p. 33). Eq. (14) varies according to two cases whose arrangement counters are listed as follows

$L = \varnothing$

$$c(\Omega : \{X_i = 1\}) = 1, c(\Omega : \{X_i = 0\}) = 0, c(\Omega) = 1.$$

$L \neq \varnothing$

$$c(\Omega : \{X_i = 1\}) = 2^{n-1} - 1, c(\Omega : \{X_i = 0\}) = 2^{n-1}, c(\Omega) = 2^n - 1.$$

The **OR-gate** inference represents prerequisite relationship satisfying OR-gate condition specified by Eq. (15) ([2], p. 157).

$$P(Y = 1 | A_i = ON \text{ for some } i) = 1 \tag{15}$$

The OR-gate condition implies

$$P(Y = 0 | A_i = ON \text{ for some } i) = 0$$

From Eq. (10), we have ([2], p. 159)

$$\begin{aligned} P(Y = 0 | X_1, X_2, \ldots, X_n) &= \sum_{A_1, A_2, \ldots, A_n} P(Y = 1 | A_1, A_2, \ldots, A_n) \prod_{i=1}^{n} P(A_i | X_i) \\ &= \prod_{i=1}^{n} P(A_i = OFF | X_i) \\ &\left(\text{due to } P(Y = 1 | A_i = ON \text{ for some } i) = 0\right) \\ &= \left(\prod_{i \in K} P(A_i = OFF | X_i = 1)\right)\left(\prod_{i \notin K} P(A_i = OFF | X_i = 0)\right) \\ &= \left(\prod_{i \in K} (1 - p_i)\right)\left(\prod_{i \notin K} 1\right) = \begin{cases} \prod_{i \in K} (1 - p_i) \text{if } K \neq \varnothing \\ 1 \text{ if } K = \varnothing \end{cases} \end{aligned}$$

(Due to Eq. (9))

In general, Eq. (16) specifies OR-gate inference.

$$\begin{gathered} P(X_1 \oplus X_2 \oplus \ldots \oplus X_n) = 1 - P(Y = 0 | X_1, X_2, \ldots, X_n) = \begin{cases} 1 - \prod_{i \in K} (1 - p_i) \text{ if } K \neq \varnothing \\ 0 \text{ if } K = \varnothing \end{cases} \\ P(Y = 0 | X_1, X_2, \ldots, X_n) = \begin{cases} \prod_{i \in K} (1 - p_i) \text{ if } K \neq \varnothing \\ 1 \text{ if } K = \varnothing \end{cases} \end{gathered} \tag{16}$$

where K is the set of X_is whose values are 1. The OR-gate inference was mentioned in Refs. ([2], p. 158) and ([3], p. 20). Eq. (16) varies according to two cases whose arrangement counters are listed as follows

$$K \neq \varnothing$$

$$c(\Omega : \{X_i = 1\}) = 2^{n-1}, c(\Omega : \{X_i = 0\}) = 2^{n-1} - 1, c(\Omega) = 2^n - 1.$$

$$K = \varnothing$$

$$c(\Omega : \{X_i = 1\}) = 0, c(\Omega : \{X_i = 0\}) = 1, c(\Omega) = 1.$$

According to De Morgan's rule with regard to AND-gate and OR-gate, we have

$$P\Big(\text{not}(X_1 \odot X_2 \odot \ldots \odot X_n)\Big) = P\Big(\Big(\text{not}(X_1)\Big) \oplus \Big(\text{not}(X_2)\Big) \oplus \ldots \oplus \Big(\text{not}(X_n)\Big)\Big)$$
$$= \begin{cases} 1 - \prod_{i \in L}\Big(1 - (1 - p_i)\Big) \text{if } L \neq \varnothing \\ 0 \text{ if } L = \varnothing \end{cases}$$

(Due to Eq. (16))

According to Eq. (14), we also have

$$P\Big(\text{not}(X_1 \oplus X_2 \oplus \ldots \oplus X_n)\Big) = P\Big(\Big(\text{not}(X_1)\Big) \odot \Big(\text{not}(X_2)\Big) \odot \ldots \odot \Big(\text{not}(X_n)\Big)\Big)$$
$$= \begin{cases} \prod_{i=1}^{n} P\Big(\text{not}(X_i)\Big) \text{ if all not } (X_i)(s) \text{ are } 1 \\ 0 \text{ if there exists at least one not } (X_i) = 0 \end{cases}$$
$$= \begin{cases} \prod_{i=1}^{n} (1 - p_i) \text{ if all } X_i(s) \text{ are } 0 \\ 0 \text{ if there exists at least one } X_i = 1 \end{cases}$$

In general, Eq. (17) specifies NAND-gate inference and NOR-gate inference derived from AND-gate and OR-gate

$$P\Big(\text{not}(X_1 \odot X_2 \odot \ldots \odot X_n)\Big) = \begin{cases} 1 - \prod_{i \in L} p_i \text{ if } L \neq \varnothing \\ 0 \text{ if } L = \varnothing \end{cases}$$
$$P\Big(\text{not}(X_1 \oplus X_2 \oplus \ldots \oplus X_n)\Big) = \begin{cases} \prod_{i=1}^{n} q_i \text{ if } K = \varnothing \\ 0 \text{ if } K \neq \varnothing \end{cases} \quad (17)$$

where K and L are the sets of X_is whose values are 1 and 0, respectively.

Suppose the number of sources X_is is even. Let O be the set of X_is whose indices are odd. Let O_1 and O_2 be subsets of O, in which all X_is are 1 and 0, respectively. Let E be the set of X_is whose indices are even. Let E_1 and E_2 be the subsets of E, in which all X_is are 1 and 0, respectively.

$$\begin{cases} E = \{2, 4, 6, \ldots, n\} \\ E_1 \subseteq E \\ E_2 \subseteq E \\ E_1 \cup E_2 = E \\ E_1 \cap E_2 = \varnothing \\ X_i = 1, \forall i \in E_1 \\ X_i = 0, \forall i \in E_2 \end{cases} \quad \text{and} \quad \begin{cases} O = \{1, 3, 5, \ldots, n-1\} \\ O_1 \subseteq O \\ O_2 \subseteq O \\ O_1 \cup O_2 = O \\ O_1 \cap O_2 = \varnothing \\ X_i = 1, \forall i \in O_1 \\ X_i = 0, \forall i \in O_2 \end{cases}$$

Thus, O_1 and E_1 are the subsets of K. Sources X_is and target Y follow **XOR-gate** if one of two XOR-gate conditions specified by Eq. (18) is satisfied.

$$P\left(Y=1 \middle| \begin{cases} A_i = ON \text{ for } i \in O \\ A_i = OFF \text{ for } i \notin O \end{cases}\right) = P(Y=1|A_1 = ON, A_2 = OFF, \ldots, A_{n-1} = ON, A_n = OFF) = 1$$
$$P\left(Y=1 \middle| \begin{cases} A_i = ON \text{ for } i \in E \\ A_i = OFF \text{ for } i \notin E \end{cases}\right) = P(Y=1|A_1 = OFF, A_2 = ON, \ldots, A_{n-1} = OFF, A_n = ON) = 1 \tag{18}$$

From Eq. (10), we have

$$P(Y = 1|X_1, X_2, \ldots, X_n) = \sum_{A_1, A_2, \ldots, A_n} P(Y = 1|A_1, A_2, \ldots, A_n) \prod_{i=1}^{n} P(A_i|X_i)$$

If both XOR-gate conditions are not satisfied then,

$$P(Y = 1|X_1, X_2, \ldots, X_n) = 0$$

If the first XOR-gate condition is satisfied, we have

$$\begin{aligned} &P(Y = 1|X_1, X_2, \ldots, X_n) \\ &= P(Y = 1|A_1 = ON, A_2 = OFF, \ldots, A_{n-1} = ON, A_n = OFF) \prod_{i=1}^{n} P(A_i|X_i) \\ &= \left(\prod_{i \in O} P(A_i = ON|X_i)\right) \left(\prod_{i \in E} P(A_i = OFF|X_i)\right) \end{aligned}$$

We have

$$\begin{aligned} &\prod_{i \in O} P(A_i = ON|X_i) \\ &= \left(\prod_{i \in O_1} P(A_i = ON|X_i = 1)\right) * \left(\prod_{i \in O_2} P(A_i = ON|X_i = 0)\right) \\ &= \left(\prod_{i \in O_1} p_i\right) * \left(\prod_{i \in O_2} 0\right) = \begin{cases} \prod_{i \in O_1} p_i \text{ if } O_2 = \varnothing \\ 0 \text{ if } O_2 \neq \varnothing \end{cases} \end{aligned}$$

http://dx.doi.org/10.5772/intechopen.70057

(Due to Eq. (9))

We also have

$$\prod_{i\in E} P(A_i = OFF|X_i)$$
$$= \left(\prod_{i\in E_1} P(A_i = OFF|X_i = 1)\right) * \left(\prod_{i\in E_2} P(A_i = OFF|X_i = 0)\right)$$
$$= \left(\prod_{i\in E_1} (1-p_i)\right)\left(\prod_{i\in E_2} 1\right) = \begin{cases} \prod_{i\in E_1} (1-p_i) \text{ if } E_1 \neq \varnothing \\ 1 \text{ if } E_1 = \varnothing \end{cases}$$

(Due to Eq. (9))

Given the first XOR-gate condition, it implies

$$P(Y=1|X_1, X_2, \ldots, X_n) = \left(\prod_{i\in O} P(A_i = ON|X_i)\right)\left(\prod_{i\in E} P(A_i = OFF|X_i)\right)$$
$$= \begin{cases} \left(\prod_{i\in O_1} p_i\right)\left(\prod_{i\in E_1} (1-p_i)\right) \text{if } O_2 = \varnothing \text{ and } E_1 \neq \varnothing \\ \prod_{i\in O_1} p_i \text{if } O_2 = \varnothing \text{ and } E_1 = \varnothing \\ 0 \text{ if } O_2 \neq \varnothing \end{cases}$$

Similarly, given the second XOR-gate condition, we have

$$P(Y=1|X_1, X_2, \ldots, X_n) = \left(\prod_{i\in E} P(A_i = ON|X_i)\right)\left(\prod_{i\in O} P(A_i = OFF|X_i)\right)$$
$$= \begin{cases} \left(\prod_{i\in E_1} p_i\right)\left(\prod_{i\in O_1} (1-p_i)\right) \text{if } E_2 = \varnothing \text{ and } O_1 \neq \varnothing \\ \prod_{i\in E_1} p_i \text{ if } E_2 = \varnothing \text{ and } O_1 = \varnothing \\ 0 \text{ if } E_2 \neq \varnothing \end{cases}$$

If one of XOR-gate conditions is satisfied then,

$$P(Y=1|X_1, X_2, \ldots, X_n)$$
$$= \left(\prod_{i\in O} P(A_i = ON|X_i)\right)\left(\prod_{i\in E} P(A_i = OFF|X_i)\right) + \left(\prod_{i\in E} P(A_i = ON|X_i)\right)\left(\prod_{i\in O} P(A_i = OFF|X_i)\right)$$

This implies Eq. (19) to specify XOR-gate inference.

$$P(X_1 \otimes X_2 \otimes \ldots \otimes X_n) = P(Y = 1 | X_1, X_2, \ldots, X_n)$$
$$= \begin{cases} \left(\prod_{i \in O_1} p_i\right)\left(\prod_{i \in E_1} (1 - p_i)\right) + \left(\prod_{i \in E_1} p_i\right)\left(\prod_{i \in O_1} (1 - p_i)\right) \text{ if } O_2 = \varnothing \textit{ and } E_2 = \varnothing \\ \left(\prod_{i \in O_1} p_i\right)\left(\prod_{i \in E_1} (1 - p_i)\right) \text{ if } O_2 = \varnothing \textit{ and } E_1 \neq \varnothing \textit{ and } E_2 \neq \varnothing \\ \prod_{i \in O_1} p_i \text{ if } O_2 = \varnothing \textit{ and } E_1 = \varnothing \\ \left(\prod_{i \in E_1} p_i\right)\left(\prod_{i \in O_1} (1 - p_i)\right) \text{ if } E_2 = \varnothing \textit{ and } O_1 \neq \varnothing \textit{ and } O_2 \neq \varnothing \\ \prod_{i \in E_1} p_i \text{ if } E_2 = \varnothing \textit{ and } O_1 = \varnothing \\ 0 \text{ if } O_2 \neq \varnothing \textit{ and } E_2 \neq \varnothing \\ 0 \text{ if } n < 2 \textit{ or } n \text{ is odd} \end{cases} \tag{19}$$

where

$$\begin{cases} O = \{1, 3, 5, \ldots, n-1\} \\ O_1 \subseteq O \\ O_2 \subseteq O \\ O_1 \cup O_2 = O \\ O_1 \cap O_2 = \varnothing \\ X_i = 1, \forall i \in O_1 \\ X_i = 0, \forall i \in O_2 \end{cases} \quad \text{and} \quad \begin{cases} E = \{2, 4, 6, \ldots, n\} \\ E_1 \subseteq E \\ E_2 \subseteq E \\ E_1 \cup E_2 = E \\ E_1 \cap E_2 = \varnothing \\ X_i = 1, \forall i \in E_1 \\ X_i = 0, \forall i \in E_2 \end{cases}$$

Where,

Given $n \geq 2$ and n is even, Eq. (19) varies according to six cases whose arrangement counters are listed as follows

$$O_2 = \varnothing \text{ and } E_2 = \varnothing$$

$$c(\Omega : \{X_i = 1\}) = 1, c(\Omega : \{X_i = 0\}) = 0, c(\Omega) = 1.$$

$$O_2 = \varnothing \text{ and } E_1 \neq \varnothing \text{ and } E_2 \neq \varnothing$$

$$c(\Omega : \{X_i = 1\}) = 2^{\frac{n}{2}} - 2, c(\Omega : \{X_i = 0\}) = 0, c(\Omega) = 2^{\frac{n}{2}} - 2.$$

$$O_2 = \varnothing \text{ and } E_1 = \varnothing$$

$$c(\Omega : \{X_i = 1\}) = 1, c(\Omega : \{X_i = 0\}) = 0, c(\Omega) = 1.$$

$$E_2 = \varnothing \text{ and } O_1 \neq \varnothing \text{ and } O_2 \neq \varnothing$$

$$c(\Omega : \{X_i = 1\}) = 2^{\frac{n}{2}-1} - 1, c(\Omega : \{X_i = 0\}) = 2^{\frac{n}{2}-1} - 1, c(\Omega) = 2^{\frac{n}{2}} - 2.$$

$$E_2 = \varnothing \text{ and } O_1 = \varnothing$$

$$c(\Omega : \{X_i = 1\}) = 0, c(\Omega : \{X_i = 0\}) = 1, c(\Omega) = 1.$$

http://dx.doi.org/10.5772/intechopen.70057

$$O_2 \neq \varnothing \text{ and } E_2 \neq \varnothing$$

$$c(\Omega : \{X_i = 1\}) = \left(2^{\frac{n}{2}-1} - 1\right)\left(2^{\frac{n}{2}} - 1\right), c(\Omega : \{X_i = 0\}) = 2^{\frac{n}{2}-1}\left(2^{\frac{n}{2}} - 1\right), c(\Omega) = \left(2^{\frac{n}{2}} - 1\right)^2.$$

Suppose the number of sources X_is is even. According to **XNOR-gate** inference [1], the output is on if all inputs get the same value 1 (or 0). Sources X_i (s) and target Y follow XNOR-gate if one of two XNOR-gate conditions specified by Eq. (20) is satisfied.

$$\begin{aligned} &P(Y = 1 | A_i = ON, \forall i) = 1 \\ &P(Y = 1 | A_i = OFF, \forall i) = 1 \end{aligned} \tag{20}$$

From Eq. (10), we have

$$P(Y = 1 | X_1, X_2, \ldots, X_n) = \sum_{A_1, A_2, \ldots, A_n} P(Y = 1 | A_1, A_2, \ldots, A_n) \prod_{i=1}^{n} P(A_i | X_i)$$

If both XNOR-gate conditions are not satisfied then,

$$P(Y = 1 | X_1, X_2, \ldots, X_n) = 0$$

If $A_i = ON$ for all i, we have

$$\begin{aligned} P(Y = 1 | X_1, X_2, \ldots, X_n) &= P(Y = 1 | A_i = ON, \forall i) \prod_{i=1}^{n} P(A_i = ON | X_i) \\ &= \prod_{i=1}^{n} P(A_i = ON | X_i) = \begin{cases} \prod_{i=1}^{n} p_i \text{if } L = \varnothing \\ 0 \text{ if } L \neq \varnothing \end{cases} \end{aligned}$$

(Please see similar proof in AND-gate inference)

If $A_i = OFF$ for all i, we have

$$P(Y = 1 | X_1, X_2, \ldots, X_n) = \prod_{i=1}^{n} P(A_i = OFF | X_i) = \begin{cases} \prod_{i \in K} (1 - p_i) \text{ if } K \neq \varnothing \\ 1 \text{ if } K = \varnothing \end{cases}$$

(Please see similar proof in OR-gate inference)

If one of XNOR-gate conditions is satisfied then,

$$P(Y = 1 | X_1, X_2, \ldots, X_n) = \prod_{i=1}^{n} P(A_i = ON | X_i) + \prod_{i=1}^{n} P(A_i = OFF | X_i)$$

This implies Eq. (21) to specify XNOR-gate inference.

$$P\Big(\text{not}(X_1 \otimes X_2 \otimes \ldots \otimes X_n)\Big) = P(Y=1|X_1, X_2, \ldots, X_n) = \begin{cases} \prod_{i=1}^{n} p_i + \prod_{i=1}^{n} (1-p_i) \text{ if } L = \varnothing \\ \prod_{i \in K} (1-p_i) \text{ if } L \neq \varnothing \text{ and } K \neq \varnothing \\ 1 \text{ if } L \neq \varnothing \text{ and } K = \varnothing \end{cases} \tag{21}$$

where K and L are the sets of X_is whose values are 1 and 0, respectively. Eq. (21) varies according to three cases whose arrangement counters are listed as follows

$$L = \varnothing$$

$$c(\Omega : \{X_i = 1\}) = 1, c(\Omega : \{X_i = 0\}) = 0, c(\Omega) = 1.$$

$$L \neq \varnothing \text{ and } K \neq \varnothing$$

$$c(\Omega : \{X_i = 1\}) = 2^{n-1} - 1, c(\Omega : \{X_i = 0\}) = 2^{n-1} - 1, c(\Omega) = 2^n - 2.$$

$$L \neq \varnothing \text{ and } K = \varnothing$$

$$c(\Omega : \{X_i = 1\}) = 0, c(\Omega : \{X_i = 0\}) = 1, c(\Omega) = 1.$$

Let U be a set of indices such that $A_i = ON$ and let $\alpha \geq 0$ and $\beta \geq 0$ be predefined numbers. The U-gate inference is defined based on α, β and cardinality of U. **Table 4** specifies four common U-gate conditions.

Note that U-gate condition on $|U|$ can be arbitrary and it is only relevant to A_is (*ON* or *OFF*) and the way to combine A_is. For example, AND-gate and OR-gate are specific cases of U-gate with $|U| = n$ and $|U| \geq 1$, respectively. XOR-gate and XNOR-gate are also specific cases of U-gate with specific conditions on A_i (s). However, it must be assured that there is at least one combination of A_is satisfying the predefined U-gate condition, which causes that U-gate probability is not always equal to 0. In this research, U-gate is the most general nonlinear gate where U-gate probability contains products of weights (see **Table 5**). Later on, we will research a so-called SIGMA-gate that contains only linear combination of weights (sum of weights, see Eq. (23)). Shortly, each X-gate is a pattern owning a particular X-gate inference that is X-gate probability $P(X_1 \times X_2 \times \ldots \times X_n)$. Each X-gate inference is based on particular X-gate condition (s) relevant to only variables A_is.

From Eq. (10), we have

$$P(Y=1|X_1, X_2, \ldots, X_n) = \sum_{A_1, A_2, \ldots, A_n} P(Y=1|A_1, A_2, \ldots, A_n) \prod_{i=1}^{n} P(A_i|X_i)$$

Let $\mathcal{U}$ be the set of all possible U (s), we have

$\|U\|=\alpha$	$P(Y=1\|A_1, A_2, \ldots, A_n)=1$ if there are exactly α variables $A_i = ON$ (s). Otherwise, $P(Y=1\|A_1, A_2, \ldots, A_n)=0$.
$\|U\|\geq\alpha$	$P(Y=1\|A_1, A_2, \ldots, A_n)=1$ if there are at least α variables $A_i = ON$ (s). Otherwise, $P(Y=1\|A_1, A_2, \ldots, A_n)=0$.
$\|U\|\leq\beta$	$P(Y=1\|A_1, A_2, \ldots, A_n)=1$ if there are at most β variables $A_i = ON$ (s). Otherwise, $P(Y=1\|A_1, A_2, \ldots, A_n)=0$.
$\alpha\leq\|U\|\leq\beta$	$P(Y=1\|A_1, A_2, \ldots, A_n)=1$ if the number of $A_i = ON$ (s) is from α to β. Otherwise, $P(Y=1\|A_1, A_2, \ldots, A_n)=0$.

Table 4. U-gate conditions.

$$\begin{aligned} P(Y=1|X_1, X_2, \ldots, X_n) &= \sum_{U\in\mathcal{U}} P(Y=1|A_1, A_2, \ldots, A_n)\prod_{i=1}^{n} P(A_i|X_i) \\ &= \sum_{U\in\mathcal{U}}\prod_{i\in U} P(A_i=ON|X_i)\prod_{j\notin U} P(A_j=OFF|X_j) \end{aligned}$$

If $X_i = 0, \forall i \in U$ then,

$$P(Y=1|X_1, X_2, \ldots, X_n) = \sum_{U\in\mathcal{U}}\prod_{i\in U} 0 \prod_{j\notin U} P(A_j=OFF|X_j) = 0$$

This implies all sets U (s) must be subsets of K. The U-gate probability is rewritten as follows

$$P(Y=1|X_1, X_2, \ldots, X_n) = \sum_{U\in\mathcal{U}}\prod_{i\in U} P(A_i=ON|X_i=1)\prod_{j\notin U} P(A_j=OFF|X_j)$$

$$= \sum_{U\in\mathcal{U}}\prod_{i\in U} p_i \prod_{j\notin U} P(A_j=OFF|X_j)$$

$$= \sum_{U\in\mathcal{U}}\prod_{i\in U} p_i \prod_{j\in K\setminus U} P(A_j=OFF|X_j=1)\prod_{j\notin K} P(A_j=OFF|X_j=0)$$

$$= \sum_{U\in\mathcal{U}}\prod_{i\in U} p_i \prod_{j\in K\setminus U} (1-p_j)\prod_{j\notin K} 1 = \sum_{U\in\mathcal{U}}\prod_{i\in U} p_i \prod_{j\in K\setminus U} (1-p_j)$$

(Due to Eq. (9))

Let P_U be the U-gate probability; **Table 5** specifies U-gate inference and cardinality of $\mathcal{U}$ where $\mathcal{U}$ is the set of subsets (U) of K.

Note that the notation $\binom{n}{j}$ denotes the number of combinations of j elements taken from n elements.

$$\binom{n}{j} = \frac{n!}{j!(n-j)!}$$

Arrangement counters relevant to U-gate inference and the set K are listed as follows

Let,	$S_U = \sum_{U\in\mathcal{U}} \prod_{i\in U} p_i \prod_{j\in K\setminus U} (1-p_j)$ $P_U = P(X_1 \uplus X_2 \uplus \ldots \uplus X_n) = P(Y=1 \mid X_1, X_2, \ldots, X_n)$
As a convention,	$\prod_{i\in U} p_i = 1 \text{ if } \lvert U\rvert = 0$ $\prod_{j\in K\setminus U} (1-p_j) = 1 \text{ if } \lvert U\rvert = \lvert K\rvert$
$\lvert U\rvert=0$	$P_U = \begin{cases} \prod_{j=1}^{n}(1-p_j) \text{ if } \lvert K\rvert > 0 \\ 1 \text{ if } \lvert K\rvert = 0 \end{cases}$ $\lvert\mathcal{U}\rvert = 1$
$\lvert U\rvert\geq 0$	$P_U = \begin{cases} S_U \text{ if } \lvert K\rvert > 0 \\ 1 \text{ if } \lvert K\rvert = 0 \end{cases} \quad \lvert\mathcal{U}\rvert = 2^{\lvert K\rvert}$ The case $\lvert U\rvert\geq 0$ is the same to the case $\lvert U\rvert\leq n$
$\lvert U\rvert=n$	$P_U = \begin{cases} \prod_{i=1}^{n} p_i \text{ if } \lvert K\rvert = n \\ 0 \text{ if } \lvert K\rvert < n \end{cases}$ $\lvert\mathcal{U}\rvert = \begin{cases} 1 \text{ if } \lvert K\rvert = n \\ 0 \text{ if } \lvert K\rvert < n \end{cases}$
$\lvert U\rvert=\alpha$ $0<\alpha<n$	$P_U = \begin{cases} S_U \text{ if } \lvert K\rvert \geq \alpha \\ 0 \text{ if } \lvert K\rvert < \alpha \end{cases}$ $\lvert\mathcal{U}\rvert = \begin{cases} \binom{\lvert K\rvert}{\alpha} \text{ if } \lvert K\rvert \geq \alpha \\ 0 \text{ if } \lvert K\rvert < \alpha \end{cases}$
$\lvert U\rvert\geq\alpha$ $0<\alpha<n$	$P_U = \begin{cases} S_U \text{ if } \lvert K\rvert \geq \alpha \\ 0 \text{ if } \lvert K\rvert < \alpha \end{cases}$ $\lvert\mathcal{U}\rvert = \begin{cases} \sum_{j=\alpha}^{\lvert K\rvert} \binom{\lvert K\rvert}{j} \text{ if } \lvert K\rvert \geq \alpha \\ 0 \text{ if } \lvert K\rvert < \alpha \end{cases}$
$\lvert U\rvert\leq\beta$ $0<\beta<n$	$P_U = \begin{cases} S_U \text{ if } \lvert K\rvert > 0 \\ 1 \text{ if } \lvert K\rvert = 0 \end{cases}$ $\lvert\mathcal{U}\rvert = \begin{cases} \sum_{j=0}^{\min(\beta, \lvert K\rvert)} \binom{\lvert K\rvert}{j} \text{ if } \lvert K\rvert > 0 \\ 1 \text{ if } \lvert K\rvert = 0 \end{cases}$
$\alpha\leq\lvert U\rvert\leq\beta$ $0<\alpha<n$ $0<\beta<n$	$P_U = \begin{cases} S_U \text{ if } \lvert K\rvert \geq \alpha \\ 0 \text{ if } \lvert K\rvert < \alpha \end{cases}$ $\lvert\mathcal{U}\rvert = \begin{cases} \sum_{j=\alpha}^{\min(\beta, \lvert K\rvert)} \binom{\lvert K\rvert}{j} \text{ if } \lvert K\rvert \geq \alpha \\ 0 \text{ if } \lvert K\rvert < \alpha \end{cases}$

Table 5. U-gate inference.

$$|K| = 0$$
$$c(\Omega : \{X_i = 1\}) = 0, c(\Omega : \{X_i = 0\}) = 1, c(\Omega) = 1.$$
$$|K| = 1$$
$$c(\Omega : \{X_i = 1\}) = 1, c(\Omega : \{X_i = 0\}) = 0, c(\Omega) = 1.$$
$$|K| = \alpha \text{ and } \alpha > 0$$
$$c(\Omega : \{X_i = 1\}) = \binom{n-1}{\alpha-1}, c(\Omega : \{X_i = 0\}) = \binom{n-1}{\alpha}, c(\Omega) = \binom{n}{\alpha}.$$
$$|K| \leq \alpha \text{ and } \alpha > 0$$
$$c(\Omega : \{X_i = 1\}) = \sum_{j=1}^{\alpha} \binom{n-1}{j-1}, c(\Omega : \{X_i = 0\}) = \sum_{j=0}^{\alpha} \binom{n-1}{j}, c(\Omega) = \sum_{j=0}^{\alpha} \binom{n}{j}.$$
$$|K| \geq \alpha \text{ and } \alpha > 0$$
$$c(\Omega : \{X_i = 1\}) = \sum_{j=\alpha}^{n} \binom{n-1}{j-1}, c(\Omega : \{X_i = 0\}) = \sum_{j=\alpha}^{n-1} \binom{n-1}{j}, c(\Omega) = \sum_{j=\alpha}^{n} \binom{n}{j}.$$

The **SIGMA-gate** inference [9] represents aggregation relationship satisfying SIGMA-gate condition specified by Eq. (22).

$$P(Y) = P\left(\sum_{i=1}^{n} A_i\right)$$

where the set of A_i is complete and mutually exclusive

$$\sum_{i=1}^{n} w_i = 1$$
$$A_i \cap A_j = \varnothing, \forall i \neq j \tag{22}$$

The sigma sum $\sum_{i=1}^{n} A_i$ indicates that Y is exclusive union of A_is and here, it does not express arithmetical additions.

$$Y = \sum_{i=1}^{n} A_i = \bigcup_{i=1}^{n} A_i$$

This implies

$$P(Y) = P\left(\sum_{i=1}^{n} A_i\right) = P\left(\bigcup_{i=1}^{n} A_i\right) = \sum_{i=1}^{n} P(A_i)$$

The sigma sum $\sum_{i=1}^{n} P(A_i)$ now expresses arithmetical additions of probabilities $P(A_i)$.

SIGMA-gate inference requires the set of A_is is complete and mutually exclusive, which means that the set of X_is is complete and mutually exclusive too. The SIGMA-gate probability is [9]

$$\begin{gathered}
P(Y|X_1, X_2, \ldots, X_n) = P\left(\sum_{i=1}^{n} A_i \middle| X_1, X_2, \ldots, X_n\right) \\
\text{(due to SIGMA} - \text{gate condition)} \\
= \sum_{i=1}^{n} P(A_i|X_1, X_2, \ldots, X_n) \\
\left(\text{because } A_i\text{(s) are mutually exclusive}\right) \\
= \sum_{i=1}^{n} P(A_i|X_i) \\
\text{(because } A_i \text{ is only dependent on } X_i\text{)}
\end{gathered}$$

It implies

$$\begin{aligned}
&P(Y = 1|X_1, X_2, \ldots, X_n) \\
&= \sum_{i=1}^{n} P(A_i = ON|X_i) \\
= &\left(\sum_{i \in K} P(A_i = ON|X_i = 1)\right) + \left(\sum_{i \notin K} P(A_i = ON|X_i = 0)\right) \\
= &\sum_{i \in K} w_i + \sum_{i \notin K} 0 = \sum_{i \in K} w_i
\end{aligned}$$

(Due to Eq. (9))

In general, Eq. (23) specifies the theorem of SIGMA-gate inference [9]. The base of this theorem was mentioned by Millán and Pérez-de-la-Cruz ([4], pp. 292-295).

$$\begin{aligned}
&P(X_1 + X_2 + \ldots + X_n) = P\left(\sum_{i=1}^{n} X_i\right) = P(Y = 1|X_1, X_2, \ldots, X_n) = \sum_{i \in K} w_i \\
&P(Y = 0|X_1, X_2, \ldots, X_n) = 1 - \sum_{i \in K} w_i = \sum_{i \in L} w_i
\end{aligned}$$

where the set of X_is is complete and mutually exclusive.

$$\begin{gathered}
\sum_{i=1}^{n} w_i = 1 \\
X_i \cap X_j = \varnothing, \forall i \neq j
\end{gathered} \tag{23}$$

The arrangement counters of SIGMA-gate inference are $c(\Omega{:}\{X_i = 1\}) = c(\Omega{:}\{X_i = 0\}) = 2^{n-1}$, $c(\Omega) = 2^n$.

Eq. (9) specifies the "clockwise" strength of relationship between X_i and Y. Event $X_i = 1$ causes event $A_i = ON$ with "clockwise" weight w_i. There is a question "given $X_i = 0$, how likely the event $A_i = OFF$ is". In order to solve this problem, I define a so-called "counterclockwise" strength of relationship between X_i and Y denoted ω_i. Event $X_i = 0$ causes event $A_i = OFF$ with "counterclockwise" weight ω_i. In other words, each arc in simple graph is associated with a clockwise weight w_i and a counterclockwise weight ω_i. Such graph is called *bi-weight simple graph* shown in **Figure 4**.

With bi-weight simple graph, all X-gate inferences are extended as so-called X-gate bi-inferences. Derived from Eq. (9), Eq. (24) specifies conditional probability of accountable variables with regard to bi-weight graph.

$$\begin{aligned} P(A_i = ON|X_i = 1) &= p_i = w_i \\ P(A_i = ON|X_i = 0) &= 1 - \rho_i = 1 - \omega_i \\ P(A_i = OFF|X_i = 1) &= 1 - p_i = 1 - w_i \\ P(A_i = OFF|X_i = 0) &= \rho_i = \omega_i \end{aligned} \tag{24}$$

The probabilities $P(A_i = ON \mid X_i = 0)$ and $P(A_i = OFF \mid X_i = 1)$ are called clockwise adder d_i and counterclockwise adder δ_i. As usual, d_i and δ_i are smaller than w_i and ω_i. When $d_i = 0$, bi-weight graph becomes normal simple graph.

$$\begin{aligned} d_i &= P(A_i = ON|X_i = 0) = 1 - \rho_i = 1 - \omega_i \\ \delta_i &= P(A_i = OFF|X_i = 1) = 1 - p_i = 1 - w_i \end{aligned}$$

The total clockwise weight or total counterclockwise weight is defined as sum of clockwise weight and clockwise adder or sum of counterclockwise weight and counterclockwise adder. Eq. (25) specifies such total weights W_i and $\mathcal{W}_i$. These weights are also called relationship powers.

$$\begin{aligned} W_i &= w_i + d_i \\ \mathcal{W}_i &= \omega_i + \delta_i \end{aligned}$$

where

$$\begin{aligned} d_i &= 1 - \rho_i = 1 - \omega_i \\ \delta_i &= 1 - p_i = 1 - w_i \end{aligned} \tag{25}$$

Given Eq. (25), the set of all A_is is complete if and only if $\sum_{i=1}^{n} w_i = 1$.

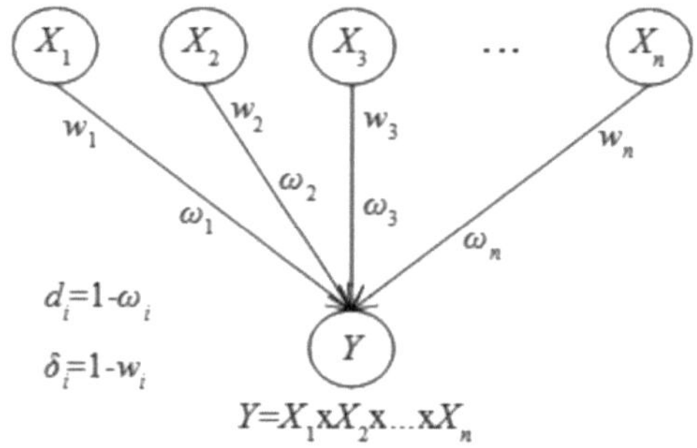

Figure 4. Bi-weight simple graph.

By extending aforementioned X-gate inferences, we get bi-inferences for AND-gate, OR-gate, NAND-gate, NOR-gate, XOR-gate, XNOR-gate, and U-gate as shown in **Table 6**.

The largest cardinalities of K (L) are 2^{n-1} and 2^n with and without fixed X_i. Thus, it is possible to calculate arrangement counters. As a convention, the product of probabilities is 1 if indices set is empty.

$$\prod_{i \in I} f_i = 1 \text{ if } I = \emptyset$$

With regard to SIGMA-gate bi-inference, the sum of all total clockwise weights must be 1 as follows

$$\sum_{i=1}^{n} W_i = \sum_{i=1}^{n} (w_i + d_i) = \sum_{i=1}^{n} (w_i + 1 - \omega_i) = 1$$

Derived from Eq. (23), the SIGMA-gate probability for bi-weight graph is

$$\begin{aligned} P(X_1 + X_2 + \ldots + X_n) &= \sum_{i=1}^{n} P(A_i = ON|X_i) \\ &= \sum_{i \in K} P(A_i = ON|X_i = 1) + \sum_{i \in L} P(A_i = ON|X_i = 0) \\ &= \sum_{i \in K} w_i + \sum_{i \in L} d_i \end{aligned}$$

Shortly, Eq. (26) specifies SIGMA-gate bi-inference.

$$P(X_1 + X_2 + \ldots + X_n) = \sum_{i \in K} w_i + \sum_{i \in L} d_i$$

where the set of X_i(s) is complete and mutually exclusive.

$$\begin{aligned} &\sum_{i=1}^{n} W_i = 1 \\ &X_i \cap X_j = \emptyset, \forall i \neq j \end{aligned} \tag{26}$$

The next section will research diagnostic relationship which adheres to X-gate inference.

4. Multihypothesis diagnostic relationship

Given a simple graph shown in **Figure 2**, if we replace the target source Y by an evidence D, we get a so-called *multihypothesis diagnostic relationship* whose property adheres to X-gate inference. Maybe there are other diagnostic relationships in which X-gate inference is not concerned. However, this research focuses on X-gate inference and so multi-hypothesis diagnostic relationship is called *X-gate diagnostic relationship*. Sources X_1, X_2,…, X_n become hypotheses. As a convention, these hypotheses have prior uniform distribution.

According to aforementioned X-gate network shown in **Figures 2** and **3**, the target variable must be binary whereas evidence D can be numeric. It is impossible to establish the evidence D as direct target variable. Thus, the solution of this problem is to add an augmented target binary variable Y and then, the evidence D is connected directly to Y. In other words, the *X-gate diagnostic network* have n sources $\{X_1, X_2, \ldots, X_n\}$, one augmented hypothesis Y, and one evidence D. As a convention, X-gate diagnostic network is called *X-D network*. The CPTs of the entire network are determined based on combination of diagnostic relationship and X-gate inference mentioned in previous sections. **Figure 5** depicts the augmented X-D network. Note that variables $X_1, X_2, \ldots, X_n$, and Y are always binary.

Appendix A3 is the proof that the augmented X-D network is equivalent to X-D network with regard to variables $X_1, X_2, \ldots, X_n$ and D. As a convention, augmented X-D network is considered as same as X-D network.

The simplest case of X-D network is NOT-D network having one hypothesis X_1 and one evidence D, equipped with NOT-gate inference. NOT-D network satisfies diagnostic condition because it essentially represents the single diagnostic relationship. Inferred from Eqs. (1) and (7), the conditional probability $P(D|X_1)$ and posterior probability $P(X_1|D)$ of NOT-D network are

$$P(D|X_1) = \begin{cases} 1 - D \text{ if } X_1 = 1 \\ D \text{ if } X_1 = 0 \end{cases}$$

$$P(X_1|D) = \frac{P(D|X_1)P(X_1)}{P(X_1)\left(P(D|X_1 = 0) + P(D|X_1 = 1)\right)}$$

(Due to Bayes' rule and uniform distribution of X_1)

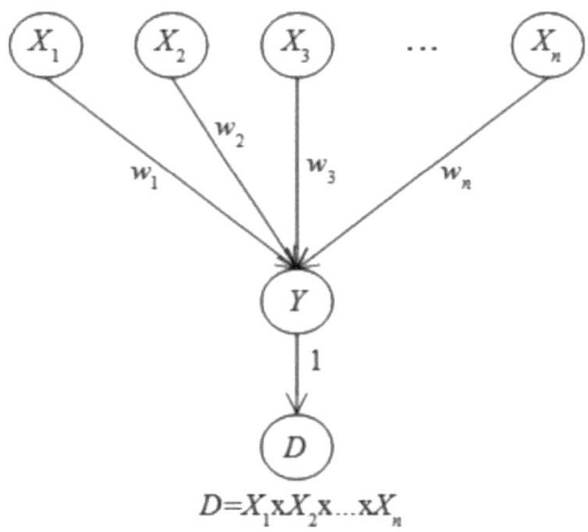

Figure 5. Augmented X-D network.

$$= \frac{P(D|X_1)}{P(D|X_1 = 0) + P(D|X_1 = 1)} = 1 * P(D|X_1)$$

$$\left(\text{due to } P(D|X_1 = 0) + P(D|X_1 = 1) = 1\right)$$

It implies NOT-D network satisfies diagnostic condition. Let

$$\Omega = \{X_1, X_2, \ldots, X_n\}$$
$$n = |\Omega|$$

We will validate whether the CPT of diagnostic relationship, $P(D|X)$ specified by Eq. (6), still satisfies diagnostic condition within general case, X-D network. In other words, X-D network is general case of single diagnostic relationship.

Recall from dependencies shown in **Figure 5**, Eq. (27) specifies the joint probability of X-D network.

$$P(\Omega, Y, D) = P(X_1, X_2, \ldots, X_n, Y, D) = P(D|Y)P(Y|X_1, X_2, \ldots, X_n)\prod_{i=1}^{n} P(X_i) \quad (27)$$
$$\text{where } \Omega = \{X1, X2, \ldots, Xn\}.$$

Eq. (28) specifies the conditional probability of D given X_i (likelihood function) and the posterior probability of X_i given D.

$$P(D|X_i) = \frac{P(X_i, D)}{P(X_i)} = \frac{\sum_{\{\Omega, Y, D\}\backslash\{X_i, D\}} P(\Omega, Y, D)}{\sum_{\{\Omega, Y, D\}\backslash\{X_i\}} P(\Omega, Y, D)}$$
$$P(X_i|D) = \frac{P(X_i, D)}{P(D)} = \frac{\sum_{\{\Omega, Y, D\}\backslash\{X_i, D\}} P(\Omega, Y, D)}{\sum_{\{\Omega, Y, D\}\backslash\{D\}} P(\Omega, Y, D)} \quad (28)$$

where $\Omega = \{X_1, X_2, \ldots, X_n\}$ and the sign "\" denotes the subtraction (excluding) operator in set theory [10]. Eq. (29) specifies the joint probability $P(X_i, D)$ and the marginal probability $P(D)$ given uniform distribution of all sources. Appendix A4 is the proof of Eq. (29).

$$P(X_i, D) = \frac{1}{2^n S}\left((2D - M)s(\Omega : \{X_i\}) + 2^{n-1}(M - D)\right)$$
$$P(D) = \frac{1}{2^n S}\left((2D - M)s(\Omega) + 2^n(M - D)\right) \quad (29)$$

where $s(\Omega)$ and $s(\Omega{:}\{X_i\})$ are specified in **Table 2**. From Eqs. (28–30) specifies conditional probability $P(D|X_i)$, posterior probability $P(X_i|D)$, and transformation coefficient for X-gate inference.

$$\begin{aligned}
&P(D|X_i=1)=\frac{P(X_i=1,D)}{P(X_i=1)}=\frac{(2D-M)s(\Omega:\{X_i=1\})+2^{n-1}(M-D)}{2^{n-1}S}\\
&P(D|X_i=0)=\frac{P(X_i=0,D)}{P(X_i=0)}=\frac{(2D-M)s(\Omega:\{X_i=0\})+2^{n-1}(M-D)}{2^{n-1}S}\\
&P(X_i=1|D)=\frac{P(X_i=1,D)}{P(D)}=\frac{(2D-M)s(\Omega:\{X_i=1\})+2^{n-1}(M-D)}{(2D-M)s(\Omega)+2^{n}(M-D)}\\
&P(X_i=0|D)=1-P(X_i=1|D)=\frac{(2D-M)s(\Omega:\{X_i=0\})+2^{n-1}(M-D)}{(2D-M)s(\Omega)+2^{n}(M-D)}\\
&k=\frac{P(X_i|D)}{P(D|X_i)}=\frac{2^{n-1}S}{(2D-M)s(\Omega)+2^{n}(M-D)}
\end{aligned} \tag{30}$$

The transformation coefficient is rewritten as follows

$$k=\frac{2^{n-1}S}{2D\left(s(\Omega)-2^{n-1}\right)+M\left(2^{n}-s(\Omega)\right)}$$

Note that S, D, and M are abstract symbols and there is no proportional connection between $2^{n-1}S$ and D for all D, specified by Eq. (6). Assuming that such proportional connection $2^{n-1}S = aD^j$ exists for all D where a is arbitrary constant. Given binary case when $D = 0$ and $S = 1$, we have

$$2^{n-1}=2^{n-1}*1=2^{n-1}S=aD^j=a*0^j=0$$

There is a contradiction, which implies that it is impossible to reduce k into the following form

$$k=\frac{aD^j}{bD^j}$$

Therefore, if k is constant with regard to D then,

$$2D\left(s(\Omega)-2^{n-1}\right)+M\left(2^{n}-s(\Omega)\right)=C\neq 0,\forall D$$

where C is constant. We have

$$\begin{aligned}
&\sum_D\left(2D\left(s(\Omega)-2^{n-1}\right)+M\left(2^{n}-s(\Omega)\right)\right)=\sum_D C\\
&\Rightarrow 2S\left(s(\Omega)-2^{n-1}\right)+NM\left(2^{n}-s(\Omega)\right)=NC\\
&\Rightarrow 2^{n}S=NC
\end{aligned}$$

It is implied that

$$P(X_1 \odot X_2 \odot \ldots \odot X_n) = \prod_{i \in K} p_i \prod_{i \in L} d_i$$

$$P(X_1 \oplus X_2 \oplus \ldots \oplus X_n) = 1 - \prod_{i \in K} \delta_i \prod_{i \in L} \rho_i$$

$$P\Big(\text{not}(X_1 \odot X_2 \odot \ldots \odot X_n)\Big) = 1 - \prod_{i \in L} \rho_i \prod_{i \in K} \delta_i$$

$$P\Big(\text{not}(X_1 \oplus X_2 \oplus \ldots \oplus X_n)\Big) = \prod_{i \in L} d_i \prod_{i \in K} p_i$$

$$P(X_1 \otimes X_2 \otimes \ldots \otimes X_n) = \prod_{i \in O_1} p_i \prod_{i \in O_2} d_i \prod_{i \in E_1} \delta_i \prod_{i \in E_2} \rho_i + \prod_{i \in E_1} p_i \prod_{i \in E_2} d_i \prod_{i \in O_1} \delta_i \prod_{i \in O_2} \rho_i$$

$$P\Big(\text{not}(X_1 \otimes X_2 \otimes \ldots \otimes X_n)\Big) = \prod_{i \in K} p_i \prod_{i \in L} d_i + \prod_{i \in K} \delta_i \prod_{i \in L} \rho_i$$

$$P(X_1 \uplus X_2 \uplus \ldots \uplus X_n) = \sum_{U \in \mathcal{U}} \left(\prod_{i \in U \cap K} p_i \prod_{i \in U \cap L} d_i \right) \left(\prod_{i \in \overline{U} \cap K} \delta_i \prod_{i \in \overline{U} \cap L} \rho_i \right)$$

There are four common conditions of U: $|U|=\alpha$, $|U|\geq\alpha$, $|U|\leq\beta$, and $\alpha\leq|U|\leq\beta$. Note that $\overline{U}$ is the complement of U,

$$\overline{U} = \{1, 2, \ldots, n\} \setminus U$$

The largest cardinality of $\mathcal{U}$ is:

$$|\mathcal{U}| = 2^n$$

Table 6. Bi-inferences for AND-gate, OR-gate, NAND-gate, NOR-gate, XOR-gate, XNOR-gate, and U-gate.

$$k = \frac{2^{n-1}S}{2D\Big(s(\Omega) - 2^{n-1}\Big) + M\Big(2^n - s(\Omega)\Big)} = \frac{NC}{2C} = \frac{N}{2}$$

This holds

$$\begin{aligned} 2^n S = N\Big(2D\Big(s(\Omega) - 2^{n-1}\Big) + M\Big(2^n - s(\Omega)\Big)\Big) &= 2ND\Big(s(\Omega) - 2^{n-1}\Big) + 2S\Big(2^n - s(\Omega)\Big) \\ \Rightarrow 2ND\Big(s(\Omega) - 2^{n-1}\Big) - 2S\Big(s(\Omega) - 2^{n-1}\Big) &= 0 \\ \Rightarrow (ND - S)\Big(s(\Omega) - 2^{n-1}\Big) &= 0 \end{aligned}$$

Assuming $ND = S$ we have

$$ND = S = 2NM \Rightarrow D = 2M$$

There is a contradiction because M is maximum value of D. Therefore, if k is constant with regard to D then $s(\Omega) = 2^{n-1}$. Inversely, if $s(\Omega) = 2^{n-1}$ then k is

$$k = \frac{2^{n-1}S}{2D(2^{n-1} - 2^{n-1}) + M(2^n - 2^{n-1})} = \frac{N}{2}$$

Given X-D network is combination of diagnostic relationship and X-gate inference:

$$P(Y = 1|X_1, X_2, \ldots, X_n) = P(X_1 x X_2 x \ldots x X_n)$$

$$P(D|Y) = \begin{cases} \frac{D}{S} \text{if } Y = 1 \\ \frac{M}{S} - \frac{D}{S} \text{if } Y = 0 \end{cases}$$

The diagnostic condition of X-D network is satisfied if and only if

$$s(\Omega) = \sum_a P\left(Y = 1|a(\Omega)\right) = 2^{|\Omega|-1}, \forall \Omega \neq \varnothing$$

At that time, the transformation coefficient becomes:

$$k = \frac{N}{2}$$

Note that weights $p_i = w_i$ and $\rho_i = \omega_i$, which are inputs of $s(\Omega)$, are abstract variables. Thus, the equality $s(\Omega) = 2^{|\Omega|-1}$ implies all abstract variables are removed and so $s(\Omega)$ does not depend on weights.

Table 7. Diagnostic theorem.

In general, the event that k is constant with regard to D is equivalent to the event $s(\Omega) = 2^{n-1}$. This implies *diagnostic theorem* stated in **Table 7**.

The diagnostic theorem is the optimal way to validate the diagnostic condition.

The Eq. (30) becomes simple with AND-gate inference. Recall that Eq. (14) specified AND-gate inference as follows

$$P(X_1 \odot X_2 \odot \ldots \odot X_n) = P(Y = 1|X_1, X_2, \ldots, X_n) = \begin{cases} \prod_{i=1}^{n} p_i \text{ if all } X_i(s) \text{ are } 1 \\ 0 \text{ if there exists at least one } X_i = 0 \end{cases}$$

Due to only one case $X_1 = X_2 = \ldots = X_n = 1$, we have

$$s(\Omega) = s(\Omega : \{X_i = 1\}) = \prod_{i=1}^{n} p_i$$

Due to $X_i = 0$, we have

$$s(\Omega : \{X_i = 0\}) = 0$$

Derived from Eq. (30), Eq. (31) specifies conditional probability $P(D|X_i)$, posterior probability $P(X_i|D)$, and transformation coefficient according to X-D network with AND-gate reference called *AND-D network*.

$$
\begin{aligned}
P(D|X_i = 1) &= \frac{(2D - M)\prod_{i=1}^{n} p_i + 2^{n-1}(M - D)}{2^{n-1}S} \\
P(D|X_i = 0) &= \frac{M - D}{S} \\
P(X_i = 1|D) &= \frac{(2D - M)\prod_{i=1}^{n} p_i + 2^{n-1}(M - D)}{(2D - M)\prod_{i=1}^{n} p_i + 2^{n}(M - D)} \\
P(X_i = 0|D) &= \frac{2^{n-1}(M - D)}{(2D - M)\prod_{i=1}^{n} p_i + 2^{n}(M - D)} \\
k &= \frac{2^{n-1}S}{(2D - M)\prod_{i=1}^{n} p_i + 2^{n}(M - D)}
\end{aligned}
\tag{31}
$$

For convenience, we validate diagnostic condition with a case of two sources $\Omega = \{X_1, X_2\}$, $p_1 = p_2 = w_1 = w_2 = 0.5$, $D \in \{0, 1, 2, 3\}$. According to diagnostic theorem stated in **Table 7**, if $s(\Omega) \neq 2$ for given X-gate then, such X-gate does not satisfy diagnostic condition.

Given AND-gate inference, by applying Eq. (14), we have

$$s(\Omega) = (0.5 * 0.5) + 0 + 0 + 0 = 0.25$$

Given OR-gate inference, by applying Eq. (16), we have

$$s(\Omega) = (1 - 0.5 * 0.5) + (1 - 0.5) + (1 - 0.5) + 0 = 3 - 3 * 0.5 * 0.5 = 1.75$$

Given XOR-gate inference, by applying Eq. (19), we have

$$s(\Omega) = (0.5 * 0.5 + 0.5 * 0.5) + 0.5 + 0.5 + 0 = 1.5$$

Given XNOR-gate inference, by applying Eq. (21), we have

$$s(\Omega) = (0.5 * 0.5 + 0.5 * 0.5) + 0.5 + 0.5 + 1 = 2.5$$

Given SIGMA-gate inference, by applying Eq. (23), we have

$$s(\Omega) = (0.5 + 0.5) + 0.5 + 0.5 + 0 = 2$$

It is asserted that AND-gate, OR-gate, XOR-gate, and XNOR-gate do not satisfy diagnostic condition and so they should not be used to assess hypotheses. However, it is not asserted if U-gate and SIGMA-gate satisfy such diagnostic condition. It is necessary to expend equation for SIGMA-gate diagnostic network (called *SIGMA-D network*) in order to validate it.

In case of SIGMA-gate inference, by applying Eq. (23), we have

$$\sum_i w_i = 1$$

$$s(\Omega) = 2^{n-1}\sum_i w_i = 2^{n-1}$$

$$s(\Omega : \{X_i = 1\}) = 2^{n-1}w_i + 2^{n-2}\sum_{j\neq i} w_j = 2^{n-1}w_i + 2^{n-2}(1 - w_i) = 2^{n-2}(1 + w_i)$$

$$s(\Omega : \{X_i = 0\}) = s(\Omega) - s(\Omega : \{X_i = 1\}) = 2^{n-2}(1 - w_i)$$

It is necessary to validate SIGMA-D network with SIGMA-gate bi-inference. By applying Eq. (26), we recalculate these quantities as follows

$$s(\Omega) = 2^{n-1}\sum_i w_i + 2^{n-1}\sum_i d_i = 2^{n-1}\sum_i (w_i + d_i) = 2^{n-1}$$

$$\left(\text{due to} \sum_i (w_i + d_i) = 1\right)$$

$$s(\Omega : \{X_i = 1\}) = 2^{n-1}w_i + 2^{n-2}\sum_{j\neq i} w_j + 2^{n-2}\sum_i d_i = 2^{n-2}w_i + 2^{n-2}\sum_i (w_i + d_i) = 2^{n-2}(1 + w_i)$$

$$s(\Omega : \{X_i = 0\}) = s(\Omega) - s(\Omega : \{X_i = 1\}) = 2^{n-2}(1 - w_i)$$

Obviously, quantities $s(\Omega)$, $s(\Omega{:}\{X_i{=}1\})$, and $s(\Omega{:}\{X_i = 0\})$ are kept intact. According to diagnostic theorem, we conclude that SIGMA-D network does satisfy diagnostic condition due to $s(\Omega){=}2^{n-1}$. Thus, SIGMA-D network can be used to assess hypotheses.

Eq. (32), an immediate consequence of Eq. (30), specifies conditional probability $P(D|X_i)$, posterior probability $P(X_i|D)$, and transformation coefficient for SIGMA-D network.

$$\begin{aligned}
P(D|X_i = 1) &= \frac{(2D - M)w_i + M}{2S} \\
P(D|X_i = 0) &= \frac{(M - 2D)w_i + M}{2S} \\
P(X_i = 1|D) &= \frac{(2D - M)w_i + M}{2M} \\
P(X_i = 0|D) &= \frac{(M - 2D)w_i + M}{2M} \\
k &= \frac{N}{2}
\end{aligned} \tag{32}$$

In case of SIGMA-gate, the augmented variable Y can be removed from X-D network. The evidence D is now established as direct target variable. **Figure 6** shows a so-called *direct SIGMA-gate diagnostic network* (direct SIGMA-D network).

Derived from Eq. (23), the CPT of direct SIGMA-D network is determined by Eq. (33).

$$P(D|X_1, X_2, \ldots, X_n) = \sum_{i \in K} \frac{D}{S} w_i + \sum_{j \in L} \frac{M-D}{S} w_j$$

where the set of X_i (s) is complete and mutually exclusive.

$$\begin{aligned} &\sum_{i=1}^{n} w_i = 1 \\ &X_i \cap X_j = \emptyset, \forall i \neq j \end{aligned} \tag{33}$$

Eq. (33) specifies valid CPT due to

$$\begin{aligned} \sum_D P(D|X_1, X_2, \ldots, X_n) &= \frac{1}{S} \sum_{i \in K} w_i \sum_D D + \frac{1}{S} \sum_{j \in L} w_j \sum_D (M-D) \\ &= \frac{1}{S} \sum_{i \in K} S w_i + \frac{1}{S} \sum_{j \in L} w_j (NM - S) = \frac{1}{S} \sum_{i \in K} S w_i + \frac{1}{S} \sum_{j \in L} S w_j = \sum_{i=1}^{n} w_i = 1 \end{aligned}$$

From dependencies shown in **Figure 6**, Eq. (34) specifies the joint probability of direct SIGMA-D network.

$$P(X_1, X_2, \ldots, X_n, Y, D) = P(D|X_1, X_2, \ldots, X_n) \prod_n^{i=1} P(X_i) \tag{34}$$

Inferred from Eq. (29), Eq. (35) specifies the joint probability $P(X_i, D)$ and the marginal probability $P(D)$ of direct SIGMA-D network, given uniform distribution of all sources.

$$\begin{aligned} P(X_i, D) &= \frac{1}{2^n} s(\Omega : \{X_i\}) \\ P(D) &= \frac{1}{2^n} s(\Omega) \end{aligned} \tag{35}$$

where $s(\Omega)$ and $s(\Omega:\{X_i\})$ are specified in **Table 2**.

By browsing all variables of direct SIGMA-D network, we have

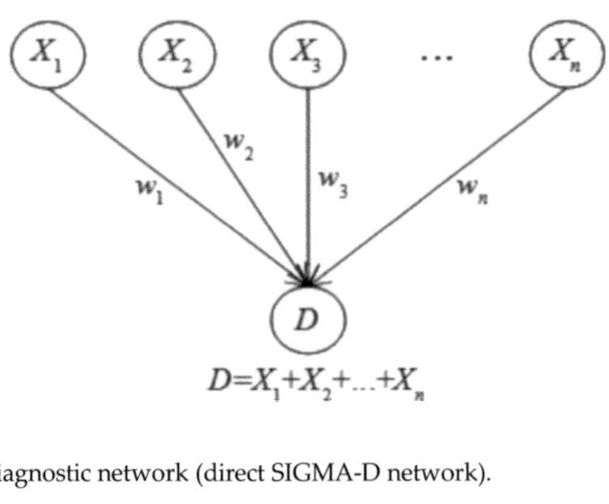

Figure 6. Direct SIGMA-gate diagnostic network (direct SIGMA-D network).

$$s(\Omega:\{X_i=1\}) = 2^{n-1}\frac{D}{S}w_i + 2^{n-2}\sum_{j\neq i}\frac{D}{S}w_j + 2^{n-2}\sum_{j\neq i}\frac{M-D}{S}w_j$$
$$= \frac{2^{n-2}}{S}(2Dw_i + M\sum_{j\neq i}w_j) = \frac{2^{n-2}}{S}\Big(2Dw_i + M(1-w_i)\Big)$$
$$\left(\text{Due to } \sum_{i=1}^{n} w_i = 1\right)$$
$$= \frac{2^{n-2}}{S}\Big((2D-M)w_i + M\Big)$$

Similarly, we have

$$s(\Omega:\{X_i=0\}) = 2^{n-1}\frac{M-D}{S}w_i + 2^{n-2}\sum_{j\neq i}\frac{M-D}{S}w_j + 2^{n-2}\sum_{j\neq i}\frac{D}{S}w_j = \frac{2^{n-2}}{S}\Big((M-2D)w_i + M\Big)$$

$$s(\Omega) = 2^{n-1}\sum_i\frac{D}{S}w_i + 2^{n-1}\sum_i\frac{M-D}{S}w_i = \frac{2^{n-1}M}{S}$$

By applying Eq. (35), $s(\Omega:\{X_i = 0\})$, $s(\Omega:\{X_i = 1\})$, and $s(\Omega)$, we get the same result with Eq. (32).

$$P(D|X_i=1) = \frac{(2D-M)w_i + M}{2S}$$
$$P(D|X_i=0) = \frac{(M-2D)w_i + M}{2S}$$
$$P(X_i=1|D) = \frac{(2D-M)w_i + M}{2M}$$
$$P(X_i=0|D) = \frac{(M-2D)w_i + M}{2M}$$

$$k = \frac{N}{2}$$

Therefore, it is possible to use direct SIGMA-D network to assess hypotheses. It is asserted that SIGMA-D network satisfy diagnostic condition when single relationship, NOT-D network, direct SIGMA-D network are specific cases of SIGMA-D network. There is a question: does an X-D network that is different from SIGMA-D network and not aforementioned exist such that it satisfies diagnostic condition?

Recall that each X-D network is a pattern owning a particular X-gate inference which in turn is based on particular X-gate condition (s) relevant to only variables A_is. The most general nonlinear X-D network is U-D network whereas SIGMA-D network is linear one. The U-gate inference given arbitrary condition on U is

$$P(X_1 \uplus X_2 \uplus \ldots \uplus X_n) = \sum_{U \in \mathcal{U}} \left(\prod_{i \in U \cap K} p_i \prod_{i \in U \cap L} (1 - \rho_i) \right) \left(\prod_{i \in \overline{U} \cap K} (1 - p_i) \prod_{i \in \overline{U} \cap L} \rho_i \right)$$

Let f be the arrangement sum of U-gate inference.

$$f(p_i, \rho_i) = \sum_{a(\Omega)} \sum_{U \in \mathcal{U}} \left(\prod_{i \in U \cap K} p_i \prod_{i \in U \cap L} (1 - \rho_i) \right) \left(\prod_{i \in \overline{U} \cap K} (1 - p_i) \prod_{i \in \overline{U} \cap L} \rho_i \right)$$

The function f is sum of many large expressions and each expression is product of four possible sub-products (Π) as follows

$$Expr = \prod_{i \in U \cap K} p_i \prod_{i \in U \cap L} (1 - \rho_i) \prod_{i \in \overline{U} \cap K} (1 - p_i) \prod_{i \in \overline{U} \cap L} \rho_i$$

In any case of degradation, there always exist expression $Expr$ (s) having at least 2 sub-products (Π), for example,

$$Expr = \prod_{i \in U \cap K} p_i \prod_{i \in U \cap L} (1 - \rho_i)$$

Consequently, there always exist $Expr$ (s) having at least 5 terms relevant to p_i and ρ_i if $n \geq 5$, for example,

$$Expr = p_1 p_2 p_3 (1 - \rho_4)(1 - \rho_5)$$

Thus, degree of f will be larger than or equal to 5 given $n \geq 5$. According to diagnostic theorem, U-gate network satisfies diagnostic condition if and only if $f(p_i, \rho_i) = 2^{n-1}$ for all $n \geq 1$ and for all abstract variables p_i and ρ_i. Without loss of generality, each p_i or ρ_i is sum of variable x and a variable a_i or b_i, respectively. Note that all p_i, ρ_i, a_i are b_i are abstract variables.

$$p_i = x + a_i$$

$$\rho_i = x + b_i$$

The equation $f - 2^{n-1} = 0$ becomes equation $g(x) = 0$ whose degree is $m \geq 5$ if $n \geq 5$.

$$g(x) = \pm x^m + C_1 x^{m-1} + \ldots + C_{m-1} x + C_m - 2^{n-1} = 0$$

where coefficients C_i s are functions of a_i and b_is. According to Abel-Ruffini theorem [11], equation $g(x) = 0$ has no algebraic solution when $m \geq 5$. Thus, abstract variables p_i and ρ_i cannot be eliminated entirely from $g(x) = 0$, which causes that there is no specification of U-gate inference $P(X_1 x X_2 x \ldots x X_n)$ so that diagnostic condition is satisfied.

It is concluded that there is no nonlinear X-D network satisfying diagnostic condition, but a new question is raised: does there exist the general linear X-D network satisfying diagnostic condition? Such linear network is called GL-D network and SIGMA-D network is specific case of GL-D network. The GL-gate probability must be linear combination of weights.

$$P(X_1 \mathrm{x} X_2 \mathrm{x} \ldots \mathrm{x} X_n) = C + \sum_{i=1}^{n} \alpha_i w_i + \sum_{i=1}^{n} \beta_i d_i$$

where C is arbitrary constant.

The GL-gate inference is singular if α_i and β_i are functions of only X_i as follows

$$P(X_1 \mathrm{x} X_2 \mathrm{x} \ldots \mathrm{x} X_n) = C + \sum_{i=1}^{n} h_i(X_i) w_i + \sum_{i=1}^{n} g_i(X_i) d_i$$

The functions h_i and g_i are not relevant to A_i because the final equation of GL-gate inference is only relevant to X_i (s) and weights (s). Because GL-D network is a pattern, we only survey singular GL-gate. Mentioned GL-gate is singular by default and it is dependent on how to define functions h_i and g_i. The arrangement sum with regard to GL-gate is

$$s(\Omega) = \sum_{a} \left(C + \sum_{i=1}^{n} h_i(X_i) w_i + \sum_{i=1}^{n} g_i(X_i) d_i \right)$$

$$= 2^n C + 2^{n-1} \sum_{i=1}^{n} \Big(h_i(X_i = 1) + h_i(X_i = 0) \Big) w_i + 2^{n-1} \sum_{i=1}^{n} \Big(g_i(X_i = 1) + g_i(X_i = 0) \Big) d_i$$

Suppose h_i and g_i are probability mass functions with regard to X_i. For all i, we have

$$0 \le h_i(X_i) \le 1$$

$$0 \le g_i(X_i) \le 1$$

$$h_i(X_i = 1) + h_i(X_i = 0) = 1$$

$$g_i(X_i = 1) + g_i(X_i = 0) = 1$$

The arrangement sum becomes

$$s(\Omega) = 2^n C + 2^{n-1} \sum_{i=1}^{n} (w_i + d_i)$$

GL-D network satisfies diagnostic condition if

$$s(\Omega) = 2^n C + 2^{n-1} \sum_{i=1}^{n} (w_i + d_i) = 2^{n-1}$$
$$\Rightarrow 2C + \sum_{i=1}^{n} (w_i + d_i) = 1$$

Suppose the set of X_is is complete.

$$\sum_{i=1}^{n} (w_i + d_i) = 1$$

This implies $C = 0$. Shortly, Eq. (36) specifies the singular GL-gate inference so that GL-D network satisfies diagnostic condition.

$$P(X_1 x X_2 x \ldots x X_n) = \sum_{i=1}^{n} h_i(X_i) w_i + \sum_{i=1}^{n} g_i(X_i) d_i$$
$$\text{where } h_i \text{ and } g_i \text{ are probability mass functions and the set of } X_i(s) \text{ is complete.} \qquad (36)$$
$$\sum_{i=1}^{n} W_i = 1$$

Functions $h_i(X_i)$ and $g_i(X_i)$ are always linear due to $X_i^m = X_i$ for all $m \geq 1$ when X_i is binary. It is easy to infer that SIGMA-D network is GL-D network with following definition of functions h_i and g_i.

$$h_i(X_i) = 1 - g_i(X_i) = X_i, \forall i$$

According to Millán and Pérez-de-la-Cruz [4], a hypothesis can have multiple evidences as seen in **Figure 7**. This is *multi-evidence diagnostic relationship* opposite to aforementioned multi-hypothesis diagnostic relationship.

Figure 7 depicts the multi-evidence diagnostic network called M-E-D network in which there are m evidences $D_1, D_2, \ldots, D_m$ and one hypothesis Y. Note that Y has uniform distribution.

In simplest case where all evidences are binary, the joint probability of M-E-D network is

$$P(Y, D_1, D_2, \ldots, D_m) = P(Y) \prod_{j=1}^{m} P(D_j|Y) = P(Y) P(D_1, D_2, \ldots, D_m|Y)$$

The product $\prod_{j=1}^{m} P(D_j|Y)$ is denoted as likelihood function as follows

$$P(D_1, D_2, \ldots, D_m|Y) = \prod_{j=1}^{m} P(D_j|Y)$$

The posterior probability $P(Y \mid D_1, D_2, \ldots, D_m)$ given uniform distribution of Y is

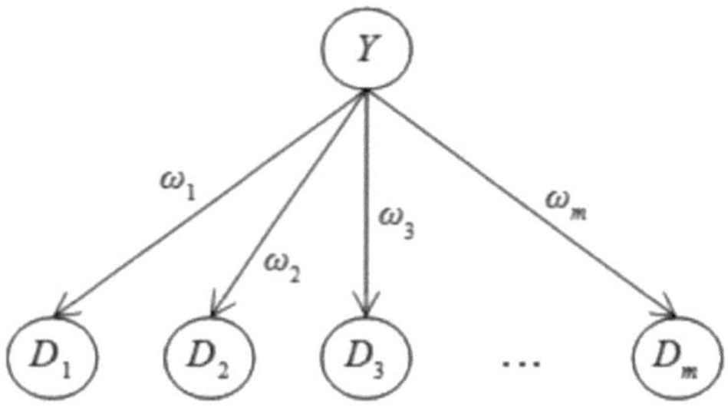

Figure 7. Diagnostic relationship with multiple evidences (M-E-D network).

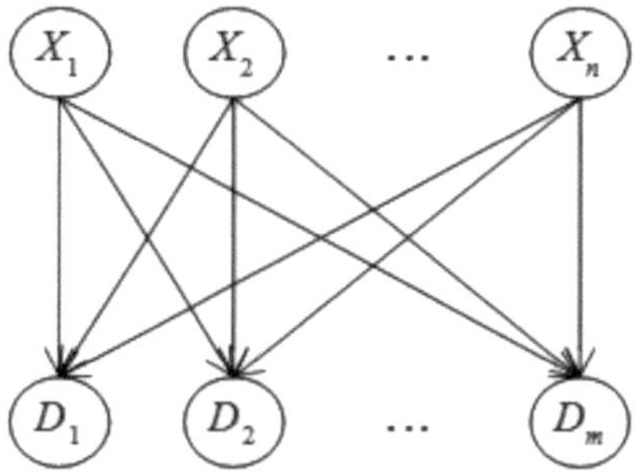

Figure 8. M-HE-D network.

$$P(Y|D_1, D_2, \ldots, D_m) = \frac{P(Y, D_1, D_2, \ldots, D_m)}{P(Y=1, D_1, D_2, \ldots, D_m) + P(Y=0, D_1, D_2, \ldots, D_m)}$$

$$= \frac{1}{\prod_{j=1}^{m} P(D_j|Y=1) + \prod_{j=1}^{m} P(D_j|Y=0)} * P(D_1, D_2, \ldots, D_m|Y)$$

The possible transformation coefficient is

$$\frac{1}{k} = \prod_{j=1}^{m} P(D_j|Y=1) + \prod_{j=1}^{m} P(D_j|Y=0)$$

M-E-D network will satisfy diagnostic condition if $k = 1$ because all hypotheses and evidence are binary, which leads that following equation specified by Eq. (37) has $2m$ real roots $P(D_j|Y)$ for all $m \geq 2$.

$$\prod_{j=1}^{m} P(D_j|Y=1) + \prod_{j=1}^{m} P(D_j|Y=0) = 1 \quad (37)$$

Eq. (37) has no real root given $m = 2$ according to following proof. Suppose Eq. (37) has 4 real roots as follows

$$a_1 = P(D_1 = 1|Y = 1)$$

$$a_2 = P(D_2 = 1|Y = 1)$$

$$b_1 = P(D_1 = 1|Y = 0)$$

$$b_2 = P(D_2 = 1|Y = 0)$$

From Eq. (37), it holds

$$\begin{cases} a_1a_2 + b_1b_2 = 1 \\ a_1(1-a_2) + b_1b_2 = 1 \\ (1-a_1)a_2 + b_1b_2 = 1 \\ a_1a_2 + b_1(1-b_2) = 1 \\ a_1a_2 + (1-b_1)b_2 = 1 \end{cases} \Rightarrow \begin{cases} a_1 = a_2 \\ b_1 = b_2 \\ a_1^2 + b_1^2 = 1 \\ a_1 + 2b_1^2 = 2 \\ b_1 + 2a_1^2 = 2 \end{cases} \Leftrightarrow \begin{cases} a_1 = a_2 = 0 \\ b_1 = b_2 \\ a_1^2 + b_1^2 = 1 \\ b_1 = 2 \end{cases} \text{or} \begin{cases} a_1 = a_2 = 0.5 \\ b_1 = b_2 \\ a_1^2 + b_1^2 = 1 \\ b_1 = 1.5 \end{cases}$$

The final equation leads a contradiction ($b_1 = 2$ or $b_1 = 1.5$) and so it is impossible to apply the sufficient diagnostic proposition into M-E-D network. Such proposition is only used for one-evidence network. Moreover, X-gate inference absorbs many sources and then produces out of one targeted result whereas the M-E-D network essentially splits one source into many results. It is impossible to model M-E-D network by X-gates. The potential solution for this problem is to group many evidences D_1, D_2,…, D_m into one representative evidence D which in turn is dependent on hypothesis Y but this solution will be inaccurate in specifying conditional probabilities because directions of dependencies become inconsistent (relationships from D_j to D and from Y to D) except that all D_js are removed and D becomes a vector. However, evidence vector does not simplify the hazardous problem and it changes the current problem into a new problem.

Another solution is to reverse the direction of relationship, in which the hypothesis is dependent on evidences so as to take advantages of X-gate inference as usual. However, the reversion method violates the viewpoint in this research where diagnostic relationship must be from hypothesis to evidence. In other words, we should change the viewpoint.

Another solution is based on a so-called *partial diagnostic condition* that is a loose case of diagnostic condition for M-E-D network, which is defined as follows

$$P(Y|D_j) = kP(D_j|Y)$$

where k is constant with regard to D_j. The joint probability is

$$P(Y, D_1, D_2, \ldots, D_m) = P(Y)\prod_{j=1}^{m} P(D_j|Y)$$

M-E-D network satisfies partial diagnostic condition. In fact, given all variables are binary, we have

$$P(Y|D_j) = \frac{\sum_{\Psi \backslash \{Y, D_j\}} P(Y, D_1, D_2, \ldots, D_m)}{\sum_{\Psi \backslash \{D_j\}} P(Y, D_1, D_2, \ldots, D_m)}$$

(Let $\Psi = \{D_1, D_2, \ldots, D_m\}$)

$$= \frac{P(D_j|Y)\prod_{k=1, k\neq j}^{m}\left(\sum_{D_k} P(D_k|Y)\right)}{\prod_{k=1, k\neq j}^{m}\left(\sum_{D_k} P(D_k|Y=1)\right) + \prod_{k=1, k\neq j}^{m}\left(\sum_{D_k} P(D_k|Y=0)\right)}$$

(Due to uniform distribution of Y)

$$= \frac{P(D_j|Y)\prod_{k=1, k\neq j}^{m} 1}{\prod_{k=1, k\neq j}^{m} 1 + \prod_{k=1, k\neq j}^{m} 1} = \frac{1}{2} P(D_j|Y)$$

$$\left(\text{Due to } \sum_{D_k} P(D_k|Y) = P(D_k = 0|Y) + P(D_k = 1|Y) = 1\right)$$

Partial diagnostic condition expresses a different viewpoint. It is not an optimal solution because we cannot test a disease based on only one symptom while ignoring other obvious symptoms, for example. The equality $P(Y|D_j) = 0.5P(D_j|Y)$ indicates the accuracy is decreased two times. However, Bayesian network provides inference mechanism based on personal belief. It is subjective. You can use partial diagnostic condition if you think that such condition is appropriate to your application.

If we are successful in specifying conditional probabilities of M-E-D network, it is possible to define an extended network which is constituted of n hypotheses $X_1, X_2, \ldots, X_n$ and m evidences $D_1, D_2, \ldots, D_m$. Such extended network represents *multi-hypothesis multi-evidence diagnostic relationship*, called M-HE-D network. **Figure 8** depicts M-HE-D network.

The M-HE-D network is the most general case of diagnostic network, which was mentioned in Ref. ([4], p. 297). We can construct any large diagnostic BN from M-HE-D networks and so the research is still open.

5. Conclusion

In short, relationship conversion is to determine conditional probabilities based on logic gates that are adhered to semantics of relationships. The weak point of logic gates is to require that all variables must be binary. For example, in learning context, it is inconvenient for expert to create an assessment BN with studying exercises (evidences) whose marks are only 0 and 1. In order to lessen the impact of such weak point, the numeric evidence is used for extending capacity of simple Bayesian network. However, combination of binary hypothesis and numeric evidence leads to errors or biases in inference. For example, given a student gets maximum grade for an exercise but the built-in inference results out that she/he has not mastered fully the associated learning concept (hypothesis). Therefore, I propose the sufficient diagnostic proposition so as to confirm that numeric evidence is adequate to make complicated inference tasks in BN. The probabilistic reasoning based on evidence is always accurate. Application of the research can go beyond learning context whenever probabilistic deduction relevant to constraints of semantic relationships is required. A large BN can be constituted of many simple BN (s). Inference in large BN is hazardous problem and there are many optimal algorithms for solving such problem. In future, I will research effective inference methods for the special BN that is constituted of X-gate BN (s) mentioned in this research because X-gate BN (s) have precise and useful features of which we should take advantages. For instance, their CPT (s) are simple in some cases and the meanings of their relationships are mandatory in many applications. Moreover, I try my best to research deeply M-E-D network and M-HE-D network whose problems I cannot solve absolutely now.

Two main documents that I referred to do this research are the book "Learning Bayesian Networks" [2] by the author Richard E. Neapolitan and the article "A Bayesian Diagnostic Algorithm for Student Modeling and its Evaluation" [4] by authors Eva Millán and José Luis Pérez-de-la-Cruz. Especially, the SIGMA-gate inference is based on and derived from the work of the Eva Millán and José Luis Pérez-de-la-Cruz. This research is originated from my PhD research "A User Modeling System for Adaptive Learning" [12]. Other references relevant to user modeling, overlay model, and Bayesian network are [13–16]. Please concern these references.

Appendices

A1. Following is the proof of Eq. (9)

$$
\begin{gathered}
P(A_i = ON|X_i) \\
= P(A_i = ON|X_i, I_i = ON)P(I_i = ON) + P(A_i = ON|X_i, I_i = OFF)P(I_i = OFF) \\
= 0 * (1 - p_i) + P(A_i = ON|X_i, I_i = OFF)p_i \\
\text{(By applying Eq. (8))} \\
= p_i P(A_i = ON|X_i, I_i = OFF)
\end{gathered}
$$

It implies

$$P(A_i = ON|X_i = 1) = p_i P(A_i = ON|X_i = 1, I_i = OFF) = p_i$$

$$P(A_i = ON|X_i = 0) = p_i P(A_i = ON|X_i = 0, I_i = OFF) = 0$$

$$P(A_i = OFF|X_i = 1) = 1 - P(A_i = ON|X_i = 1) = 1 - p_i$$

$$P(A_i = OFF|X_i = 0) = 1 - P(A_i = ON|X_i = 0) = 1 \ \blacksquare$$

A2. Following is the proof of Eq. (10)

$$P(Y|X_1, X_2, \ldots, X_n) = \frac{P(Y, X_1, X_2, \ldots, X_n)}{P(X_1, X_2, \ldots, X_n)}$$

(Due to Bayes' rule)

$$= \frac{\sum_{A_1, A_2, \ldots, A_n} P(Y, X_1, X_2, \ldots, X_n|A_1, A_2, \ldots, A_n) * P(A_1, A_2, \ldots, A_n)}{P(X_1, X_2, \ldots, X_n)}$$

(Due to total probability rule)

$$= \sum_{A_1, A_2, \ldots, A_n} P(Y, X_1, X_2, \ldots, X_n|A_1, A_2, \ldots, A_n) * \frac{P(A_1, A_2, \ldots, A_n)}{P(X_1, X_2, \ldots, X_n)}$$

$$= \sum_{A_1, A_2, \ldots, A_n} P(Y|A_1, A_2, \ldots, A_n) * P(X_1, X_2, \ldots, X_n|A_1, A_2, \ldots, A_n) * \frac{P(A_1, A_2, \ldots, A_n)}{P(X_1, X_2, \ldots, X_n)}$$

(Because Y is conditionally independent from X_is given A_is)

$$= \sum_{A_1, A_2, \ldots, A_n} P(Y|A_1, A_2, \ldots, A_n) * \frac{P(X_1, X_2, \ldots, X_n, A_1, A_2, \ldots, A_n)}{P(X_1, X_2, \ldots, X_n)}$$

$$= \sum_{A_1, A_2, \ldots, A_n} P(Y|A_1, A_2, \ldots, A_n) * P(A_1, A_2, \ldots, A_n|X_1, X_2, \ldots, X_n)$$

(Due to Bayes' rule)

$$= \sum_{A_1, A_2, \ldots, A_n} P(Y|A_1, A_2, \ldots, A_n) \prod_{i=1}^{n} P(A_i|X_1, X_2, \ldots, X_n)$$

(Because A_is are mutually independent)

$$= \sum_{A_1, A_2, \ldots, A_n} P(Y|A_1, A_2, \ldots, A_n) \prod_{i=1}^{n} P(A_i|X_i)$$

(Because each A_i is only dependent on X_i) ■

A3. Following is the proof that the augmented X-D network (shown in **Figure 5**) is equivalent to X-D network (shown in shown in **Figures 2** and **3**) with regard to variables X_1, X_2,…, X_n, and D.

The joint probability of augmented X-D network shown in **Figure 5** is

$$P(X_1, X_2, \ldots, X_n, Y, D) = P(D|Y)P(Y|X_1, X_2, \ldots, X_n) \prod_{i=1}^{n} P(X_i)$$

The joint probability of X-D network is

$$P(X_1, X_2, \ldots, X_n, D) = P(D|X_1, X_2, \ldots, X_n)\prod_{i=1}^{n}P(X_i)$$

By applying total probability rule into X-D network, we have

$$P(X_1, X_2, \ldots, X_n, D) = \frac{P(D, X_1, X_2, \ldots, X_n)}{P(X_1, X_2, \ldots, X_n)}\prod_{i=1}^{n}P(X_i)$$

(Due to Bayes' rule)

$$= \frac{\sum_Y P(D, X_1, X_2, \ldots, X_n|Y)P(Y)}{P(X_1, X_2, \ldots, X_n)}\prod_{i=1}^{n}P(X_i)$$

(Due to total probability rule)

$$= \frac{\sum_Y P(D, X_1, X_2, \ldots, X_n|Y)P(Y)}{P(X_1, X_2, \ldots, X_n)}\prod_{i=1}^{n}P(X_i)$$

$$= \left(\sum_Y P(D, X_1, X_2, \ldots, X_n|Y) * \frac{P(Y)}{P(X_1, X_2, \ldots, X_n)}\right) * \prod_{i=1}^{n}P(X_i)$$

$$= \left(\sum_Y P(D|Y) * \frac{P(X_1, X_2, \ldots, X_n|Y)P(Y)}{P(X_1, X_2, \ldots, X_n)}\right) * \prod_{i=1}^{n}P(X_i)$$

(Because D is conditionally independent from all X_i (s) given Y)

$$= \left(\sum_Y P(D|Y) * \frac{P(Y, X_1, X_2, \ldots, X_n)}{P(X_1, X_2, \ldots, X_n)}\right) * \prod_{i=1}^{n}P(X_i)$$

$$= \sum_Y P(D|Y)P(Y|X_1, X_2, \ldots, X_n)\prod_{i=1}^{n}P(X_i)$$

(Due to Bayes' rule)

$$= \sum_Y P(X_1, X_2, \ldots, X_n, Y, D) \blacksquare$$

A4. Following is the proof of Eq. (29)

Given uniform distribution of X_i (s), we have

$$P(X_1) = P(X_2) = \cdots = P(X_n) = \frac{1}{2}$$

The joint probability becomes

$$P(\Omega, Y, D) = \frac{1}{2^n}P(Y|X_1, X_2, \ldots, X_n)P(D|Y)$$

The joint probability of X_i and D is

$$
\begin{aligned}
P(X_i, D) = & \sum_{\{\Omega, Y, D\} \backslash \{X_i, D\}} P(\Omega, Y, D) \\
= & P(X_1 = 1, X_2 = 1, \dots, X_i, \dots, X_{n-1} = 1, X_n = 1, Y = 1, D) \\
& + P(X_1 = 1, X_2 = 1, \dots, X_i, \dots, X_{n-1} = 1, X_n = 0, Y = 1, D) + \cdots \\
& + P(X_1 = 0, X_2 = 0, \dots, X_i, \dots, X_{n-1} = 0, X_n = 1, Y = 1, D) \\
& + P(X_1 = 0, X_2 = 0, \dots, X_i, \dots, X_{n-1} = 0, X_n = 0, Y = 1, D) \\
& + P(X_1 = 1, X_2 = 1, \dots, X_i, \dots, X_{n-1} = 1, X_n = 1, Y = 0, D) \\
& + P(X_1 = 1, X_2 = 1, \dots, X_i, \dots, X_{n-1} = 1, X_n = 0, Y = 0, D) + \cdots \\
& + P(X_1 = 0, X_2 = 0, \dots, X_i, \dots, X_{n-1} = 0, X_n = 1, Y = 0, D) \\
& + P(X_1 = 0, X_2 = 0, \dots, X_i, \dots, X_{n-1} = 0, X_n = 0, Y = 0, D) \\
= & \frac{1}{2^n} \frac{D}{S} \Big(P(Y = 1 | X_1 = 1, X_2 = 1, \dots, X_i, \dots, X_{n-1} = 1, X_n = 1) + P(Y = 1 | X_1 = 1, X_2 \\
= & 1, \dots, X_i, \dots, X_{n-1} = 1, X_n = 0) + \cdots + P(Y = 1 | X_1 = 1, X_2 = 1, \dots, X_i, \dots, X_{n-1} = 0, X_n = 1) \\
& + P(Y = 1 | X_1 = 1, X_2 = 1, \dots, X_i, \dots, X_{n-1} = 0, X_n = 0) \Big) \\
& + \frac{1}{2^n} \frac{M - D}{S} \Big(P(Y = 0 | X_1 = 1, X_2 = 1, \dots, X_i, \dots, X_{n-1} = 1, X_n = 1) + P(Y = 0 | X_1 = 1, X_2 \\
= & 1, \dots, X_i, \dots, X_{n-1} = 1, X_n = 0) + \cdots + P(Y = 0 | X_1 = 1, X_2 = 1, \dots, X_i, \dots, X_{n-1} = 0, X_n = 1) \\
& + P(Y = 0 | X_1 = 1, X_2 = 1, \dots, X_i, \dots, X_{n-1} = 0, X_n = 0) \Big)
\end{aligned}
$$

(Due to Eq. (6))

The marginal probability of D is

$$
\begin{aligned}
P(D) = & \sum_{\{\Omega, Y, D\} \backslash \{D\}} P(\Omega, Y, D) \\
= & P(X_1 = 1, X_2 = 1, \dots, X_n = 1, Y = 1, D) + P(X_1 = 1, X_2 = 1, \dots, X_n = 0, Y = 1, D) + \cdots \\
& + P(X_1 = 0, X_2 = 0, \dots, X_n = 1, Y = 1, D) + P(X_1 = 0, X_2 = 0, \dots, X_n = 0, Y = 1, D) \\
& + P(X_1 = 1, X_2 = 1, \dots, X_n = 1, Y = 0, D) \\
= & \frac{1}{2^n} \frac{D}{S} \Big(P(Y = 1 | X_1 = 1, X_2 = 1, \dots, X_n = 1) + P(Y = 1 | X_1 = 1, X_2 = 1, \dots, X_n = 0) + \cdots \\
& + P(Y = 1 | X_1 = 1, X_2 = 1, \dots, X_n = 1) + P(Y = 1 | X_1 = 1, X_2 = 1, \dots, X_n = 0) \Big) \\
& + \frac{1}{2^n} \frac{M - D}{S} \Big(P(Y = 0 | X_1 = 1, X_2 = 1, \dots, X_n = 1) + P(Y = 0 | X_1 = 1, X_2 = 1, \dots, X_n = 0) + \cdots \\
& + P(Y = 0 | X_1 = 1, X_2 = 1, \dots, X_n = 1) + P(Y = 0 | X_1 = 1, X_2 = 1, \dots, X_n = 0) \Big) \\
& + P(X_1 = 1, X_2 = 1, \dots, X_n = 0, Y = 0, D) + \cdots
\end{aligned}
$$

By applying **Table 2**, the joint probability $P(X_i, D)$ is determined as follows

$$P(X_i, D) = \frac{1}{2^n S}\left(D\sum_a P\Big(Y = 1|a(\Omega : \{X_i\})\Big) + (M - D)\sum_a P\Big(Y = 0|a(\Omega : \{X_i\})\Big)\right)$$
$$= \frac{1}{2^n S}\left(D\sum_a P\Big(Y = 1|a(\Omega : \{X_i\})\Big) + (M - D)\sum_a \Big(1 - P\Big(Y = 1|a(\Omega : \{X_i\})\Big)\Big)\right)$$
$$= \frac{1}{2^n S}\Big((2D - M)s(\Omega : \{X_i\}) + 2^{n-1}(M - D)\Big)$$

Similarly, the marginal probability $P(D)$ is

$$P(D) = \frac{1}{2^n S}\Big((2D - M)s(\Omega) + 2^n(M - D)\Big) \blacksquare$$

Author details

Loc Nguyen

Address all correspondence to: ng_phloc@yahoo.com

Sunflower Soft Company, An Giang, Vietnam

References

[1] Wikipedia. Logic gate. Wikimedia Foundation [Internet]. 2016. [Online]. Available from: https://en.wikipedia.org/wiki/Logic_gate [Accessed June 4, 2016]

[2] Neapolitan RE. Learning Bayesian Networks. Upper Saddle River, New Jersey: Prentice Hall; 2003. p. 674

[3] Díez FJ, Druzdzel MJ. Canonical Probabilistic Models. Madrid: Research Centre on Intelligent Decision-Support Systems; 2007

[4] Millán E, Pérez-de-la-Cruz JL. A bayesian diagnostic algorithm for student modeling and its evaluation. User Modeling and User-Adapted Interaction. 2002;**12**(2-3):281–330

[5] Wikipedia. Factor graph. Wikimedia Foundation [Internet]. 2015. [Online]. Available from: https://en.wikipedia.org/wiki/Factor_graph [Accessed: February 8, 2017]

[6] Kschischang FR, Frey BJ, Loeliger HA. Factor graphs and the sum-product algorithm. IEEE Transactions on Information Theory. 2001;**47**(2):498–519

[7] Pearl J. Fusion, propagation, and structuring in belief networks. Artificial Intelligence. 1986;**29**(3):241–288

[8] Millán E, Loboda T, Pérez-de-la-Cruz JL. Bayesian networks for student model engineering. Computers & Education. 2010:**55**(4):1663–1683

[9] Nguyen L. Theorem of SIGMA-gate inference in Bayesian network. Wulfenia Journal. 2016;**23**(3):280–289

[10] Wikipedia. Set (mathematics), Wikimedia Foundation [Internet]. 2014. [Online]. Available from: http://en.wikipedia.org/wiki/Set_(mathematics) [Accessed: October 11, 2014]

[11] Wikipedia. Abel-Ruffini theorem. Wikimedia Foundation [Internet]. 2016. [Online]. Available from: https://en.wikipedia.org/wiki/Abel%E2%80%93Ruffini_theorem [Accessed: June 26, 2016]

[12] Nguyen L. A User Modeling System for Adaptive Learning. Abuja, Nigeria: Standard Research Journals; 2014

[13] Fröschl C. User modeling and user profiling in adaptive E-learning systems [master thesis]. Graz, Austria: Graz University of Technology; 2005

[14] De Bra P, Smits D, Stash N. The Design of AHA!. In Proceedings of the Seventeenth ACM Hypertext Conference on Hypertext and hypermedia (Hypertext '06); 22-25 August 2006; Odense, Denmark. New York, NY: ACM; 2006. pp. 133–134

[15] Murphy KP. A Brief Introduction to Graphical Models and Bayesian Networks. University of British Columbia; 1998. [Online]. Available from: http://www.cs.ubc.ca/~murphyk/Bayes/bnintro.html [Accessed: 2008]

[16] Heckerman D. A Tutorial on Learning With Bayesian Networks. Redmond: Microsoft Research; 1995

Section 2

Applications of Bayesian Inference in Life Sciences

Chapter 7

Bayesian Estimation of Multivariate Autoregressive Hidden Markov Model with Application to Breast Cancer Biomarker Modeling

Hamid El Maroufy, El Houcine Hibbah,
Abdelmajid Zyad and Taib Ziad

Additional information is available at the end of the chapter

http://dx.doi.org/10.5772/intechopen.70053

Abstract

In this work, a first-order autoregressive hidden Markov model (AR(1)HMM) is proposed. It is one of the suitable models to characterize a marker of breast cancer disease progression essentially the progression that follows from a reaction to a treatment or caused by natural developments. The model supposes we have observations that increase or decrease with relation to a hidden phenomenon. We would like to discover if the information about those observations can let us learn about the progression of the phenomenon and permit us to evaluate the transition between its states (supposed discrete here). The hidden states governed by the Markovian process would be the disease stages, and the marker observations would be the depending observations. The parameters of the autoregressive model are selected at the first level according to a Markov process, and at the second level, the next observation is generated from a standard autoregressive model of first order (unlike other models considering the successive observations are independents). A Markov Chain Monte Carlo (MCMC) method is used for the parameter estimation, where we develop the posterior density for each parameter and we use a joint estimation of the hidden states or block update of the states.

Keywords: autoregressive hidden Markov model, breast cancer progression marker, Gibbs sampler, hidden states joint estimation, Markov Chain Monte Carlo

1. Introduction

The main motivation behind this work is to characterize progression in breast cancer. In fact, disease progression cannot be assessed correctly without the use of biomarkers, which would effectively monitor the evolution of the patient health state, and this is the case for breast cancer. The major challenge in this matter for researchers and clinicians is to unravel the stage of the disease, so as to tailor the treatment for each patient and to monitor the response of a patient to a treatment.

Currently, studies have shown that there is a correlation between the levels of certain markers such as cancer antigen CA15-3, carcinoembryonic antigen (CEA), and serum HER2 Neu with the stage of the disease [1]. This gives an opportunity of using a hidden Markov model (HMM) to predict the stage of the disease based on biomarker data and to address the effectiveness of the treatments in their influence on the transition of the cancer from one state to another. In HMM, we have two constituents: the Markovian hidden process suitable to represent the breast cancer stage and the observation process given by the biomarker data. By the way, we can learn about the disease transition rates and how it progresses from primary breast cancer to advanced cancer stage, for example.

Indeed, HMM is a useful tool for tackling numerous concrete problems in many fields but some possible applications of HMM are in speech processing [2], in biology [3], in disease progression [4], in economics [5, 6], and in gene expression [7]. For a complete review of HMM, the reader is referred to Zucchini and MacDonald [8], in which properties and definitions of HMM are presented in a plausible way with both classical estimation by maximum likelihood method and expectation maximization (EM) algorithm and the new Bayesian inference is addressed.

The model we consider here is a variation of the regular hidden Markov model, since we will use extensions to incorporate dependence among successive observations, suggesting autoregressive dependence among continuous observations. Consequently, we have relaxed the conditional independence assumption from a standard HMM, because we would like to add some dynamics to the patient disease progression and because in reality the current patient biomarker observation is dependent on the past one. In fact, the autoregressive assumption in HMM has shown its advantage over regular HMM that cannot catch the strong dependence between successive observations (e.g., Ref. [9]). A similar model to ours can be found in Ref. [10]. This kind of models, which were first proposed in Ref. [11] to describe econometrics time series, is generalization of both HMM and autoregressive models, will be effective in representing multiple heterogeneous dynamics such as the disease progression dynamics, and can be even generalized to a regime-switching ARMA models such as in Ref. [12].

Moreover, Our model can also be viewed as an extension of the multivariate double-chain Markov model (DCMM) developed by Ref. [13], where there are two discrete Markov chains of first order: the first Markov chain is observed and the second one is hidden. In contrast to this DCMM, our multivariate first-order autoregressive hidden Markov model (MAR(1)HMM) will lead to continuous observations, where each observation conditional on the hidden process will depend on the previous observation according to an autoregressive process of first order. This dynamic is promising for continuous observed disease biomarkers.

Parameter estimation is very challenging for HMM family models since the likelihood is not available in a closed form most of the time. Thus, we call for a Markov Chain Monte Carlo (MCMC) procedure instead of a maximum likelihood-based approach. This choice rises from the fact that the Bayesian analysis uses prior knowledge about the process being measured, and it allows direct probability statements and an approximation of posterior distributions for the parameters. Instead in the maximum likelihood approach, we cannot have declared prior or have exact distribution for the parameters when the likelihood is untractable or when we have missing data (e.g., Refs. [14–16]).

Since the realization of HMM includes two separate entities: the parameters and the hidden states, the Bayesian computation is carried out after augmenting the likelihood by the missing hidden states [17]. The hidden states are sampled using a Gibbs sampler adopting a joint estimation of the hidden states or block update of the states (instead of a single update of each state separately) by means of a forward filtering/backward smoothing algorithm. Given the hidden states, we can compute the autoregressive parameters and the transition probabilities of the Markov chain by Gibbs sampler from their posterior densities after specifying conjugate priors for the parameters. Hence, the MCMC algorithm will alternate between simulating the hidden states and the parameters. Finally, we can obtain posteriors statistics such as the means, standard deviations and confidence intervals after assessing the convergence of the MCMC algorithm.

This chapter is organized as follows: after a preliminary on HMM, a description of the model is given in Section 3. In Section 4, we will give the Bayesian estimation of the parameters and the hidden states and provide the details of the MCMC algorithm, before presenting the results of a simulation studies in Section 5 and we will finish by a conclusion.

2. Preliminary

Since the model suggested is of the HMM type, we will describe HMM in more detail: an HMM is a stochastic process $\{X_t, Y_t\}_{t=0}^{T}$, where $\{X_t\}_{t=0}^{T}$ is a hidden Markov chain (unobservable) and $\{Y_t\}_{t=0}^{T}$ is a sequence of observable independent random variables such that Y_t depends only on X_t for the time $t = 0,1,\ldots T$. Here the process $\{X_t\}_{t=0}^{T}$ evolves independently of $\{Y_t\}_{t=0}^{T}$ and is supposed to be a homogeneous finite Markov chain with probability transition matrix Π of dimension $a \times a$, where a indicates the number of the hidden states and $\Pi_0 = (\Pi_{01},\ldots,\Pi_{0a})$ is the initial state distribution.

We denote the probability density function of $Y_t = y_t$ given $X_t = k$ for $k \in \{1, \ldots, a\}$ with $P_{x_t}(y_t, \theta_k)$, where θ_k refers to the parameters of P when $X_t = k$. We suppose further that the processes $Y_t | X_t$ and $Y_{t'} | X_{t'}$ are independent for $t \neq t'$. Let $\Theta = (\theta_1,\ldots, \theta_a)$ and $\theta = (\Pi_0, \Pi, \Theta)$, and then, the HMM can be described as follows: First, the likelihood of the observations and the hidden states can be decomposed to $P(y_0, \ldots, y_T, x_0, \ldots, x_T, \theta) = P(y_0, \ldots, y_T|x_0, \ldots, x_T, \theta)P(x_0, \ldots, x_T, \theta)$. Since $\{X_t\}_{t=0}^{T}$ is a Markov chain, $P(x_0, \ldots, x_T, \theta) = \Pi_0(x_0)\prod_{t=1}^{T}\Pi(x_t|x_{t-1})$. Under the conditional independence of the observations given the hidden states, $P(y_0, \ldots, y_T|x_0, \ldots, x_T, \theta) = P_{x_0}(y_0|\theta_{x_0})\prod_{t=1}^{T}P_{x_t}(y_t|\theta_{x_t})$. Consequently, the likelihood function for the hidden states and the observations is given by $P(y_0, y_1, \ldots, y_T, x_0, x_1, \ldots, x_T, \theta) = \Pi(x_0)P_{x_0}(y_0|\theta_{x_0})\prod_{t=0}^{T}\Pi(x_t|x_{t-1})P_{x_t}(y_t|\theta_{x_t})$.

3. Model description and specification

The MAR(1)HMM model we consider in this work is a hidden Markov model, where conditionally on the latent states, the observations are not independent like it is the case for a regular hidden Markov model. Instead, the current observation is allowed to depend on the previous observation according to an autoregressive model of first order. As in an HMM model, the latent states evolve according to a discrete first-order time homogeneous Markov model. We consider data of n continuous random variables observed over time, each of potentially different lengths, i.e., for each individual $i = 1,2, \ldots, n$, we observe a vector $y_{i,.} = (y_{i,u_i}, \ldots, y_{i,m_i})^T$, with $u_i < m_i$.

Define $u_0 = \min_{1 \le i \le n}\{u_i\}$ and $M = \max_{1 \le i \le n}\{m_i\}$ and note that the times u_i and m_i may vary over the entire observation period from u_0 to M with the restriction that $u_i - m_i \ge 1$, for $i = 1,2,\ldots,n$.

We assume, for $i = 1,2,\ldots,n$ for integer time $t = u_i,\ldots,m_i$, that the random variable $Y_{i,t}$ taking nonnegative values depends only on the states X_t and the previous observation $Y_{i,t-1}$, and based on the model developed by Farcomeni and Arima [10], we get the following model:

$$Y_{i,t}|_{X_t=x_t} = \beta^{(x_t)} Y_{i,t-1} + \mu^{(x_t)} + \varepsilon_{i,t}. \tag{1}$$

The choice of the autoregressive part of the model is motivated by the fact that successive biomarker observations are most of the time correlated from many diseases unlike the hypothesis of independence between observations in HMMs.

We interpret x as the vector of the hidden health states of the patients; in the case of breast cancer, those states would be localized or advanced metastatic breast cancer for example, while y is the vector of the biomarkers observed and measured for the patients. The $\varepsilon_{i,t}$ are normal variables with mean 0 and variance σ^2 such that $\varepsilon_{i,t}$ and $\varepsilon_{i',t'}$ are uncorrelated, $(i, t) \neq (i', t')$.

The parameters $\beta^{(x_t)}$ and $\mu^{(x_t)}$ are parameters taking values in $\mathbb{R}$ for each hidden state and $\sigma^2 \in \mathbb{R}^+$.

Similar to Ref. [13], the transition matrix of the Markov chain Π is time homogeneous with dimension $a \times a$ where a is the number of hidden states, and $\Pi = (\Pi_{gh}, g = 1,\ldots,a; h = 1,\ldots,a)$ where $\Pi_{gh} = P(X_t = h \mid X_{t-1} = g)$, for $g = 1,2,\ldots,a; h = 1,2,\ldots, a$; and $t = u_{0+1},\ldots, M$. We let the first state X_{u_0} to be selected from a discrete distribution with vector of probabilities $r = (r_1,\ldots,r_a)$. Also we consider the time of initial observation u_i, the initial observed state y_{i,u_i}, and the number of consecutive time points that were observed $m_i - u_i + 1$. Let $\mu = (\mu^{(1)},\ldots, \mu^{(a)})$, $\beta = (\beta^{(1)},\ldots,\beta^{(a)})$, and $\theta = (\mu, \beta, \sigma^2, r, \Pi)$ be the set of all parameters in the model. We suppose that the individuals, i.e., $Y_{i,t}$, behave independently conditionally on X. Therefore, for $i = 1,\ldots,n$, $P(y_{i,.}|y_{i,u_i}, x, \theta) = \prod_{t=u_i+1}^{m_i} P(y_{i,t}|y_{i,t-1}, x_t, \Theta)$ and $P(x|\theta) = P(x_{u_0}) \prod_{t=u_0+1}^{M} P(x_t|x_{t-1}, \Pi)$, where $P(x_t|x_{t-1}, \Pi) = P(X_t = x_t|X_{t-1} = x_{t-1}, \Pi) = \Pi_{x_{t-1},x_t}$. Then, the likelihood density for the observations of all individuals $y = (y_1,\ldots,y_n)$ given first time vector of observations $y_0 = (y_{1,u_1}, \ldots, y_{n,u_n})$, x, and θ is

$$P(y|y_0, x, \theta) = \prod_{i=1}^{n} P(y_{i,.}|y_{i,u_i}, x, \theta),$$

This is due to the conditional independence of the y_i, given x and θ. The joint mass of each $y_{i,\cdot}$ and x given y_{u_i} and θ can be written as follows: $P(y_{i,\cdot}, x|y_{i,u_i}, \theta) = P(y_{i,\cdot}|y_{i,u_i}, x, \theta) \times P(x|y_{u_i}, \theta)$. Using the Markov property of the hidden process, we have after simplification $P(x|y_{i,u_i}, \theta) \propto P(y_{i,u_i}|x, \theta)P(x|\theta) = P(y_{i,u_i}|x_{u_i}, \theta) r_{x_{u_0}} \Pi_{x_{u_0}, x_{u_0+1}} \times \cdots \times \Pi_{x_{M-1}, x_M}$. In addition, $P(y_{i,\cdot}|y_{i,u_i}, x, \theta) = \prod_{t=u_i+1}^{m_i} P(y_{i,t}|y_{i,t-1}, x, \theta)$, and consequently,

$P(y_{i,\cdot}, x|y_{i,u_i}, \theta) \propto r_{x_{u_0}} P(y_{i,u_i}|x_{u_i}, \theta) \prod_{t=u_0+1}^{M} \Pi_{x_{t-1}, x_t} \prod_{t=u_i+1}^{m_i} P(y_{i,t}|y_{i,t-1}, x, \theta)$. Finally, under the hypothesis of normal error distribution for the autoregressive parameters of the model (Eq. (1)) and the Chapman-Kolmogorov property, the joint distribution of $y_{i,\cdot}$ and x given y_{i,u_i} and θ can be simplified to:

$$P(y_{i,\cdot}, x|y_{i,u_i} \theta) \propto P(y_{i,u_i}|x_{u_i}, \theta) \prod_{h=1}^{a} r_h^{\chi_{\{x_{u_0}\}}(h)} \prod_{t=u_0+1}^{M} \prod_{g=1}^{a} \prod_{h=1}^{a} \Pi_{g,h}^{\chi_{\{x_t, x_{t-1}\}}(g,h)}$$
$$\times \prod_{t=u_i+1}^{m_i} \prod_{h=1}^{a} \left[\frac{1}{\sigma} \phi\left(\frac{y_{i,t} - \mu^{(h)} - \beta^{(h)} y_{i,t-1}}{\sigma} \right) \right]^{\chi_{\{x_t\}}(h)},$$

where ϕ denotes the density of a standard normal distribution $\mathcal{N}(0,1)$ and $\chi_{\{A\}}(x)$ is the usual indicator function of a set A. Finally, the joint distribution of y and x has the following form:

$$P(y, x|y_0, \theta) \propto \prod_{h=1}^{a} r_h^{\chi_{\{x_{u_0}\}}(h)} \prod_{t=u_0+1}^{M} \prod_{g=1}^{a} \prod_{h=1}^{a} \Pi_{g,h}^{\chi_{\{x_t, x_{t-1}\}}(g,h)}$$
$$\times \prod_{i=1}^{n} \prod_{l=1}^{a} \left[\frac{1}{\sigma} \phi\left(\frac{y_{i,u_i} - \mu^{(l)}}{\sigma} \right) \right]^{\chi_{\{x_t\}}(l)} \times \prod_{i=1}^{n} \prod_{t=u_i+1}^{m_i} \prod_{h=1}^{a} \left[\frac{1}{\sigma} \phi\left(\frac{y_{i,t} - \mu^{(h)} - \beta^{(h)} y_{i,t-1}}{\sigma} \right) \right]^{\chi_{\{x_t\}}(h)}. \quad (2)$$

4. Bayesian estimation of the model parameters

We will use a Bayesian approach to estimate the model parameters. Inference in the Bayesian framework is obtained through the posterior density, which is proportional to the prior multiplied by the likelihood. The posterior distribution for our model, as in most cases, cannot be derived analytically, and we will approximate it through MCMC methods specifically designed for working with the augmented likelihood with the hidden states. In fact, MCMC methods start by specifying the prior density $\Pi(\theta)$ for the parameters. Since the data Y are available, the general sampling methods work recursively by alternating between simulating the full conditional distributions X given y and θ given x and y.

4.1. Prior distributions

Under the assumption of independence between the parameters $\theta = (\mu, \beta, \sigma^2, r, \Pi)$, the prior density could be written as $P(\theta) = P(r)P(\Pi)P(\mu)P(\beta)P(\sigma^2)$. r is the parameters of a multinomial distribution; hence, the natural choice for the prior would be a Dirichlet distribution

$r \sim \mathbb{D}(\alpha_{01}, \ldots, \alpha_{0a})$. Later on, $\sum_{j=1}^{a} \Pi_{ij} = 1$, and we assume that $\Pi_i \sim \mathbb{D}(\delta_{i1}, \ldots, \delta_{ia})$ for each row i of the transition matrix. This choice of the Dirichlet prior can be even the default $\mathbb{D}(1,\ldots,1)$ as recently discussed in Ref. [18]. In fact, a Dirichlet prior is justified because the posterior density of each row of the transition matrix is proportional to the density of a Dirichlet distribution, and hence, choosing a Dirichlet prior would give a posterior Dirichlet. This can be justified as follows for a given set of parameters $\lambda = (\lambda_1, \ldots, \lambda_a)$ from a discrete or from a multinomial density:

$\pi(x_1, \ldots, x_a, \lambda_1, \ldots, \lambda_a) = \frac{n!}{x_1!\ldots x_a!} \lambda_1^{x_1} \ldots \lambda_a^{x_a}$ for the nonnegative integers $x_1, \ldots, x_a$, with $\sum_{i=1}^{a} x_i = n$.

This probability mass function can be expressed, using the gamma function Γ, as $\pi(x_1, \ldots, x_a, \lambda_1, \ldots, \lambda_a) = \frac{\Gamma\left(\sum_{i=1}^{a} x_i + 1\right)}{\prod_{i=1}^{a} \Gamma(x_i + 1)} \prod_{i=1}^{a} \lambda_i^{x_i}$. This form shows its resemblance to the Dirichlet distribution, and by starting from supposing the prior $\lambda \propto \mathbb{D}(\alpha_0, \ldots, \alpha_a)$, the posterior is $P(\lambda|x) \propto P(\lambda)P(x|\lambda) \propto \prod_i \lambda_i^{x_i} \prod_i \lambda_i^{\alpha_i - 1} \propto \prod_i \lambda_i^{x_i + \alpha_i - 1} \propto \mathbb{D}(x_1 + \alpha_1, \ldots, x_a + \alpha_a)$.

Furthermore, concerning the priors for parameters of the autoregressive model, we suppose for $h = 1,\ldots,a$: $\mu^{(h)} \sim \mathcal{N}(\alpha_h, \tau_h)$, $\beta^{(h)} \sim \mathcal{N}(b_h, c_h)$, and inverse gamma ($\mathbb{IG}$) prior for $\sigma^2 \sim \mathbb{IG}(\varepsilon, \zeta)$. $\alpha_h, \tau_h, b_h, c_h, \epsilon, \zeta$ are hyperparameters to be specified. For more details on Bayesian inference and prior selection in HMM, the reader is referred to Ref. [19]. In our case, prior distributions for the autoregressive parameters were proposed by Ref. [20] for a mixture autoregressive model, who points out that they are conventional prior choices for mixture models.

4.2. Sampling the posterior distribution for the hidden states

Chib [21] developed a method for the simulation of the hidden states from the full joint distribution for the univariate hidden Markov model case. We will describe his full Bayesian algorithm for the univariate hidden Markov model before a generalization to our MAR(1) HMM.

4.2.1. *Chib's algorithm for the univariate hidden Markov model for estimation of the states*

Suppose we have an observed process $Y_n = (y_1, \ldots, y_n)$ and the hidden states $X_n = (x_1, \ldots, x_n)$, θ are the parameters of the model. We adopt for simplicity $X_t = (x_1, \ldots, x_t)$ the history of the states up to time t and $X^{t+1} = (x_{t+1}, \ldots, x_n)$ the future from $t + 1$ to n. We use the same notation for Y_t and Y^{t+1}.

For each state $x_t \in \{1, 2, \ldots, a\}$ for $t = 1, 2, \ldots, n$, the hidden model can be described by a conditional density given the hidden states $\pi(y_t|Y_{t-1}, x_t = k) = \pi(y_t|Y_{t-1}, \theta_k)$, $k = 1,\ldots,a$, with x_t

depending only on x_{t-1} and having transition matrix Π and initial distribution Π_0, and the parameters for $\pi(.)$ are $\theta = (\theta_1, \ldots, \theta_a)$.

Chib [21] shows that it is preferable to simulate the full latent data $X_n = (x_1, \ldots, x_n)$ from the joint distribution of $x_1, \ldots, x_n | Y_n, \theta$, in order to improve the convergence property of the MCMC algorithm because instead of n additional blocks if each state is simulated separately, only one additional block is required. First, we write the joint conditional density as

$$P(X_n|Y_n, \theta, \Pi) = P(x_n|Y_n, \theta) P(x_{n-1}|Y_n, x_n, \theta, \Pi) \times \cdots \times P(x_1|Y_n, X^2, \theta, \Pi).$$

For sampling, it is sufficient to consider the sampling of x_t from $P(x_t|Y_n, X^{t+1}, \theta, \Pi)$. Moreover, $P(x_t|Y_n, X^{t+1}, \theta, \Pi) \propto P(x_t|Y_t, \theta, \Pi) P(x_{t+1}|x_t, \Pi)$. This expression has two ingredients: the first is $P(x_{t+1}|x_t, \Pi)$, which is the transition matrix from the Markov chain. The second is $P(x_t|Y_t, \theta, \Pi)$ that would be obtained by recursively starting at $t = 1$.

The mass function $P(x_{t-1}|Y_{t-1}, \theta, \Pi)$ is transformed into $P(x_t|Y_t, \theta, \Pi)$, which is in turn transformed into $P(x_{t+1}|Y_{t+1}, \theta, \Pi)$ and so on. The update is as follows: for $k = 1, \ldots, a$, we could write

$$P(x_t = k|Y_t, \theta, \Pi) = \frac{P(x_t = k|Y_{t-1}, \theta, \Pi) \pi(y_t|y_{t-1}, \theta_k)}{\sum_{l=1}^{a} P(x_t = l|Y_{t-1}, \theta, \Pi) \pi(y_t|y_{t-1}, \theta_l)}.$$

These calculations are initialized at $t = 0$, by setting $P(x_1|Y_0, \theta)$ to be the stationary distribution of the Markov chain. Precisely, the simulation proceeds for $k = 1, \ldots, a$, recursively by first simulating $P(x_1 = k|Y_0, \theta)$, from the initial distribution $\Pi_0(k)$ and $P(x_1 = k|Y_1, \theta, \Pi) \propto P(x_1 = k|Y_0, \theta, \Pi) \pi(y_1|Y_0, \theta_k)$. Then, we get by forward calculation $P(x_t = k|Y_{t-1}, \theta) = \sum_{l=1}^{a} \Pi_{lk} P(x_{t-1} = l|Y_{t-1}, \theta)$, for each $t = 2, \ldots, n$, where Π_{lk} is the transition probability and $P(x_t = k|Y_t, \theta) \propto P(x_t = k|Y_{t-1}, \theta, P) \Pi(y_t|Y_{t-1}, \theta_k)$. The last term in the forward computation $P(x_n = k|Y_n, \theta)$ would serve as a start for the backward pass, and we get recursively for each $t = n-1, \ldots, 1.$; $P(x_t = k|Y_n, X^{t+1}, \theta) \propto P(x_t|Y_{t-1}, \theta, \Pi) P(x_{t+1}|x_t = k, \Pi)$, which permits the obtention of $X_n = (x_1, \ldots, x_n)$.

4.2.2. *Simulating the hidden states for the MAR(1)HMM*

Returning to our model, and adopting notations and algorithm developed by Fitzpatrick and Marchev, f will denote the observation density for the MAR(1)HMM, and for $u_0 < t < M$; $x_{-t} = (x_{u_0}, \ldots, x_t)$, $x^t = (x_t, \ldots, x_M)$, $y(t) = (y_{i,t}, i = 1, 2, \ldots, n)$, $y_{,t} = \bigcup_{i:u_i<t} \left\{y_{i,u_i}, \ldots, y_{i,\min\{t,m_i\}}\right\}$, and $y^t = \bigcup_{i:t<m_i} \left\{y_{i,\max\{t+1,u_i\}}, \ldots, y_{i,m_i}\right\}$. The posterior distribution of the hidden state could be written as: $P(x_{-M}|y_{,M}, \theta) = P(x_M|y_{,M}, \theta) \times \cdots \times P(x_{u_0}|y_{,M}, x^{u_0+1}, \theta)$. So we could sample the whole sequence of states by sampling from $P(x_t|y_{,M}, x^{t+1}, \theta)$. Hence, the estimation of the hidden states is performed recursively by first initializing

$$P(x_{u_0}|y_{,u_0},\theta) \propto P(y_{,u_0}|x_{u_0})P(x_{u_0}|r); y_{,u_0} = \left\{y_{i,u_i,}, u_i = u_0, i = 1,\dots,n\right\}.$$

$$P(x_{u_0+1} = k|y_{,u_0},\theta) = \sum_{l=1}^{a}\Pi_{lk}P(x_{u_0} = l|Y_{,u_0}); k = 1,\dots,a.$$

$$P(x_{u_0+1} = k|y_{,u_0+1},\theta) \propto P(x_{u_0+1} = k|y_{,u_0},\theta)f(y(u_0)|y_{,u_0},\theta_k).$$

We perform a similar calculation for every state at time t, and we conclude by calculating $P(x_M = k|y_{,M-1},\theta) = \sum_{l=1}^{a}\Pi_{lk}P(x_{M-1} = l|Y_{,M-1},\theta)$, and $P(x_M = k|y_{,M},\theta) \propto P(x_M = k|y_{,M-1},\theta$ $)f(y(M)|y_{,M-1},\theta_k)$. Later on, we get $P(x_M = k|y_{,M},\theta)$, which permits the simulation of $P(x_M|y_{,M},\theta)$. Finally, by backward calculation, we simulate from the probabilities $P(x_t|y_{,M},x^{t+1},\theta) \propto P(x_{t+1}|x_t,\Pi)P(x_t|y_{,t},\theta)$ for each time $t = M-1,\dots,u_0$. Those backward probabilities would permit the simulation of the latent states.

4.3. Sampling from $P(\theta\,|\,x, y)$

4.3.1. Sampling Π

Under the prior assumption of Dirichlet prior for each row of the transition matrix $P(\Pi_i) \propto \mathbb{D}(\delta_{i1},\dots,\delta_{ia})$, and the independence assumption between those rows, the posterior distribution for Π_i can be developed using Eq. (2) as follows: Let n_{ij} denote the number of single transitions from state i to state j, so

$$P(\Pi_i|y,x) \propto P(\Pi_i)\prod_{t=u_0+1}^{M}\prod_{j=1}^{a}\Pi^{\chi_{\{x_{t-1},x_t\}}(i,j)} \propto P(\Pi_i)\prod_{j=1}^{a}\Pi_{ij}^{n_{ij}} \propto \prod_{j=1}^{a}\Pi_{ij}^{\delta_{ij}+n_{ij}-1}$$
$$\propto \mathbb{D}(\delta_{i1}+n_{i1},\dots,\delta_{ia}+n_{ia}).$$

4.3.2. Sampling posterior distribution for initial distribution

Let $n_{0l} = \chi_{x_{u_0}}(l)$, for $l = 1,\dots,a$. Using (2), under Dirichlet prior $\mathbb{D}(\delta_{01},\dots,\delta_{0a})$ for the parameter r, we obtain $P(r|x,y) \propto P(r)\prod_{l=1}^{a}r_l^{\chi_{\{x_{u_0}\}}(l)} \propto \prod_{l=1}^{a}r_l^{\delta_{0l}+n_{0l}-1} \propto \mathbb{D}(\delta_{01}+n_{01},\dots,\delta_{0a}+n_{0a})$.

4.3.3. Sampling posterior distribution for the autoregressive parameters μ, β, σ^2

When a complete conditional distribution is known such as the normal distribution or beta distribution, we use the Gibbs sampler to draw the random variable. This is the case for our model. Let us define $n_{u_i}(l) = \sum_{i=1}^{n}\chi_{\{x_{u_i}=l\}}$, $n_l = \sum_{i=1}^{n}\sum_{t=u_i+1}^{m_i}\chi_{\{x_t=l\}}$, $N = \sum_{l=1}^{a}n_l$, $n_{0l} = \chi_{\{x_{u_0}\}}(l)$. So for $l = 1,2,\dots,a$; by supposing $\mathcal{N}(\alpha_l,\tau_l)$ as prior distribution and using Eq. (2), the conditional posterior distribution of $\mu^{(l)}$ is:

$$P(\mu^{(l)}|y,x) \propto P(\mu^{(l)})\prod_{i=1}^{n}\left[\frac{1}{\sigma}\phi\left(\frac{y_{i,u_i}-\mu^{(l)}}{\sigma}\right)\right]^{\chi_{\{x_{u_i}\}}(l)} \times \prod_{i=1}^{n}\prod_{t=u_i+1}^{m_i}\left(\frac{1}{\sigma}\phi\left(\frac{y_{i,t}-\mu^{(l)}-\beta^{(l)}y_{i,t-1}}{\sigma}\right)\right)^{\chi_{\{x_t\}}(l)}.$$

$$\propto \exp\frac{-1}{2}\left\{\frac{(\mu^{(l)}-\alpha_l)^2}{\tau_l} + \sum_{i=1,x_{u_i}=l}^{n}\left(\frac{y_{i,u_i}-\mu^{(l)}}{\sigma}\right)^2 + \sum_{i=1}^{n}\sum_{t=u_i+1,x_t=l}^{m_i}\left(\frac{y_{i,t}-\mu^{(l)}-\beta^{(l)}y_{i,t-1}}{\sigma}\right)^2\right\}.$$

then $\mu^{(l)}/y,x \sim \mathcal{N}(\tilde{\tau}_l,\tilde{\alpha}_l)$ with inverse mean $\tilde{\tau}_l{}^{-1} = \frac{n_{u_i}(l)+n_l}{\sigma^2} + \frac{1}{\tau_l}$ and variance

$$\tilde{\alpha}_l = \tilde{\tau}_l\left(\frac{\sum_{i=1,x_{u_i}=l}^{n} y_{i,u_i} + \sum_{i=1}^{n}\sum_{t=u_i+1,x_t=l}^{m_i}(y_{i,t}-\beta^{(l)}y_{i,t-1})}{\sigma^2} + \frac{\alpha_l}{\tau_l}\right).$$

For $\beta^{(l)}, l = 1,\dots, a$, and similar to $\mu^{(l)}$, $\mathcal{N}(b_l, c_l)$ was proposed as prior choice to obtain:

$$P(\beta^{(l)}|y,x) \propto P(\beta^{(l)})\prod_{i=1}^{n}\prod_{t=u_i+1}^{m_i}\left[\frac{1}{\sigma}\phi\left(\frac{y_{i,t}-\mu^{(l)}-\beta^{(l)}y_{i,t-1}}{\sigma}\right)\right]^{\chi_{\{x_t\}}(l)},$$

and therefore, $\beta^{(l)}/y,x \sim \mathcal{N}(\tilde{c}_l,\tilde{b}_l)$ with inverse mean $\tilde{c}_l{}^{-1} = \frac{1}{c_l} + \frac{\sum_{i=1}^{n}\sum_{t=u_i+1,x_t=l}^{m_i} y_{i,t-1}^2}{\sigma^2}$ and variance

$$\tilde{b}_l = \tilde{c}_l\left(\frac{b_l}{c_l} + \frac{\sum_{i=1}^{n}\sum_{t=u_i+1,x_t=l}^{m_i}(y_{i,t}-\mu^{(l)})y_{i,t-1}}{\sigma^2}\right).$$

For the posterior distribution of σ^2, by supposing $\mathbb{IG}(\varepsilon,\zeta)$ as prior, we deduce from Eq. (2)

$$P(\sigma^2|y,x) \propto (\sigma^2)^{-(\varepsilon+1)}\exp\left(-\frac{\zeta}{\sigma^2}\right)\prod_{i=1}^{n}\left[\frac{1}{\sigma}\phi\left(\frac{y_{i,u_i}-\mu^{(x_{u_i})}}{\sigma}\right)\right]$$
$$\times\prod_{i=1}^{n}\prod_{t=u_i+1}^{m_i}\left[\frac{1}{\sigma}\phi\left(\frac{y_{i,t}-\mu^{(x_t)}-\beta^{(x_t)}y_{i,t-1}}{\sigma}\right)\right],$$

consequently $\sigma^2/y,x \sim \mathbb{IG}(\tilde{\varepsilon},\tilde{\zeta})$ *with parameters* $\tilde{\varepsilon} = \frac{n_{u_i}+N}{2} + \varepsilon$ and

$$\tilde{\zeta} = \frac{\sum_{i=1}^{n}(y_{i,u_i}-\mu^{(x_{u_i})})^2 + \sum_{i=1}^{n}\sum_{t=u_i+1}^{m_i}(y_{i,t}-\mu^{(x_t)}-\beta^{(x_t)}y_{i,t-1})^2}{2} + \zeta.$$

Finally, the algorithm is ran for $d = 1,\ldots,D$ iterations by alternating between the following steps, where in each step we compute a conditional posterior for the given parameter:

The MCMC algorithm:

1. For $h = 1, 2, \ldots., a$, give reference values for the hyperparameters α_h, τ_h, a_h, b_h, δ_{0h}, and δ_{ih} for $i = 1, 2, \ldots., a$.

2. Initialization (Step $d = 1$ of the MCMC iterations): Initialize $\Pi^{(1)}$, $r^{(1)}$, $\mu^{(1)}$, $\beta^{(1)}$, and $\sigma^{2(1)}$.

3. Simulation of the hidden states:

   a. Initialization of forward simulation: $P(x_{u_0}^{(d)}|y_{,u_0}, \theta) \propto P(y_{,u_0}|x_{u_0}^{(d)})P(x_{u_0}^{(d)}|r^{(d)})$, with $y_{,u_0} = \left\{y_{i,u_i}, u_i = u_0, i = 1, \ldots, n\right\}$.

   b. Forward simulation: For $k = 1, \ldots.,a$ and $t = u_0 + 1, \ldots., M$:
   $P(x_t^{(d)} = k|y_{,t-1}, \theta) = \sum_{l=1}^{a} \Pi_{lk}^{(d)} P(x_{t-1}^{(d)} = l|Y_{,t-1}, \theta)$ and

   $$P(x_t^{(d)} = k|y_{,t}, \theta) = \frac{P(x_t^{(d)}=k|y_{,t-1}, \theta_k)f(y(t)|y_{,t-1}, \theta_k)}{\sum_{l=1}^{a} P(x_t^d=l|y_{,t-1}, \theta)f(y(t)|y_{,t-1}, \theta_l)}.$$

   c. Initialization of backward simulation: For $k = 1, \ldots.,a$, given

   $P(x_M^{(d)} = k|y_{,M}, \theta)$ from forward simulation, we get $P(x_M^{(d)}|y_{,M}, \theta)$.

   d. Backward simulation: For $k = 1, \ldots.,a$ and $t = M - 1, \ldots,u_0$:

   $$P(x_t^{(d)}|y_{,M}, x^{t+1(d)}, \theta) \propto P(x_{t+1}^{(d)}|x_t^{(d)}, \pi)P(x_t^{(d)}|y_{,t}, \theta).$$

4. Estimation of the initial distribution and the transition distribution

   a. for $l = 1, \ldots.,a$, $k = 1, \ldots.,a$. Calculate $n_{0l} = \chi_{\{x_{u_0}^{(d)}\}}(l)$ and $n_{kl} = \Sigma_{t=u_0+1}^{M} \chi_{\{x_{t-1}^{(d)}, x_t^{(d)}\}}(k, l)$.

   b. Sample $(r_1^{(d+1)}, \ldots., r_a^{(d+1)}) \propto \mathbb{D}(\delta_{01} + n_{01}, \ldots., \delta_{0a} + n_{0a})$.

   c. For $i = 1, \ldots., a$; sample $(\Pi_{i1}^{(d+1)}, \ldots., \Pi_{ia}^{(d+1)}) \propto \mathbb{D}(\delta_i 1 + n_{i1}, \ldots., \delta_i a + n_{ia})$.

5. Simulation of μ: For $l = 1, \ldots., a$,

   a. $\tilde{\tau}_l^{-1} = \frac{n_{u_i}(l) + n_l}{\sigma_{(d)}^2} + \frac{1}{\tau_l}$.

   b. $$\tilde{\alpha}_l = \tilde{\tau}_l \left(\frac{\sum_{i=1, x_{u_i}=l}^{n} y_{i,u_i} + \sum_{i=1}^{n} \sum_{t=u_i+1, x_t=l}^{m_i} (y_{i,t} - \beta_{(d)}^{(l)} y_{i,t-1})}{\sigma_{(d)}^2} + \frac{\alpha_l}{\tau_l} \right).$$

   c. Simulate $\mu_{(d+1)}^{(l)}/y, x \sim \mathcal{N}(\tilde{\alpha}_l, \tilde{\tau}_l)$.

6. Simulation of β: For $l = 1, \ldots, a$,

a. $\tilde{c}_l^{-1} = \frac{1}{c_l} + \frac{\sum_{i=1}^{n} \sum_{t=u_i+1, x_t=l}^{m_i} y_{i,t-1}^2}{\sigma_{(d)}^2}$.

b. $\tilde{b}_l = \tilde{c}_l \left(\frac{b_l}{c_l} + \frac{\sum_{i=1}^{n} \sum_{t=u_i+1, x_t=l}^{m_i} (y_{i,t} - \mu_{d+1}^{(l)}) y_{i,t-1}}{\sigma_{(d)}^2} \right)$.

c. Simulate $\beta_{(d+1)}^{(l)} / y, x \sim \mathcal{N}(\tilde{b}_l, \tilde{c}_l)$.

7. - Simulation of σ^2:

a. $\tilde{\epsilon} = \frac{n_{u_i} + N}{2} + \epsilon$.

b. $\tilde{\zeta} = \frac{\sum_{i=1}^{n} (y_{i,u_i} - \mu_{(d+1)}^{(x_{u_i})})^2 + \sum_{i=1}^{n} \sum_{t=u_i+1}^{m_i} (y_{i,t} - \mu_{(d+1)}^{(x_t)} - \beta_{(d+1)}^{(x_t)} y_{i,t-1})^2}{2} + \zeta$.

c. Simulate $\sigma_{(d+1)}^2 / y, x \sim \mathbb{IG}(\epsilon, \tilde{\zeta})$.

---

5. Simulation study

In this section, we apply our results to the breast cancer model discussed earlier. The main reason behind our work is that the progression of breast cancer cannot be seen directly unless we use observations related to the disease that could characterize its progression; those observations here are quantities which could be measured; they are called biomarkers, where the word biomarker is used to designate any objective indication of a biological process or disease condition including during treatment and should be measurable. Furthermore, biomarkers are increasingly used in the management of breast cancer patients. One example is reported in Ref. [22], stating that there is correlation between elevation of CEA and/or CA15-3 and disease progression, in breast cancer patients. Also we use the autoregressive dependence among the observations to add more dynamics to the model unlike conventional HMMs where the successive observations given the Markov process are independent. We used the classification of breast cancer in three states: local where the disease is confined within the breast, the regional phase when the lymph nodes are involved, and the distant stage where the cancer is found in other parts of the body. We restrict ourselves to these three stages unlike other stage classifications that divide the progression in more than three stages such as the TNM (tumor, node, and metastasis) system. By lack of finding data about breast cancer biomarkers, we will confine ourselves to simulate an MAR(1)HMM model for observation time M = 24, and a number of individuals n = 210, a = 3 for Markov states number, with the length observation time for each individual selected uniformly between 2 and M. The simulation process supposes we have for the autoregressive means $\mu = (\mu^{(1)}, \mu^{(2)}, \mu^{(3)}) = (12, 24, 36)$, since markers such as CA15–3 increase as the disease advances toward metastatic breast cancer. In addition, CA15–3 increase rapidly between successive observations, and thus, we take in the simulation the parameters $\beta = (\beta^{(1)}, \beta^{(2)}, \beta^{(3)}) = (0.2, 0.4, 0.8)$.

The algorithm of simulation works as follows:

1. For each individual $i = 1, \ldots, n$, choose m_i the length of observation for that individual i.
2. Generate each discrete disease state x_t using transition matrix$\Pi = (0.7, 0.2, 0.1; 0.1, 0.6, 0.3; 0.2, 0.3, 0.5)$ for $t = u_{0+1}, \ldots, M$.
3. Generate the observations $y_{i,t}$ for all individuals using our model 1.

We choose a prior $\sigma^2 \sim \mathbb{IG}(0.001, 0.001)$, a $\mathbb{D}(1, \ldots, 1)$ prior for each row of Π, and Gaussian noninformative priors for the μs and the βs. Having the hidden states and the observations, we ran our algorithm for 8000 MCMC iterations. MCMC algorithm convergence was assessed by analyzing MCMC iterations mixing plots that are shown in **Figure 1**, autocorrelation sample graphs checking as illustrated in **Figure 2**, and inspecting histograms of posterior densities for the parameters of the models in **Figure 3**. All parameters show good mixing of chains, autocorrelations that decay immediately after a few lags, and perfect posterior densities fitting. Also the Gelman [23] potential scale reduction factor (PSRF) was plot. The PSRF is measured for more than two MCMC chains (three chains in this works are considered), and it is measured for each parameter of the model; it should show how the chains have forgotten their initial values and that the output from all chains is indistinguishable. It is based on a comparison of within-chain and between-chains variances and is similar to a classical analysis of variance; when the PSRF is high (perhaps greater than 1.1 or 1.2), then we should run our chains out longer to improve convergence to the stationary distribution. Each PSRF declines to 1 as the number of iterations approaches infinity to confirm convergence. All the parameters have shown a PSRF less than 1.1 as the number of iteration increases and by the way a good sign of convergence (**Figure 4**). Moreover, we should point out that in the family of Markov switching model there is the so-called label switching problem (e.g., Ref. [24]) which arises identifiability problem, and hence, we would not estimate perfectly the parameters. In addition, the posterior densities could show evidence of multimodality. some authors postprocess the output of the MCMC to deal with the issue (e.g., [25]), while other uses a random permutation of the parameters in each iteration of the MCMC algorithm (e.g., [26]) or one can call for an invariant loss function method (e.g., [27]). In our case, no identifiability issue is noticed since we used well-separated prior hyperparameters. Even when we start from different initial values for the parameters, our algorithm converges immediately after a few iterations.

Finally and before giving our results, we should report that the simulation of the Dirichlet posterior was carried out following ([28, p. 22], [29, p. 155]) who reported that the posterior Dirichlet parameters should be simulated using the beta distribution approach. **Table 1** shows how the posterior values estimated from algorithm are very close to the true ones.

http://dx.doi.org/10.5772/intechopen.70053

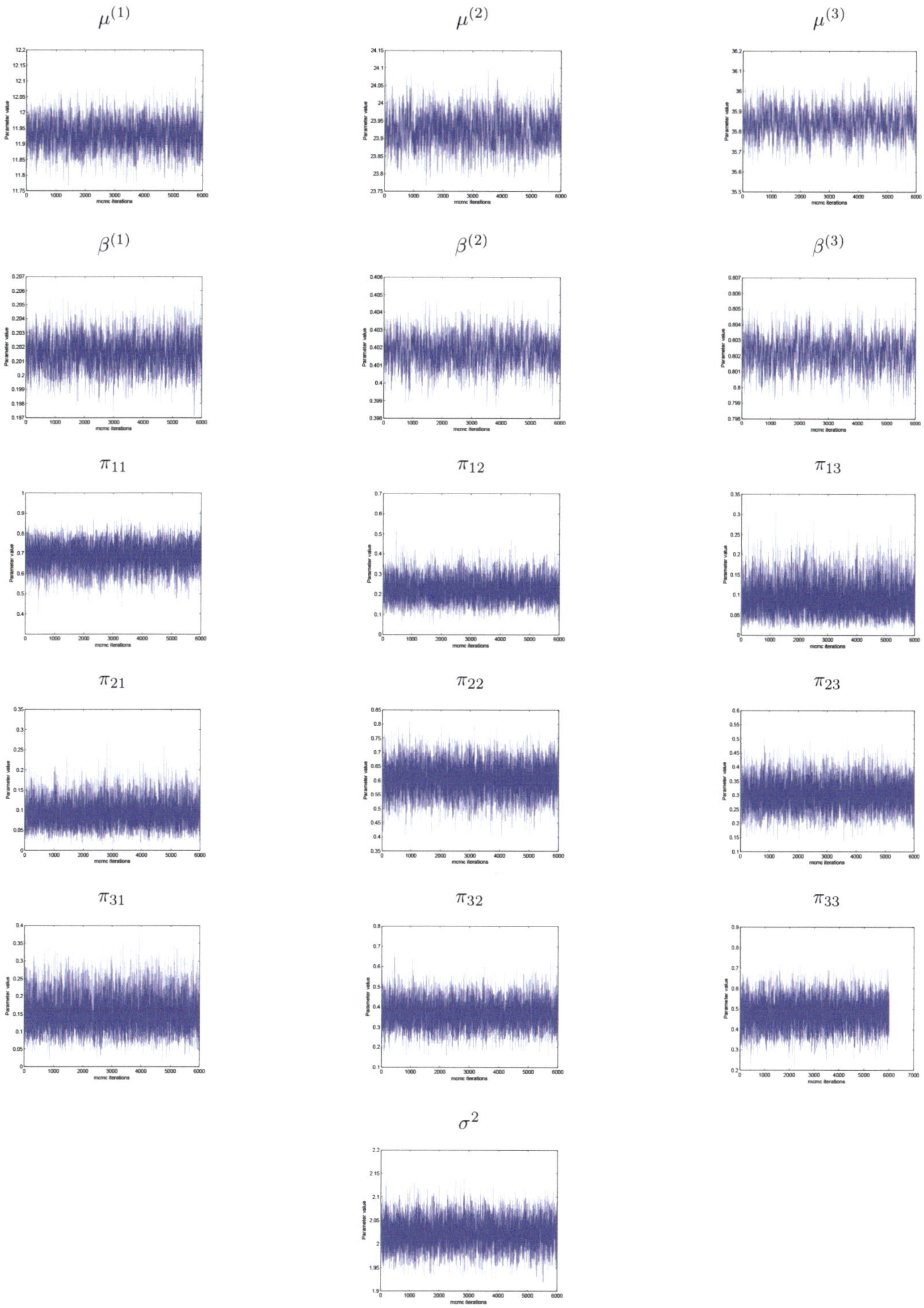

Figure 1. Markov chain mixing for each parameter through MCMC algorithm simulation.

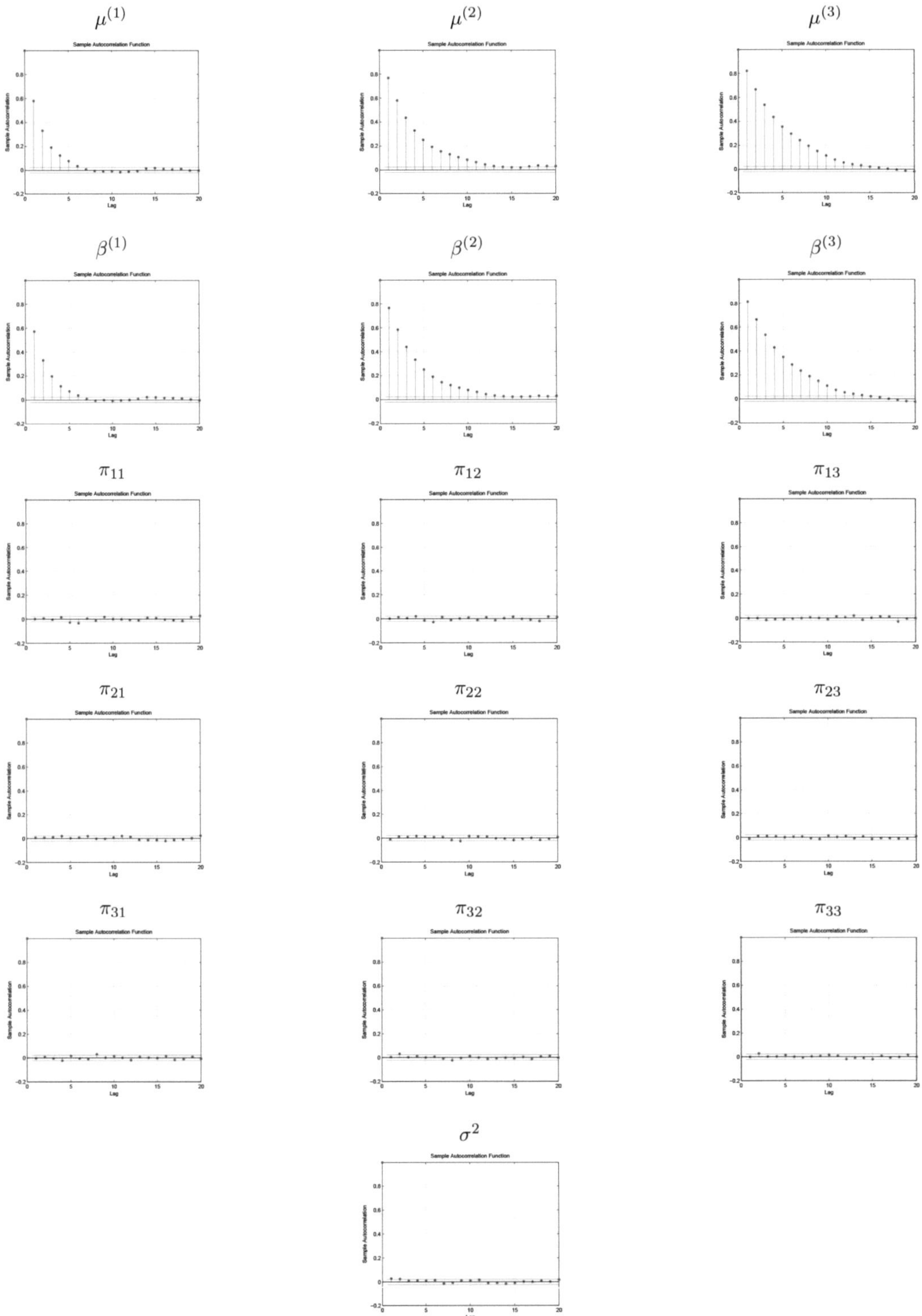

Figure 2. Autocorrelation sample plots for parameters of the model.

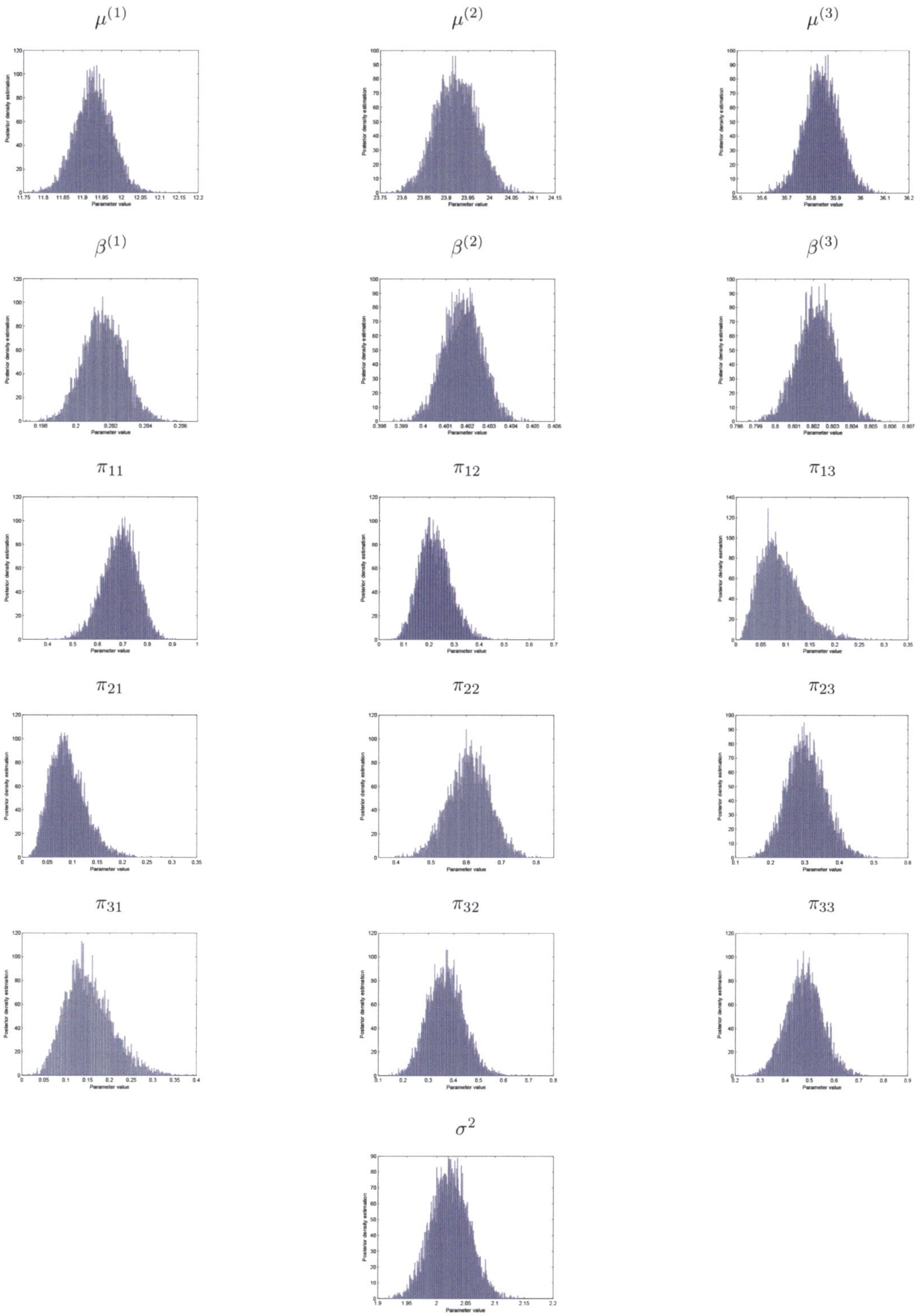

Figure 3. Posterior densities for the parameters of the model (after 8000 iterations).

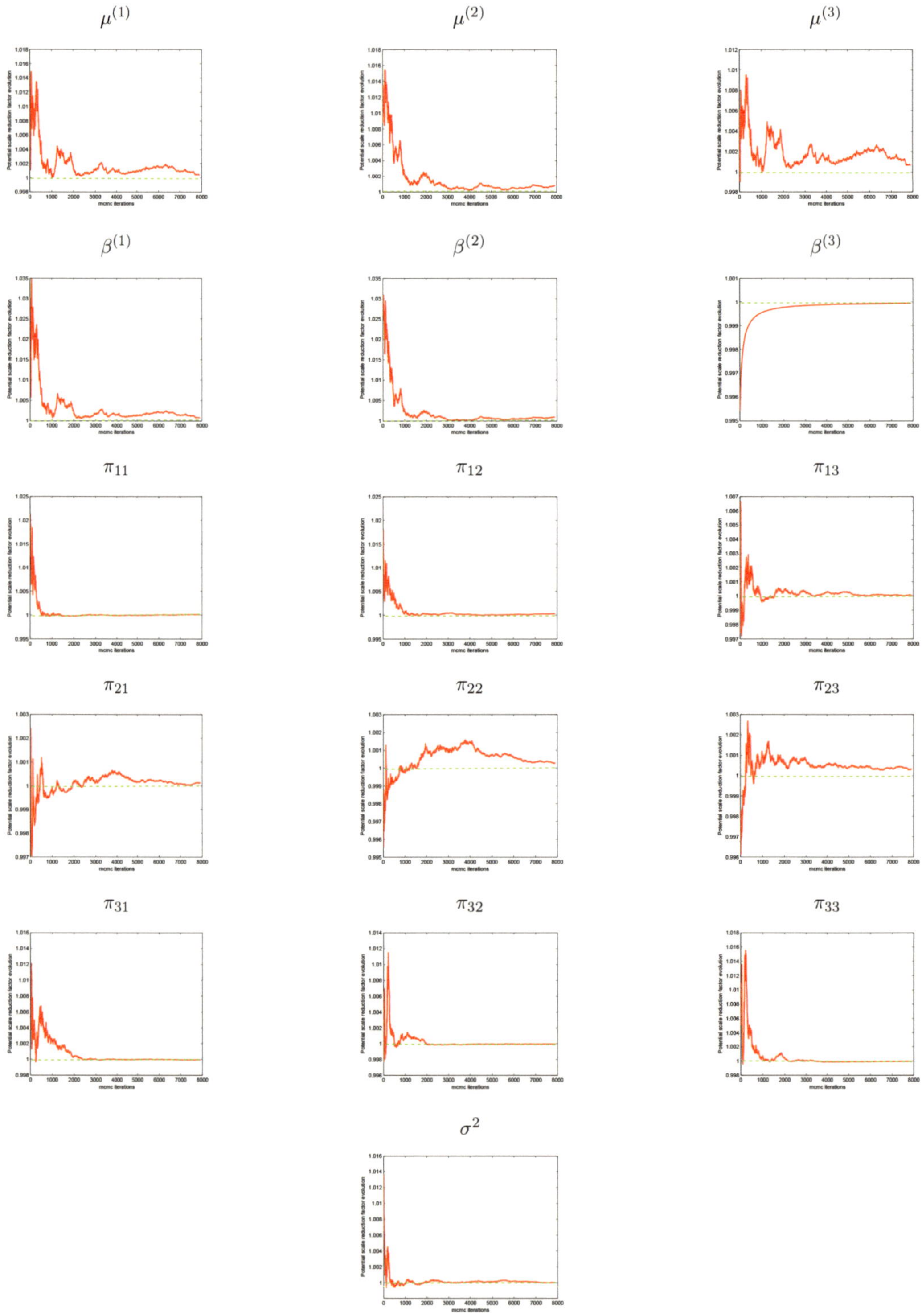

Figure 4. Potential scale reduction factor convergence to less than 1.02 with more iterations.

http://dx.doi.org/10.5772/intechopen.70053

Parameter	True value	Posterior statistics		
		Mean	Standard deviation	Confidence interval (5%)
μ^1	12	11.929	0.047	(11.851–12.005)
μ^2	24	23.923	0.047	(23.847–24.000)
μ^3	36	35.843	0.070	(35.729–35.959)
β^1	0.2	0.2016	0.0012	(0.1997–0.2035)
β^2	0.4	0.4018	0.0009	(0.4004–0.4032)
β^3	0.8	0.8022	0.0010	(0.8005–0.8038)
π_{11}	0.7	0.688	0.068	(0.5715–0.797)
π_{12}	0.2	0.223	0.062	(0.129–0.332)
π_{13}	0.1	0.090	0.042	(0.032–0.17)
π_{21}	0.1	0.091	0.035	(0.041–0.154)
π_{22}	0.6	0.607	0.059	(0.507–0.701)
π_{23}	0.3	0.302	0.055	(0.214–0.397)
π_{31}	0.2	0.153	0.053	(0.075–0.250)
π_{32}	0.3	0.368	0.071	(0.257–0.488)
π_{33}	0.5	0.479	0.073	(0.358–0.599)
σ^2	2	2.023	0.032	(1.970–2.077)

Table 1. Posterior inference for the parameters of the MAR(1)HMM model.

6. Conclusion

We have extended the method of Chib [21] for block update estimation of the states to a MAR (1)HMM model. Furthermore, we would like to point out that our model can easily be extended to include missing observations, as we should only add an extra step in each MCMC iteration to estimate the missing observations. Also, we can estimate the autoregressive model for different values of the autoregressive order, $p \geq 1$, by evaluating the Bayesian information criterion to select the best order that fits the observations of the model. Our model would capture the complexity and the dynamics of the evolution of breast cancer by introducing the latent states; the probabilities of transition between the latent states allow to compare among the effects of treatments on slowing or accelerating the transition of the disease from one health stage to another the autoregressive parameter mean values corresponding to different stages of the disease would guide medical doctors and scientists to monitor patients in different phases of the disease. The model incorporates individual observations with different lengths.

Last but not least, we like to mention the utilities of switching diffusion processes in addressing and analyzing many complicated applications such as in finance and risk management. Our future work would be to apply these processes to explore disease progression, because they are characterized by the coexistence of continuous dynamics and discrete events as well as their interactions.

Acknowledgements

We would like to thank the editorial staff for the comments that helped in improving this work. Also, we would like to thank the supporters of this work: The Lalla Salma Foundation Prevention and Treatment of Cancer, Rabat, Morocco; and the Germano-Morrocan Program for Scientific Research PMARS 2015-060.

Author details

Hamid El Maroufy[1*†], El Houcine Hibbah[1†], Abdelmajid Zyad[2†] and Taib Ziad[3]

*Address all correspondence to: h_elmaroufy@hotmail.com

1 Department of Mathematics, Faculty of Sciences and Technics, Sultan Moulay Slimane University, Béni Mellal, Morocco

2 Biological Engineering Laboratory, Team of Natural Substances, Cell and Molecular Immuno-Pharmacology, Sultan Moulay Slimane University, Morocco

3 Early Clinical Development, Astra Zeneca RD, Gothenburg, Mölndal, Sweden

†The three first authors acknowledge the financial support of the Lalla Salma Fondation of Cancer: Prevention and Treatment, Project 09/2013.

References

[1] Samy N, Ragab HM, El Maksoud NA, Shaalan M. Prognostic significance of serum Her2/neu, BCL2, CA15-3 and CEA in breast cancer patients: A short follow up. Cancer Biomarkers. 2009;**6**:63-72

[2] Benmiloud B, Piczunski W. Estimation des parametres dans les chaines de markov cachees et segmentation. Traitement du Signal. 1995;**12**:433–454

[3] Boys R, Handerson D. A Bayesian approach to DNA sequence segmentation (with discussion). Biometrics. 2004;**60**:573–588

[4] Guihenneuc-Jouyaux C, Richardson S, Longini IM Jr. Modeling disease progression by a hidden Markov process: Application to characterizing CD4 cell decline. Biometrics. 2000;**56**:733–741

[5] Albert J, Chib S. Bayes inference via Gibbs sampling of autoregressive time series subject to Markov mean and variance shifts. Journal of Business and Economic Statistics. 1993;**11**:1–15

[6] Korolkiewickz M, Elliot J. A hidden Markov model of credit quality. Journal of Economic Dynamics and Control. 2008;**32**:3807–3819

[7] Zeng Y, Frias J. A novel HMM-based clustering algorithm for the analysis of gene expression time-course data. Computational Statistics and Data Analysis. 2006;**50**:2472–2494

[8] Zucchini W, MacDonald I. Hidden Markov Models for Time Series: An Introduction Using R. New York: Springer; 2009

[9] Ailliot P, Monbet V. Markov-switching autoregressive models for wind time series. Environmental Modelling & Software. 2012;**30**:92–101

[10] Farcomeni A, Arima S. A Bayesian autoregressive three state hidden Markov model for identifying switching monotonic regimes in microarray time course data. Statistical Applications in Genetics and Molecular Biology. 2013;**23**:467–480

[11] Hamilton J. A new approach to the economic analysis of nonstationary time series and the business cycle. Econometrica. 1989;**57**:357–384

[12] Kim C, Kim J. Bayesian inference in regime-switching ARMA models with absorbing states: The dynamics of the ex-ante real interest rate under regime shifts. Journal of Business and Economic Statistics. 2015;**33**:566–578

[13] Fitzpatrick, M. and Marchev, D. Efficient bayesian estimation of the multivariate double chain markov model. Statistics and Computing, 2013;**23**(4):467-480

[14] Lindley D. The philosophy of statistics. The Statistician. 2000;**49**:293–337

[15] Bolstad, W. M. (2007). Introduction to Bayesian Statistics. John Wiley and Sons, Inc., Hoboken, New Jersey; 2007

[16] Gelman A, Shalizi CR. Philosophy and the practice of Bayesian statistics in the social sciences. British Journal of Mathematical and Statistical Psychology. 2013;**66**:8–38

[17] Hobert, J.P. The data augmentation algorithm: Theory and methodology. In: Brooks S, Gelman A, Jones GL, Meng X-L, editors. The Handbook of Markov Chain Monte Carlo. Boca Raton, FL: Chapman and Hall/CRC; 2011. p. 253

[18] Tuyl F, Gerlach R, Mengersen K. Posterior predictive arguments in favor of the Bayes-Laplace priors as the consensus prior for the binomial and multinomial parameters. Bayesian Analysis. 2013;**4**:151–158

[19] Cappe O, Moulines E, Ryden T. Inference in Hidden Markov Models. New York: Springer-Verlag; 2005

[20] Sampietro S. Mixture of autoregressive components for modeling financial market volatility. LIUC Papers. Serie Metodi quantitativi 16. 2005

[21] Chib S. Calculating posterior distributions and model estimates in Markov mixtures models. Journal of Econometrics. 1996;**75**:79–98

[22] Laessig D, Nagel D, Heinemann V, Untch M, Kahlert S, Bauerfeind I, Stieber P. Importance of CEA and CA15-3 during disease progression in metastatic breast cancer patients. Anticancer Research. 2007;**27**:1963–1968

[23] Gelman A. Inference and monitoring convergence. In: Gilks W, Richardson S, Spiegelhalter D, editors. Markov Chain Monte Carlo in Practice. London: Chapman and Hall/CRC; 1995. p. 131

[24] Fruhwirth-Schnatter S. Finite Mixture and Markov Switching Models. New York: Springer; 2006

[25] Celoux G. Bayesian inference for mixture: The label switching problem. In: Payne R, Green P, editors. Proceedings in Computational Statistics. Heidelberg: Physica; 1998. pp. 227–232

[26] Fruhwirth-Schnatter S. Markov Chain Monte Carlo estimation of classical and dynamic switching and mixture models. Journal of the American Statistical Association. 2001; **96**:194–209

[27] Hurn M, Justel A, Rober C. Estimating mixtures of regressions. Journal of Computational and Graphical Statistics. 2003;**79**:55–79

[28] Kim C, Nelson C. State-Space Models with Regime Switching: Classical and Gibbs Sampling Approaches with Applications. Cambridge, MA: MIT Press; 1999

[29] Krozlig H. Markov-Switching Vector Autoregressions. Berlin, Heidelberg: Springer-Verlag; 1997

Chapter 8

Bayesian Model Averaging and Compromising in Dose-Response Studies

Steven B. Kim

Additional information is available at the end of the chapter

http://dx.doi.org/10.5772/intechopen.68786

Abstract

Dose-response models are applied to animal-based cancer risk assessments and human-based clinical trials usually with small samples. For sparse data, we rely on a parametric model for efficiency, but posterior inference can be sensitive to an assumed model. In addition, when we utilize prior information, multiple experts may have different prior knowledge about the parameter of interest. When we make sequential decisions to allocate experimental units in an experiment, an outcome may depend on decision rules, and each decision rule has its own perspective. In this chapter, we address the three practical issues in small-sample dose-response studies: (i) model-sensitivity, (ii) disagreement in prior knowledge and (iii) conflicting perspective in decision rules.

Keywords: dose-response models, model-sensitivity, model-averaging, prior-sensitivity, consensus prior, Bayesian decision theory, individual-level ethics, population-level ethics, Bayesian adaptive designs, sequential decisions, continual reassessment method, c-optimal design, Phase I clinical trials

1. Introduction

Dose-response modeling is often used to learn about the effect of an agent on a particular outcome with respect to dose. It is widely applied to animal-based cancer risk assessments and human-based clinical trials. A sample size is typically small; so many statistical issues can arise from a limited amount of data. The issues include the impact of a misspecified model, prior-sensitivity, and conflicting ethical perspectives in clinical trials. In this chapter, we focus on cases when an outcome variable of interest is binary (a predefined event happened or not) when an experimental unit is exposed to a dose. Main ideas are preserved for cases when an outcome variable is continuous or discrete.

There are two different approaches to statistical inference. One approach is called frequentist inference. In this framework, we often rely on the sampling distribution of a statistic and large-sample theories. Another approach is called Bayesian inference. It is founded on Bayes' Theorem, and it allows researchers to express prior knowledge independent of data. In a small-sample study, Bayesian inference can be more useful than frequentist inference because we can incorporate both researcher's prior knowledge and observed data to make inference for the parameter of interest. Bayesian ideas are briefly introduced for dose-response modeling with a binary outcome in Section 2.

In a small-sample study, we often rely on a parametric model to gain statistical efficiency (i.e., less variance in parameter estimation), but our inference can be severely biased by the use of a wrong model. To account for model uncertainty, it is reasonable to specify multiple models and make inference based on "averaged-inference." In this regard, Bayesian model averaging (BMA) is a useful method to gain robustness [1]. The BMA method has a wide range of application, and we focus its application to animal-based cancer risk assessments in Section 3.

In clinical trials, study participants are real patients, and therefore, we need to carefully consider ethics. There are conflicting perspectives of individual- and population-level ethics in early phase clinical trials. Individual-level ethics focuses on the benefit of trial participants, whereas population-level focuses on the benefit of future patients, which may require some level of sacrifice from trial participants. We compare the two conflicting perspectives in clinical trials based on Bayesian decision theory, and we discuss a compromising method in Section 4 [2, 3].

A sample size for an early phase (Phase I) clinical trial is often less than 30 subjects. Dose allocations for first few patients and statistical inference for future patients heavily depend on researcher's prior knowledge in sparse data. When multiple researchers have different prior knowledge about a parameter of interest, one compromising approach is to combine their prior elicitations and average them (i.e., consensus prior) [4, 5]. When we average the prior elicitations, there are two different approaches to determine the weight of each prior elicitation, weights determined before observing data and after observing data. We discuss operating characteristics of the two different weighting methods in the context of Phase I clinical trials in Section 5.

2. Bayesian inference

In statistics, we address a research question by a parameter, which is often denoted by θ. We begin Bayesian inference by modeling the prior knowledge about θ. A function, which models the prior knowledge about θ, is called the prior density function of θ, and we denote it by $f(\theta)$. It is a non-negative function, which satisfies $\int_{\Omega} f(\theta)\, d\theta = 1$, where Ω is the set of all possible values of θ (i.e., parameter space). We then model data $\vec{y} = (y_1, \ldots, y_n)$ given θ. The likelihood function, denoted by $f(\vec{y}|\theta)$, quantifies the likelihood of observing a particular sample $\vec{y} = (y_1, \ldots, y_n)$ under an assumed probability model. By Bayes' Theorem, we update our knowledge about θ after observing data $\vec{y}$ as

$$f(\theta|\vec{y}) = \frac{f(\vec{y}|\theta) f(\theta)}{f(\vec{y})}. \tag{1}$$

The function $f(\theta|\vec{y})$ is called the posterior density function of θ given data $\vec{y}$. Since we treat observed data $\vec{y} = (y_1, \ldots, y_n)$ as fixed numbers, we often express Eq. (1) as follows

$$f(\theta|\vec{y}) \propto f(\vec{y}|\theta)\, f(\theta) = k\, f(\vec{y}|\theta)\, f(\theta), \tag{2}$$

where k is the normalizing constant which makes $\int_\Omega f(\theta|\vec{y})\, d\theta = 1$. We can often realize $f(\theta|\vec{y})$ based on the prior density function $f(\theta)$ and the likelihood function $f(\vec{y}|\theta)$ without considering the denominator $f(\vec{y}) = \int f(y|\theta) f(\theta)\, d\theta$ in Eq. (1) which is called the marginal likelihood.

2.1. Example

Suppose we observe n = 20 rats for 2 years. Let π be the parameter of interest, which is interpreted as the probability of developing some type of tumor. Suppose a researcher models the prior knowledge about π using the prior density function

$$f(\pi) = \frac{\Gamma(a+b)}{\Gamma(a)\,\Gamma(b)}\, \pi^{a-1}\,(1-\pi)^{b-1},\ \ 0 < \pi < 1. \tag{3}$$

It is known as the beta distribution with shape parameters $a > 0$ and $b > 0$. We often denote the beta distribution by $\pi \sim \text{Beta}(a, b)$, and the values of a and b must be specified by the researcher independent from data. Let $\vec{y} = (y_1, \ldots, y_n)$ denote observed data, where y_i = 1 if the $_i^{\text{th}}$ rat developed tumor and y_i = 0 otherwise. Assuming $y_1, \ldots, y_n$ are independent observations, the likelihood function is as follows

$$f(\vec{y}\,|\pi) = \prod_{i=1}^{n} \pi^{y_i}\,(1-\pi)^{1-y_i} = \pi^{s}\,(1-\pi)^{n-s}, \tag{4}$$

where $s = \sum_{i=1}^{n} y_i$ is the total number of rats developed tumor. By Eq. (2), the posterior density function of π is as follows

$$f(\pi|\vec{y}) = k\,\pi^{a+s-1}\,(1-\pi)^{b+n-s-1}, \tag{5}$$

where $k = \frac{\Gamma(a+b+n)}{\Gamma(a+s)\,\Gamma(b+n-s)}$ is the normalizing constant, which makes $\int_0^1 f(\pi|\vec{y})\, d\pi = 1$. We can recognize that $\pi|\vec{y} \sim \text{Beta}(a+s, b+n-s)$.

If the researcher fixed $a = 2$ and $b = 3$ and observed $s = 9$ from a sample of size $n = 20$, the prior density function is $f(\pi) = k\pi(1-\pi)^2$ with $k = \frac{\Gamma(5)}{\Gamma(2)\Gamma(3)} = 12$, and the posterior density function is $f(\pi|\vec{y}) = k\pi^{10}(1-\pi)^{13}$ with $k = \frac{\Gamma(25)}{\Gamma(11)\Gamma(14)} = 27457584$. The prior and posterior distributions are shown in **Figure 1**. The knowledge about π becomes more certain (less variance) after observing the data.

2.2. Example

This example is simplified from Shao and Small [6]. In dose-response studies, we model π as a function of dose x. There are many link functions between π and x used in practice. In this example, we focus on a link function

$$\pi_x = \frac{e^{\beta_0+\beta_1 x}}{1+e^{\beta_0+\beta_1 x}} \tag{6}$$

which is known as a logistic regression model. It is commonly assumed that a dose-response curve increases with respect to dose, so we assume $\beta_1 > 0$ (and β_0 can be any real number). There are two regression parameters in Eq. (6), β_0 and β_1, and we denote them as $\vec{\beta} = (\beta_0, \beta_1)$. **Figure 2** presents two dose-response curves. The solid curve is generated by $\vec{\beta} = (-1, 2)$, and the dotted curve is generated by $\vec{\beta} = (-2, 5)$. As β_0 increases, the background risk $\pi_0 = \frac{e^{\beta_0}}{1+e^{\beta_0}}$ increases, where π_0 is interpreted as the probability of tumor development at dose $x = 0$. The dose-response curve increases when $\beta_1 > 0$, and it decreases when $\beta_1 < 0$. The rate of change in the dose-response curve is determined by $|\beta_1|$.

To express prior knowledge about $\vec{\beta}$, we need to find an appropriate prior density function $f(\vec{\beta})$. It is not simple because it is difficult to express one's knowledge on the two-dimensional parameters $\vec{\beta} = (\beta_0, \beta_1)$. For mathematical convenience, some practitioners use a flat prior density function $f(\vec{\beta}) \propto 1$. Another way of expressing a lack of prior knowledge about $\vec{\beta}$ is as follows

$$f(\vec{\beta}) \propto \frac{1}{2\pi\sigma^2} e^{-\frac{\beta_0^2+\beta_1^2}{2\sigma^2}} \mathrm{I}_{\beta_1>0} \tag{7}$$

with an arbitrarily large value of σ [6]. When a reliable source of prior information is available, there is a practical method, which is known as the conditional mean prior [7], and it will be discussed in a later section (see Section 4.2). In an experiment, the experimental doses $\vec{x} = (x_1, \ldots, x_n)$ are fixed, and we observe random binary outcomes $\vec{y} = (y_1, \ldots, y_n)$. Given $\vec{y}$ (and fixed $\vec{x}$), the likelihood function is as follows

$$f(\vec{y}\,|\vec{\beta}) = \prod_{i=1}^{n} \left(\frac{e^{\beta_0+\beta_1 x_i}}{1+e^{\beta_0+\beta_1 x_i}}\right)^{y_i} \left(\frac{1}{1+e^{\beta_0+\beta_1 x_i}}\right)^{1-y_i} = \frac{e^{\beta_0 s_1+\beta_1 s_2}}{\prod_{i=1}^{n}(1+e^{\beta_0+\beta_1 x_i})}, \tag{8}$$

http://dx.doi.org/10.5772/intechopen.68786

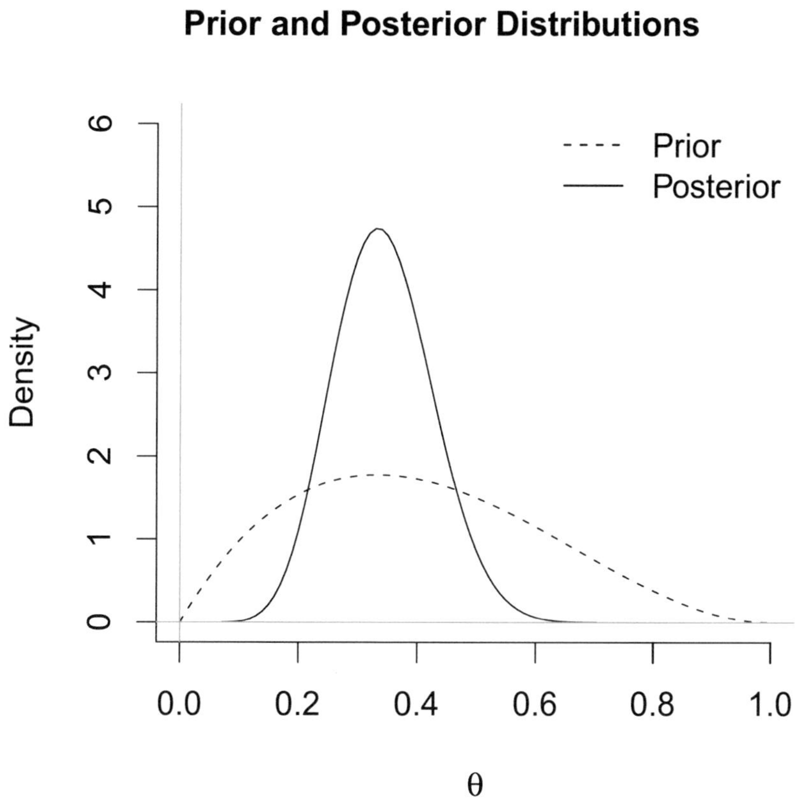

Figure 1. The prior $f(\pi)$ in the dotted curve and the posterior $f(\pi|\vec{y})$ in the solid curve.

where $s_1 = \sum_{i=1}^{n} y_i$ and $s_2 = \sum_{i=1}^{n} x_i y_i$. By incorporating both prior and data, the posterior density function is as follows

$$f(\vec{\beta}\,|\vec{y}) \propto f(\vec{\beta})\,\frac{e^{\beta_0 s_1 + \beta_1 s_2}}{\prod_{i=1}^{n}\left(1 + e^{\beta_0 + \beta_1 x_i}\right)}. \tag{9}$$

In an animal-based studies, one parameter of interest is the median effective dose, which is denoted by ED_{50}. It is the dose, which satisfies

$$\pi_{ED_{50}} = \frac{e^{\beta_0 + \beta_1 ED_{50}}}{1 + e^{\beta_0 + \beta_1 ED_{50}}} = .5, \tag{10}$$

and it can be shown that $ED_{50} = -\frac{\beta_0}{\beta_1}$ by algebra. In the case of $\beta_0 = -2$ and $\beta_1 = 5$, we have $ED_{50} = .4$ as describe in the figure with the dotted curve. In the case of $\beta_0 = -1$ and $\beta_2 = 2$, we have $ED_{50} = .5$ as described in the figure with the solid curve.

In 1997, International Agency for Research on Cancer classified 2,3,7,8-Tetrachlorodibenzo-p-dioxin (known as TCDD) as a carcinogen for humans based on various empirical evidence [8].

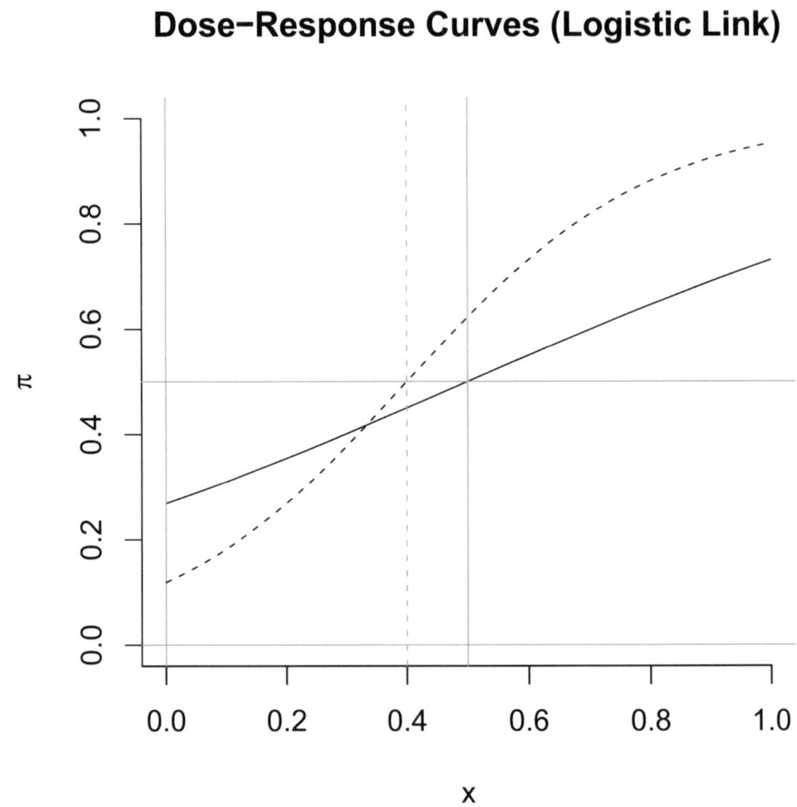

Figure 2. Two dose-response curves using the logistic link.

In 1978, Kociba et al. presented the data on male Sprague-Dawley rats at four experimental doses 0, 1, 10 and 100 nanograms per kilogram per day (ng/kg/day) [9]. In the control dose group, nine of 86 rats developed tumor (known as hepatocellular carcinoma); three of 50 rats developed the tumor at dose 1; 18 of 50 rats developed the tumor at dose 10; and 34 of 48 rats developed the tumor at dose 100 [6]. Without loss of generosity, we let $x_i = 0$ for $i = 1,\ldots, 86$; $x_i = 1$ for $i = 87,\ldots 136$; $x_i = 10$ for $i = 137,\ldots, 186$; and $x_i = 100$ for $i = 187,\ldots, 234$. The given information is sufficient to calculate $s_1 = \sum_{i=1}^{n} y_i = 64$ and $s_2 = \sum_{i=1}^{n} x_i y_i = 3583$. By the use of the flat prior $f(\vec{\beta}) \propto 1$ with the restriction $\beta_1 > 0$, given the observed sample of size $n = 234$, we can generate random numbers of $\vec{\beta} = (\beta_0, \beta_1)$ from the posterior density function

$$f(\vec{\beta}\,|\vec{y}) \propto \frac{e^{\beta_0 s_1 + \beta_1 s_2}}{\prod_{i=1}^{n}(1 + e^{\beta_0 + \beta_1 x_i})} \mathrm{I}_{\beta_1 > 0}, \tag{11}$$

where $\mathrm{I}_{\beta_1 > 0} = 1$ if $\beta_1 > 0$ and $\mathrm{I}_{\beta_1 > 0} = 0$ otherwise. Using a method of Markov Chain Monte Carlo (MCMC), we can approximate the posterior distribution of $\vec{\beta}$ as shown in the left panel of **Figure 3**. By transforming (β_0, β_1) to $\mathrm{ED}_{50} = -\frac{\beta_0}{\beta_1}$, we can approximate the posterior

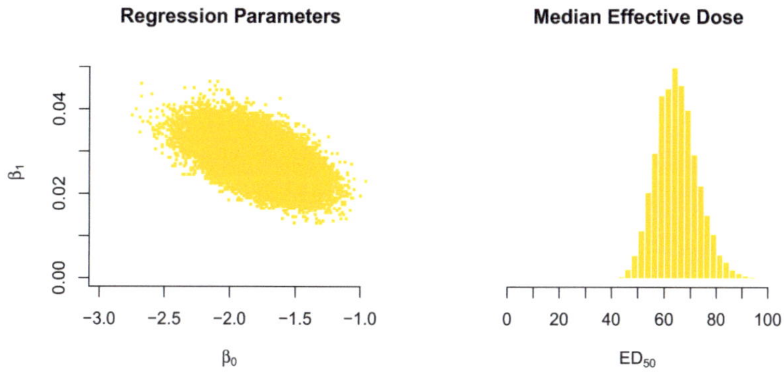

Figure 3. Approximate posterior distributions of (β_0, β_1) and $\mathrm{ED}_{50} = -\frac{\beta_0}{\beta_1}$.

distribution of the median effective dose ED_{50} as shown in the right panel. The posterior mean of ED_{50} is $E(\mathrm{ED}_{50}|\vec{y}) = 64.9$ with 95% credible interval (50.8, 82.5), the 2.5th percentile and the 97.5th percentile of the posterior distribution.

3. Bayesian model averaging

In a small sample, we borrow the strength of a parametric model to gain efficiency in parameter estimation. However, an assumed model may not describe the true dose-response relationship adequately. The impact of model misspecification is not negligible particularly in a poor experimental design. In such a limited practical situation, Bayesian model averaging (BMA) can be a useful method to account for model uncertainty. It is widely applied in practice, and in this section, we focus on the application to cancer risk assessment for the estimation of a benchmark dose [1, 6, 10, 11].

Let θ denote a parameter of interest. Suppose we have a set of K candidate models denoted by $\mathcal{M} = \{M_1, \ldots, M_K\}$. Let $\vec{\beta}_k$ denote the vector of regression parameters under model M_k for k =1, …, K. Suppose θ is a function of $\vec{\beta}_k$, and the interpretation of θ must be common across all models. Let $f(\vec{\beta}_k|M_k)$ and $f(\vec{y}|\vec{\beta}_k, M_k)$ denote the prior density function and the likelihood function, respectively, under M_k. By the Law of Total Probability, the posterior density function of θ is as follows

$$f(\theta|\vec{y}) = \sum_{k=1}^{K} f(\theta|M_k, \vec{y})\, P(M_k|\vec{y}). \tag{12}$$

In Eq. (12), the posterior density function $f(\theta|M_k, \vec{y})$ depends on model M_k, and the posterior model probability $P(M_k|\vec{y})$ quantifies the plausibility of model M_k after observing data, which is given by

$$P(M_k|\vec{y}\,) = \frac{f(\vec{y}|M_k)\,P(M_k)}{\sum_{j=1}^{K} f(\vec{y}|M_j)\,P(M_j)}. \tag{13}$$

In Eq. (13), the prior model probability $P(M_k)$ is determined before observing data such that $P(M_k) > 0$ for $k = 1, \ldots, K$ and $\sum_{k=1}^{K} P(M_k) = 1$. The marginal likelihood under M_k requires the integration

$$f(\vec{y}\,|M_k) = \int f(\vec{y}|\vec{\beta}_k, M_k) f(\vec{\beta}_k|M_k)\, d\vec{\beta}_k. \tag{14}$$

In the BMA method, all K models contribute to inference of θ through the averaged posterior density function in Eq. (12), and the weight of contribution is determined by Bayes' Theorem in Eq. (13).

3.1. Example

This example is continued from the example in Section 2.2. Recall π_x is interpreted as the probability of a toxic event (tumor development) at dose x. In many cancer risk assessments, a parameter of interest is θ_γ at a fixed risk level γ, which is defined as follows

$$\gamma = \frac{\pi_{\theta_\gamma} - \pi_0}{1 - \pi_0} \tag{15}$$

or equivalently $\pi_{\theta_\gamma} = \pi_0 + (1 - \pi_0)\,\gamma$. In words, θ_γ is a dose corresponding to a fixed increase in the risk level. In frequentist framework, Crump defined a benchmark dose as a lower confidence limit for θ_γ [12]. In Bayesian framework, an analogous definition would be a lower credible bound (i.e., a fixed low percentile of the posterior distribution of θ_γ). The definition is widely applied to the public health protection [13].

In practice, γ is fixed between 0.01 and 0.1. Often, the estimation of θ_γ is highly sensitive to an assumed dose-response model because we have a lack of information at low doses. Shao and Small fixed $\gamma = 0.1$ and applied BMA with $K = 2$ models, logistic model and quantal-linear model [6]. In the quantal-linear model, the probability of tumor development is modeled by

$$\pi_x = \beta_0 + (1 - \beta_0)(1 - e^{-\beta_1 x}). \tag{16}$$

with the restrictions $0 < \beta_0 < 1$ and $\beta_1 > 0$ under the monotonic assumption. The logistic model was given in Eq. (6) of Section 2.2.

Let M_1 denote the logistic model, and let M_2 denote the quantal-linear model. Assume the uniform prior model probabilities $P(M_1) = P(M_2) = .5$ and flat priors on the regression parameters. By posterior sampling, we can approximate the posterior model probabilities $P(M_1|\vec{y}) = .049$ and $P(M_2|\vec{y}) = .951$. Under M_1, the posterior mean of $\theta_{0.1}$ is 20.95 with the 5th percentile 16.74. Under M_2, the posterior mean is 8.25 with the 5th percentile 5.95. These

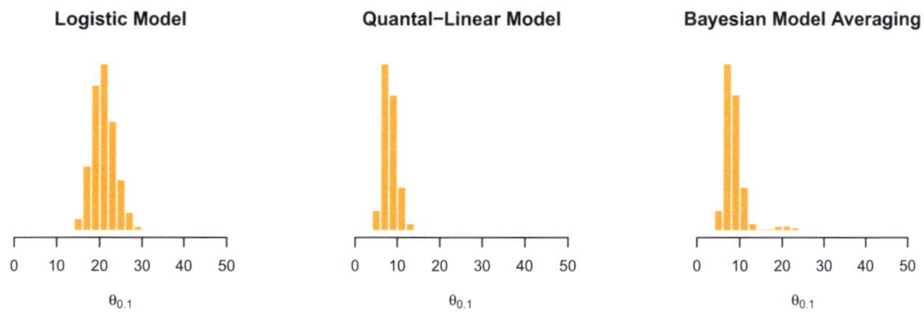

Figure 4. Posterior distributions of $\theta_{0.1}$ from the logistic model (left panel), the quantal-linear model (middle panel), and the Bayesian model averaging (right panel).

results are very similar to the results reported by Shao and Small [6]. From these model-specific statistics, we can calculate the model-averaged posterior mean

$$E(\theta_{0.1}|\vec{y}\,) = \sum_{k=1}^{2} E(\theta_{0.1}|M_k, \vec{y}\,)\, P(M_k|\vec{y}\,) = 20.95\,(.049) + 8.25\,(.951) = 8.87\,. \tag{17}$$

However, we are not able to calculate the 5th percentile of the model-averaged posterior distribution based on the given statistics. In fact, we need to approximate the posterior distribution $f(\theta_{0.1}|\vec{y}\,)$, which is a mixture of $f(\theta_{0.1}|M_1, \vec{y}\,)$ and $f(\theta_{0.1}|M_2, \vec{y}\,)$ weighted by $P(M_1|\vec{y}\,) = .049$ and $P(M_2|\vec{y}\,) = .951$, respectively, as shown in **Figure 4**. In the figure, the left panel shows an approximation of $f(\theta_{0.1}|M_1, \vec{y}\,)$, the middle panel shows an approximation of $f(\theta_{0.1}|M_2, \vec{y}\,)$, and the right panel shows an approximation of the averaged posterior $f(\theta_{0.1}|\vec{y}\,)$. The averaged posterior density $f(\theta_{0.1}|\vec{y}\,)$ is bimodal, but it is very close to $f(\theta_{0.1}|M_2, \vec{y}\,)$ because the quantal-linear model M_2 fits the data better than the logistic model M_1 by a Bayes factor of $\frac{P(M_2|\vec{y}\,)}{P(M_1|\vec{y}\,)} = \frac{.951}{.049} = 19.4$. The 5th percentile of the model-averaged posterior distribution is approximately 5.97, and it is a BMA-BMD based on the BMA method proposed by Raftery et al. [1] and the BMD estimation method suggested by Crump [12].

4. Application of Bayesian decision theory to Phase I trials

In a Phase I cancer trial, the main objectives are to study the safety of a new chemotherapy and to determine an appropriate dose for future patients. Since trial participants are cancer patients, dose allocations require ethical considerations. Whitehead and Williams discussed several Bayesian approaches to dose allocations [14]. One decision rule is devised from the perspective of trial participants (individual-level ethics), and another decision rule is devised from the perspective of future patients (population-level ethics). However, a decision rule, which is devised from the population-level ethics, is not widely accepted in current practice [15]. Instead, there are some proposed decision rules, which compromise between the individual- and population-level perspectives [3, 16]. In this section, we discuss the two conflicting perspectives in Phase I clinical trials and a compromising method based on Bayesian decision theory.

Assume a dose-response relationship follows a logistic model

$$\pi_x = \frac{e^{\beta_0+\beta_1 x}}{1+e^{\beta_0+\beta_1 x}}, \tag{18}$$

where x is a dose in the logarithmic scale (base e) and π_x is the probability of observing an adverse event due to the toxicity of a new chemotherapy at dose x. The logarithmic transformation on the dose is to satisfy $\pi_x \to 0$ as $x \to 0$. Let $\vec{x}_n = (x_1, \dots, x_n)$ denote a series of decisions for n patients (i.e., allocated doses) and $\vec{y}_n = (y_1, \dots, y_n)$ denote a series of observed responses, where $y_i = 1$ indicates an adverse event and $y_i = 0$ otherwise. Let $L(\vec{\beta}, x_{n+1})$ denote a loss by allocating the next patient at x_{n+1}. Based on Bayesian Decision Theory, we want to find x_{n+1} which minimizes the posterior mean of $L(\vec{\beta}, x_{n+1})$. If we let $\mathcal{A}$ denote an action space, a set of all possible dose allocations for the next patients, the decision rule can be written as follows:

$$x^*_{n+1} = \text{argmin}_{x_{n+1} \in \mathcal{A}} E\left(L(\vec{\beta}, x_{n+1}) \mid \vec{y}_n\right). \tag{19}$$

A choice of L has a substantial impact on the operating characteristics of a Phase I trial including (i) the degree of under- and over-dosing in trial, (ii) the observed number of adverse events at the end of a trial, and (iii) the quality of estimation at the end of a trial.

4.1. Parameter of interest: maximum tolerable dose

Let N denote an available sample size for a Phase I clinical trial. A typical sample size is $N \leq 30$. Let γ denote a target risk level, the probability of an adverse event. In a cancer study, a typical target risk level γ is fixed between .15 and .35 depending on the severity of an adverse event. Then, the dose corresponding to γ is called a maximum tolerable dose (MTD) at level γ, and we denote it by θ_γ in the logarithmic scale. Under the logistic model in Eq. (18), it is defined as follows

$$\theta_\gamma = \frac{\log\left(\frac{\gamma}{1-\gamma}\right) - \beta_0}{\beta_1}. \tag{20}$$

At the end of a trial (observing N responses), we estimate θ_γ by the posterior mean $\hat{\theta}_{\gamma,N} = E(\theta_\gamma | \vec{y}_N)$ for future patients.

4.2. Prior density function: conditional mean priors

A consequence of sequential decisions heavily depends on a prior density function $f(\vec{\beta})$. In particular, the first decision x_1 must be made based on prior knowledge only because empirical evidence is not observed yet. In addition, the later decisions $x_2, x_3, \dots$ and the final inference of

θ_γ are substantially affected by $f(\vec{\beta})$ as a Phase I study is typically based on a small sample. In this regard, we want to carefully utilize researchers' prior knowledge about $\vec{\beta}$, but it may be difficult to express their prior knowledge directly through $f(\vec{\beta})$. In this section, we discuss a method of eliciting prior knowledge, which is more tractable than prior elicitation directly on $\vec{\beta}$.

Suppose a researcher selects two arbitrarily doses, say $x_{-1} < x_0$. Then, the researcher may express their prior knowledge by two independent beta distributions

$$\pi_{x_i} = \frac{e^{\beta_0+\beta_1 x_i}}{1+e^{\beta_0+\beta_1 x_i}} \sim \text{Beta}(a_i, b_i), \ \ j = -1, 0. \tag{21}$$

Using the Jacobian transformation from $(\pi_{x_{-1}}, \pi_{x_0})$ to $\vec{\beta} = (\beta_0, \beta_1)$, it can be shown that the prior density function of $\vec{\beta}$ is given by

$$f(\vec{\beta}) \propto (x_0 - x_{-1}) \prod_{i=-1}^{0} \left(\frac{e^{\beta_0+\beta_1 x_i}}{1+e^{\beta_0+\beta_1 x_i}}\right)^{a_i} \left(\frac{1}{1+e^{\beta_0+\beta_1 x_i}}\right)^{b_i}. \tag{22}$$

It is known as conditional mean priors under the logistic model [7].

4.3. Posterior density function: conjugacy

For notational convenience, we let $y_i = a_i$ and $n_i = a_i + b_i$ for $i = -1,0$. By conjugacy, the posterior density function of $\vec{\beta}$ can be concisely written as follows

$$f(\vec{\beta}\,|\vec{y}_n) \propto \frac{e^{\beta_0 s_1+\beta_1 s_2}}{\prod_{i=-1}^{n}(1+e^{\beta_0+\beta_1 x_i})}, \tag{23}$$

where $s_1 = \sum_{i=-1}^{n} y_i$ and $s_2 = \sum_{i=-1}^{n} x_i\, y_i$. After observing n responses, the decision rule for the next patient is as follows

$$x_{n+1}^* = \text{argmin}_{x_{n+1}\in\mathcal{A}} \int L(\vec{\beta}, x_{n+1}) f(\vec{\beta}|\vec{y}_n)\, d\vec{\beta}. \tag{24}$$

4.4. Loss functions for individual- and population-level ethics

A loss function, which reflects the perspective of individual-level ethics, is as follows:

$$L_I(\vec{\beta}, x_{n+1}) = (x_{n+1} - \theta_\gamma)^2. \tag{25}$$

This loss function is analogous to the original continual reassessment method proposed by O'Quigley et al. [17]. The square error loss attempts to treat a trial participant at θ_γ, and the expected square error loss is minimized by the posterior mean of θ_γ.

From the perspective of population-level ethics, Whitehead and Brunier proposed a loss function, which is equal to the asymptotic variance of the maximum likelihood estimator for θ_γ [18]. The Fisher expected information matrix with a sample of size $n + 1$ is given by

$$\mathcal{I}(\vec{\beta}) = \begin{pmatrix} \sum_{i=1}^{n+1} \tau_i & \sum_{i=1}^{n+1} \tau_i x_i \\ \sum_{i=1}^{n+1} \tau_i x_i & \sum_{i=1}^{n+1} \tau_i x_i^2 \end{pmatrix}, \quad (26)$$

where $\tau_i = \pi_{x_i}(1 - \pi_{x_i})$. Then, the loss function (the asymptotic variance) is given by

$$L_P(\vec{\beta}, x_{n+1}) = \left[\nabla \vec{h}(\vec{\beta})\right]^T \left[\mathcal{I}(\vec{\beta})\right]^{-1} \left[\nabla \vec{h}(\vec{\beta})\right], \quad (27)$$

where

$$\nabla \vec{h}(\vec{\beta}) = \begin{pmatrix} \dfrac{\partial \theta_\gamma}{\partial \beta_0} \\ \dfrac{\partial \theta_\gamma}{\partial \beta_1} \end{pmatrix} = -\frac{1}{\beta_1}\begin{pmatrix} 1 \\ \theta_\gamma \end{pmatrix} \quad (28)$$

is the gradient vector, the partial derivatives of θ_γ with respect to β_0 and β_1. Kim and Gillen decomposed the population-level loss function as follows

$$L_P(\vec{\beta}, x_{n+1}) = \frac{\tau_{n+1}(x_{n+1} - \theta_\gamma)^2 + s_n^{(0)}\left[(\theta_\gamma - \mu_n)^2 + \sigma_n^2\right]}{\left[s_n^{(0)} s_n^{(2)} - s_n^{(1)} s_n^{(1)}\right] + s_n^{(0)} \tau_{n+1}\left[(x_{n+1} - \mu_n)^2 + \sigma_n^2\right]}, \quad (29)$$

where

$$\begin{aligned} s_n^{(m)} &= \sum_{i=1}^{n} \tau_i x_i^m, \quad m = 0, 1, 2, \\ \mu_n &= \sum_{i=1}^{n} w_i x_i, \\ \sigma_n^2 &= \sum_{i=1}^{n} w_i x_i^2 - \left(\sum_{i=1}^{n} w_i x_i\right)^2 \end{aligned} \quad (30)$$

with the weight defined as $w_i = \frac{\tau_i}{\sum_{i=1}^{n} \tau_j}$ [3]. Eq. (29) has the following important remarks. In fact, $L_P(\vec{\beta}, x_{n+1})$ considers individual-level ethics by including $L_I(\vec{\beta}, x_{n+1}) = (x_{n+1} - \theta_\gamma)^2$ in the numerator. By including $(x_{n+1} - \mu_n)^2$ in the denominator, where $\mu_n = \sum_{i=1}^{n} w_i x_i$, the population-level loss function reduces a loss by allocating the next patient further away from the weighted average of previously allocated doses (i.e., devised from information gain). In long run, $L_P(\vec{\beta}, x_{n+1})$ is devised from a compromise between individual- and population-level ethics, but the compromising process is rather too slow to be implemented in a small-sample Phase I clinical trial [3].

4.5. Loss function for compromising the two perspectives

Kim and Gillen proposed to accelerate the compromising process by modifying $L_P(\vec{\beta}, x_{n+1})$ of Eq. (29) as follows

$$L_{B,\lambda}(\vec{\beta}, x_{n+1}) = \frac{a_n(\lambda)\, \tau_{n+1}(x_{n+1} - \theta_\gamma)^2 + s_n^{(0)}\left[(\theta_\gamma - \mu_n)^2 + \sigma_n^2\right]}{\left[s_n^{(0)} s_n^{(2)} - s_n^{(1)}\right] + s_n^{(0)} \tau_{n+1}\left[(x_{n+1} - \mu_n)^2 + \sigma_n^2\right]}, \tag{31}$$

where

$$a_n(\lambda) = \left(1 + \tfrac{n}{N}\right)^{\lambda\left(1 + \frac{\sum_{i=1}^{n} y_i}{N\gamma}\right)} \tag{32}$$

is an accelerating factor [3]. It has two implications. First, the compromising process is accelerated toward the individual-level ethics as the trial proceeds (i.e., n increases). Second, the compromising process toward the individual-level ethics is accelerated at a faster rate when an adverse event is observed (i.e., $\sum_{i=1}^{n} y_i$ increases). The tuning parameter λ controls the rate of acceleration. It imposes more emphasis on population-level ethics as $\lambda \to 0$ and more emphasis on individual-level ethics as $\lambda \to \infty$. The choice of λ shall depend on the severity level of an adverse event.

4.6. Simulation

To study the operating characteristics of $L_{B,\lambda}$ with respect to λ, we assume the logistic model with $\beta_0 = -3$ and $\beta_1 = .8$ as a true dose-response relationship as shown in **Figure 5** in the left panel. The target risk level is fixed at $\gamma = .2$, so the true MTD is given by $\theta_{.2} = 2.02$ in the logarithmic scale. We consider three different priors based on the conditional mean priors given in Eq. (22). For simplicity, we set $a_{-1} = 1$, $b_{-1} = 3$, $a_0 = 3$ and $b_0 = 1$ for all three priors. Then, we let $x_{-1} = -4$ and $x_0 = 4$ for Prior 1; $x_{-1} = 0$ and $x_0 = 8$ for Prior 2; and $x_{-1} = 4$ and $x_0 = 12$ for Prior 3. **Figure 5** in the right panel shows an approximated $f(\theta_{.2})$ for each prior. Prior 1 significantly underestimates the true $\theta_{.2} = 2.02$ with prior mean $E(\theta_{.2}) = -1.70$, Prior 3 overestimates the truth with $E(\theta_{.2}) = 5.38$, and Prior 2 has a prior estimate relatively close to the truth with $E(\theta_{.2}) = 1.40$.

Let $N = 20$ be a fixed sample size. Let $Y_i = 1$ denote an adverse event observed from the ith patient ($Y_i = 0$ otherwise), so $\sum_{i=1}^{N} Y_i$ denotes the total number of adverse events observed at the end of a trial. The sum $\sum_{i=1}^{n} Y_i$ is random from a trial to another trial, and we want $\sum_{i=1}^{n} Y_i$ to behave like Binomial(20, .2) which is the case when we treat $N = 20$ to the true MTD $\theta_{.2}$. **Figure 6** shows three simulated trials under the loss function $L_{B,\lambda}$ with $\lambda = 0,1,5$. When $\lambda = 0$,

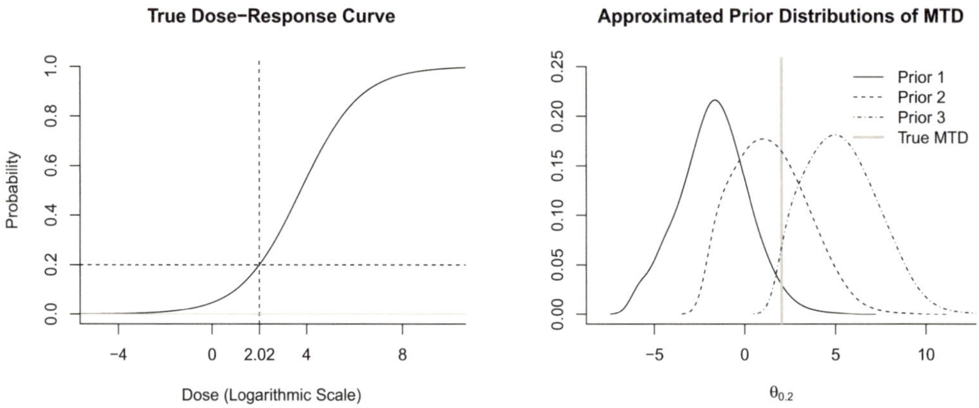

Figure 5. The true dose-response relationship $\pi_x = \frac{e^{\beta_0+\beta_1 x}}{1+e^{\beta_0+\beta_1 x}}$ with $\beta_0 = -3$ and $\beta_1 = .8$ (where x is the dose in the logarithmic scale) in the simulation (left panel) and the three prior distributions of $\theta_{.2}$ approximated by kernel density (right panel).

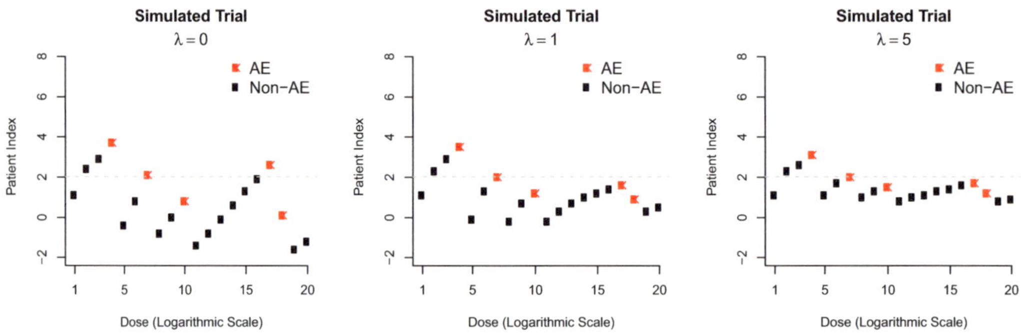

Figure 6. Three simulated trials using the loss function $L_{B,\lambda}$ with $\lambda = 0$ (left), $\lambda = 1$ (middle) and $\lambda = 5$ (right) with a sample of size $N = 20$ and assumed parameter values $\beta_0 = -3$, $\beta_1 = 8$ and $\theta_{.2} = 2.02$.

the up-and-down scheme has a high degree of fluctuation in order to maximize information about $\theta_{.2}$. When $\lambda = 1$, the up-and-down scheme is stabilized after the first few adverse events, and the stabilization occurs quickly when $\lambda = 5$ to treat trial participants near an estimated $\theta_{.2}$.

Let $\hat{\theta}_{.2} = E(\theta_{.2}|\vec{y}_N)$, the posterior estimate of $\theta_{.2}$ at the end of a trial, so $\pi_{\hat{\theta}_{.2}}$ implies the true probability of an adverse event at the estimated MTD. We focus on the following criteria: (i) $E(\pi_{\hat{\theta}_{.2}})$ which we desire to be close to $\gamma = .2$ for future patients, (ii) $V(\pi_{\hat{\theta}_{.2}})$ which we desire to be as low as possible for future patients, (iii) $E[(\pi_{\hat{\theta}_{.2}} - .2)^2]$ which we desire to be as low as possible for future patients, (iv) $E(\sum_{i=1}^{20} Y_i)$ which we desire to be close to $N\gamma = 4$ for trial participants and (v) $P(3 \le \sum_{i=1}^{20} Y_i \le 5)$ which we desire to be close to one for trial participants.

Prior	λ	$E(\pi_{\hat{\theta}.2})$	$V(\pi_{\hat{\theta}.2})$	$E[(\pi_{\hat{\theta}.2} - .2)^2]$	$E(\sum_{i=1}^{20} Y_i)$	$P(3 \le \sum_{i=1}^{20} Y_i \le 5)$
1	0	0.0964	0.0019	0.0126	2.4353	0.4318
	.5	0.1034	0.0024	0.0118	2.0997	0.2298
	1	0.1082	0.0028	0.0113	1.8969	0.1714
	2	0.1100	0.0031	0.0112	1.6929	0.1211
	5	0.1157	0.0035	0.0106	1.3128	0.0596
2	0	0.1665	0.0054	0.0065	4.1217	0.9889
	.5	0.1705	0.0056	0.0065	3.9598	0.9877
	1	0.1727	0.0060	0.0068	3.9025	0.9670
	2	0.1751	0.0066	0.0072	3.8707	0.9291
	5	0.1763	0.0067	0.0073	3.8442	0.9068
3	0	0.2743	0.0048	0.0103	6.1875	0.1600
	.5	0.2673	0.0048	0.0093	6.3954	0.1430
	1	0.2606	0.0046	0.0083	6.6194	0.1165
	2	0.2562	0.0045	0.0077	6.8035	0.1020
	5	0.2499	0.0044	0.0068	7.0274	0.0760

Table 1. Simulation results of 10,000 replicates for λ = 0, .5, 1, 2, 5 and each prior.

Table 1 summarizes simulation results of 10,000 replicates for each prior. For all three priors, we observe similar tendencies. First, $E(\pi_{\hat{\theta}.2})$ gets closer to $\theta = .2$ as λ increases. Second, $V(\pi_{\hat{\theta}.2})$ decreases as λ decreases to zero. The average square distance between $\pi_{\hat{\theta}.2}$ and $\gamma = .2$ measures a balance between $|E(\pi_{\hat{\theta}.2}) - .2|$ and $V(\pi_{\hat{\theta}.2})$, and the superiority depends on priors. Lastly, as $\lambda \to 0$, we have larger $P(3 \le \sum_{i=1}^{20} Y_i \le 5)$ and more robust $E(\sum_{i=1}^{20} Y_i)$ to prior elicitation.

In summary, when we emphasize more on population-level ethics, we have a smaller variance in the estimation for future patients (with a greater absolute bias, potentially due to Jensen's Inequality), and the distribution of $\sum_{i=1}^{n} Y_i$ becomes more robust to prior elicitations. When we emphasize more on individual-level ethics, we have a larger variance in the estimation, and the distribution of $\sum_{i=1}^{n} Y_i$ becomes more sensitive to prior elicitations.

5. Consensus prior

In Bayesian inference, researchers are able to utilize information, which is independent of observed data. It allows researchers to incorporate any form of information, such as one's experience and existing literature, which may be particularly useful in a small-sample study. On the other hand, we concern subjectivity and prior sensitivity in sparse data. Furthermore, it is possible to have disagreement among multiple researchers' prior elicitations about a parameter θ.

Suppose there are K researchers with their own prior density functions, say $f(\theta|Q_k)$ for $k = 1,\ldots, K$, and they have the same likelihood function $f(\vec{y}|\theta)$. Each prior elicitation leads to a unique Bayes estimator

$$\hat{\theta}_k = E(\theta|\vec{y}, Q_k) = \int \theta f(\theta|\vec{y}, Q_k)\, d\theta, \tag{33}$$

where $f(\theta|\vec{y}, Q_k) \propto f(\vec{y}|\theta)f(\theta|Q_k)$ is the posterior density function of θ given data $\vec{y}$ and the kth prior elicitation Q_k. For posterior estimation, one reasonable approach to compromise is a weighted average $\sum_{k=1}^{K} w_k\hat{\theta}_k$, where $w_k > 0$ for $k = 1,\ldots,K$ and $\sum_{k=1}^{K} w_k = 1$. In this section, we discuss two different weighting methods. The first method is to fix w_k before observing data (referred to as prior weighting scheme). The second method is to determine $w_k(\vec{y})$ after observing data $\vec{y}$ so that $w_k(\vec{y})$ increases when the kth prior elicitation Q_k is better supported by the observed data $\vec{y}$ (referred to as posterior weighting scheme) [5].

For a prior weighting scheme, we denote $w_k = P(Q_k)$ which quantifies the credibility of the kth prior elicitation. For a posterior weighting scheme, we consider

$$w_k(\vec{y}) = P(Q_k|\vec{y}) = \frac{f(\vec{y}|Q_k)\, P(Q_k)}{\sum_{j=1}^{K} f(\vec{y}|Q_j)\, P(Q_j)} = \frac{w_k f(\vec{y}|Q_k)}{\sum_{j=1}^{K} w_j\, f(\vec{y}|Q_j)}, \tag{34}$$

where $f(\vec{y}|Q_k) = \int f(\vec{y}|\theta)\, f(\theta|Q_k)\, d\theta$ is the marginal likelihood from the kth prior elicitation. This formulation is similar to the BMA method discussed in Section 3. It can be shown that $\sum_{k=1}^{K} w_k(\vec{y})\, \hat{\theta}_k$ is the Bayes estimator (the posterior mean of θ) when a consensus prior $f(\theta) = \sum_{k=1}^{K} w_k f(\theta|Q_k)$ is used with $w_k = P(Q_k)$ [5].

Samaniego discussed self-consistency when compromised inference is used through the prior weighting scheme $\sum_{k=1}^{K} w_k\hat{\theta}_k$ [4]. Let θ denote a parameter of interest and

$$E(\theta) = \int \theta f(\theta)\, d\theta = \theta^* \tag{35}$$

be the prior expectation, the mean of the prior density function $f(\theta)$. Let $\tilde{\theta}$ denote a sufficient statistic, which serves as an unbiased estimator for θ. When we satisfy $E(\theta|\tilde{\theta} = \theta^*) = \theta^*$, it is called self-consistency [4].

Self-consistency can be achieved under simple models. For example, let $\vec{Y} = (Y_1, \ldots, Y_n)$ be a random sample, where $Y_i \sim \text{Bernoulli}(\theta)$, and assume $\theta \sim \text{Beta}(a, b)$ for prior. It can be shown

http://dx.doi.org/10.5772/intechopen.68786

that the maximum likelihood estimator $\tilde{\theta} = \frac{1}{n}\sum_{i=1}^{n} Y_i$ is a sufficient statistic and an unbiased estimator for θ. The posterior mean is a weighted average between θ^* and $\tilde{\theta}$ as follows

$$E(\theta|\tilde{\theta} = \theta^*) = c\,\theta^* + (1-c)\,\tilde{\theta}, \tag{36}$$

where $c = \frac{a+b}{a+b+n}$. If we observe $\tilde{\theta} = \theta^*$, we can achieve the self-consistency because $E(\theta|\hat{\theta} = \theta^*) = \theta^*$. In words, when prior estimate and maximum likelihood estimate are identical, the posterior estimate must be consistent with the prior estimate and the maximum likelihood estimate. The self-consistency can be also achieved in the prior weighting scheme under certain conditions as illustrated in the following example.

5.1. Binomial experiment

Let $Y_i \sim \text{Bernoulli}(\pi)$ for $i = 1,\ldots,n$ and assume $Y_1, \ldots, Y_n$ are independent. Suppose the kth researcher specifies the prior distribution $\pi|Q_k \sim \text{Beta}(a_k, b_k)$ for $k = 1,\ldots,K$. For the prior weighting scheme, let $w_k = P(Q_k)$, the prior probability for the kth prior elicitation (fixed before observing data). Since $E(\pi|Q_k) = \frac{a_k}{a_k+b_k}$ and the expectation $E\,(\cdot)$ is a linear operator, the average of "consensus prior" is

$$E(\pi) = \int_0^1 \pi\, f(\pi)\, d\pi = \int_0^1 \pi\Big(\sum_{k=1}^{K} f(\pi|Q_k)\, P(Q_k)\Big)\, d\pi = \sum_{k=1}^{K} w_k\Big(\int_0^1 \pi\, f(\pi|Q_k)\, d\pi\Big) = \sum_{k=1}^{K} w_k\, E(\pi|Q_k). \tag{37}$$

Let $E(\pi) = \pi^*$ and suppose the K researchers observed the consistent result $\tilde{\pi} = \frac{1}{n}\sum_{i=1}^{n} Y_i = \pi^*$. The individual-specific Bayes estimator is as follows

$$\hat{\pi}_k = E(\pi|\tilde{\pi} = \pi^*, Q_k) = c_k\, E(\pi|Q_k) + (1-c_k)\,\pi^*, \tag{38}$$

for the kth researcher, where $c_k = \frac{a_k+b_k}{a_k+b_k+n}$. The compromised Bayes estimator is as follows

$$E(\pi|\tilde{\pi} = \pi^*) = \sum_{k=1}^{K} w_k\, \hat{\pi}_k = \sum_{k=1}^{K} w_k\,[c_k\, E(\pi|Q_k) + (1-c_k)\,\pi^*]. \tag{39}$$

If we allow individual-specific prior elicitation a_k and b_k with the restriction $a_k + b_k = m$ for all K researchers (i.e., the same strength of prior elicitation), the value $c_k = \frac{m}{m+n}$ is constant over all researcher. By letting the constant $c_k = c$,

$$E(\pi|\tilde{\pi} = \pi^*) = c\left(\sum_{k=1}^{K} w_k\, E(\pi|Q_k)\right) + (1-c)\,\pi^*\left(\sum_{k=1}^{K} w_k\right) = c\,E(\pi) + (1-c)\,\pi^* = \pi^*, \tag{40}$$

so the self-consistency is satisfied.

For the posterior weighting scheme given data $\vec{y} = (y_1, \ldots, y_n)$, the marginal likelihood from the kth prior elicitation is as follows

$$f(\vec{y}\,|Q_k) = \int_0^1 f(\vec{y}|\pi) f(\pi|Q_k)\, d\pi = \frac{\Gamma(a_k + b_k)}{\Gamma(a_k)\Gamma(b_k)} \frac{\Gamma(a_k + s)\Gamma(b_k + n - s)}{\Gamma(a_k + b_k + n)}, \tag{41}$$

where $s = \sum_{i=1}^{n} y_i$ is an observed sufficient statistic. Then, the posterior weighting scheme becomes $\sum_{k=1}^{K} w_k(\vec{y})\, \hat{\pi}_k$ with

$$\begin{aligned} w_k(\vec{y}) &= \frac{w_k f(\vec{y}|Q_k)}{\sum_{j=1}^{K} w_j f(\vec{y}|Q_j)}, \\ \hat{\pi}_k &= \frac{a_k + s}{a_k + b_k + n}. \end{aligned} \tag{42}$$

If we desire an equal strength from each researcher's prior elicitation, we may fix $a_k + b_k = m$ and $w_k = \frac{1}{K}$. In the posterior weighting scheme, it is difficult to achieve the self-consistency.

Whether self-consistency is satisfied, the practical concern is the quality of estimation such as bias, variance and mean square error. Assuming K = 2 researchers have disagreeing prior knowledge and a sample of size n = 10, let us consider three cases. Suppose two researchers express relatively mild disagreement as $(a_1, b_1) = (1, 3)$ and $(a_2, b_2) = (3, 1)$ in Case 1, relatively strong disagreement as $(a_1, b_1) = (2, 6)$ and $(a_2, b_2) = (6, 2)$ in Case 2, and even stronger disagreement as $(a_1, b_1) = (3, 9)$ and $(a_2, b_2) = (9, 3)$ in Case 3. For each case, **Figure 7** provides the relative bias, variance and mean square error (MSE) for comparing the posterior weighting scheme $\sum_{k=1}^{3} w_k(\vec{y})\, \hat{\pi}_k$ to the prior weighting scheme $\sum_{k=1}^{3} w_k\, \hat{\pi}_k$. When a relative MSE is smaller than one, it implies a smaller MSE for the posterior weighting scheme. As the true value of π is well between the two prior guesses $E(\pi|Q_1) = .25$ and $E(\pi|Q_2) = .75$, the posterior weighting scheme shows a greater MSE due to greater variance. When the true value of π deviates away from either prior guess, the posterior weighting schemes show a smaller MSE due to smaller bias. The tendency is stronger when the two disagreeing prior elicitations are stronger (i.e., stronger prior disagreement). The bottom line is a clear bias-variance tradeoff

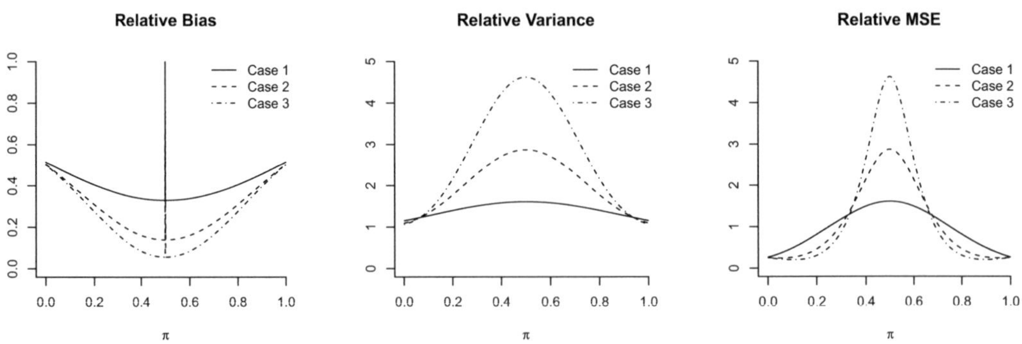

Figure 7. Comparing prior and posterior weighting schemes for different degrees of disagreements.

when we compare the two weighting schemes. $\sum_{k=1}^{3} w_k(\vec{y})\, \hat{\pi}_k$ is able to reduce bias when there is strong discrepancy between "consensus prior" and data, but it has larger variance than $\sum_{k=1}^{3} w_k\, \hat{\pi}_k$ because $w_k(\vec{y})$ depends on random data.

5.2. Applications to Phase I trials under logistic regression model

In this section, we apply the prior weighting scheme and the posterior weighting scheme to Phase I clinical trials under the logistic regression model. We consider the three priors considered in Section 4.6. We denote Prior 1, 2 and 3 by Q_1, Q_2 and Q_3, respectively. The three priors had the same hyper-parameters $a_{-1,k} = 1$, $b_{-1,k} = 3$, $a_{0,k} = 3$, $b_{0,k} = 1$, but they were different by $x_{-1,k} = -4, 0, 4$ and $x_{0,k} = 4, 8, 12$ for $k = 1, 2, 3$, respectively. By the use of the conditional mean prior in Eq. (22), the prior density function of $\vec{\beta}$ for prior Q_k is given by

$$f(\vec{\beta}\,|Q_k) \propto (x_{0,k} - x_{-1,k}) \prod_{i=-1}^{0} \left(\frac{e^{\beta_0+\beta_1 x_i}}{1+e^{\beta_0+\beta_1 x_i}} \right)^{a_{i,k}} \left(\frac{1}{1+e^{\beta_0+\beta_1 x_i}} \right)^{b_{i,k}}. \tag{43}$$

The prior means were $E(\theta_{.2}|Q_1) = -1.70$, $E(\theta_{.2}|Q_2) = 1.40$ and $E(\theta_{.2}|Q_3) = 5.38$ for Priors 1, 2 and 3, respectively.

For simulation study, we consider three simulation scenarios with sample size $N = 20$. In Scenario 1, we assume $\beta_0 = -5$ and $\beta_1 = .6$, so the true MTD is $\theta_{.2} = 6.02$, which deviates significantly from all of the three prior means. In Scenario 2, we assume $\beta_0 = -3$ and $\beta_1 = .8$ as in Section 4.6, so $\theta_{.2} = 2.02$ is well surrounded by the three prior means. In Scenario 3, we assume $\beta_0 = -1$ and $\beta_1 = 1.2$, so $\theta_{.2} = -.32$ is close to the most conservative prior mean $E(\theta_{.2}|Q_1) = -1.70$. We consider the loss function $L_I(\vec{\beta}, x_{n+1}) = (x_{n+1} - \theta_{.2})^2$ discussed in Section 4.4, which focuses on individual-level ethics. We use the uniform prior probabilities $w_k = P(Q_k) = 1/3$ for $k = 1, 2, 3$ for implementing both prior and posterior weighting scheme.

Table 2 provides the simulation results of 10,000 replicates for each scenario under the prior weighting scheme and under the posterior weighting scheme. Since the posterior weighting scheme adaptively updates $w_k(\vec{y})$ based on empirical evidence, it can reduce bias, but it has greater variance in the estimation of θ_2. As a consequence, when the true MTD was close to one extreme prior estimate (Scenarios 1 and 3), the use of the posterior weighting scheme yields a smaller $E\left[(\pi_{\hat{\theta}_{.2}} - .2)^2\right]$, $E(\sum_{i=1}^{20} Y_i)$ closer to $N\gamma = 4$, and $P(3 \leq \sum_{i=1}^{20} Y_i \leq 5)$ closer to one when compared to the use of the prior weighting scheme. In Scenario 3, the average number of adverse events was 4.6 for the posterior weighting scheme, but it was as high as 7.1 in the prior weighting scheme. On the other hand, when the true MTD was well surrounded by the three prior estimates (Scenario 2), the use of the prior weighting scheme yielded more plausible results.

The simulation results are analogous to the simpler model in Section 5.1. When the true parameter is not well surrounded by prior guesses, the posterior weighting scheme is preferable with respect to mean square error due to smaller bias. When the true parameter is well surrounded by prior guesses, the prior weighting scheme is beneficial with respect to mean square error due to smaller variance.

Scenario	Method	$E(\pi_{\hat{\theta}.2})$	$V(\pi_{\hat{\theta}.2})$	$E[(\pi_{\hat{\theta}.2}-.2)^2]$	$E(\sum_{i=1}^{20} Y_i)$	$P(3\leq\sum_{i=1}^{20} Y_i\leq 5)$
1	Prior weighting	0.0967	0.0014	0.0121	1.1090	0.0398
	Posterior weighting	0.1853	0.0073	0.0075	2.7304	0.5900
2	Prior weighting	0.2018	0.0059	0.0059	3.8432	0.9042
	Posterior weighting	0.2048	0.0110	0.0110	4.2848	0.8920
3	Prior weighting	0.2929	0.0071	0.0157	7.1090	0.0568
	Posterior weighting	0.1951	0.0133	0.0133	4.6036	0.8646

Table 2. Simulation results of 10,000 replicates for the prior and posterior weighting schemes.

As a final comment, we shall be careful about the strength of individual prior elicitations when we implement the posterior weighting scheme in Phase I clinical trials. The strength of individual prior elicitations depends on (i) the hyper-parameters $a_{i,k}$ and $b_{i,k}$, (ii) the prior weight $w_k = P(Q_k)$ as well as (iii) the distance between the two arbitrarily chosen doses $x_{0,k} - x_{1,k}$. It can be seen through the expression

$$f(\vec{\beta}) = \sum_{k=1}^{K} f(\vec{\beta}|Q_k)\,P(Q_k) \propto \sum_{k=1}^{K} w_k\,(x_{0,k} - x_{-1,k}) \prod_{i=-1}^{0} \left(\frac{e^{\beta_0+\beta_1 x_i}}{1+e^{\beta_0+\beta_1 x_i}}\right)^{a_{i,k}} \left(\frac{1}{1+e^{\beta_0+\beta_1 x_i}}\right)^{b_{i,k}}. \quad (44)$$

When researchers determine consensus prior elicitations before initiating a trial, the multiplicative term $w_k\,(x_{0,k} - x_{-1,k})$ shall be carefully considered together with the hyper-parameters $a_{i,k}$ and $b_{i,k}$ [5].

6. Concluding remarks

In this chapter, we have discussed Bayesian inference with averaging, balancing, and compromising in sparse data. In the cancer risk assessment, we have observed that low-dose inference can be very sensitive to an assumed parametric model (Section 3.1). In this case, the Bayesian model averaging can be a useful method. It provides robustness by using multiple models and posterior model probabilities to account for model uncertainty. In the application of Bayesian decision theory to Phase I clinical trials, we have observed that the sequential sampling scheme heavily depends on a loss function. A loss function, which is devised from individual-level ethics, focuses on the benefit of trial participants, and a loss function, which is devised from population-level ethics, focuses on the benefit of future patients. It is possible to balance between the two conflicting perspectives, and we can adjust a focusing point by the tuning parameter (Sections 4.5 and 4.6). Finally, the use of a weighted posterior estimate can be a compromising method when two or more researchers have prior disagreement. We have compared the prior and posterior weighting schemes in a small-sample binomial problem (Section 5.1) and in a small-sample Phase I clinical trial (Section 5.2). The prior weighting scheme (data-independent weights) outperforms when prior estimates surround the truth, and the posterior weighting scheme (data-dependent weights) outperforms when the truth is not well surrounded by prior estimates. One method does not outperform the other method for all parameter values, so it is important to be aware of their bias-variance tradeoff.

Author details

Steven B. Kim

Address all correspondence to: stkim@csumb.edu

Department of Mathematics and Statistics, California State University, CA, United States

References

[1] Raftery AE, Madigan D, Hoeting JA. Bayesian model averaging for linear regression models. Journal of the American Statistical Association. 1997;**92**:171-191

[2] Whitehead J, Williamson D. Bayesian decision procedures based on logistic regression models for dose-finding studies. Journal of Biopharmaceutical Statistics. 1998;**8**:445-467

[3] Kim SB, Gillen DL. A Bayesian adaptive dose-finding algorithm for balancing individual- and population-level ethics in Phase I clinical trials, Sequential Analysis. 2016;**35**(4):423-439

[4] Samaniego FJ. A Comparison of the Bayesian and Frequentist Approaches to Estimation. New York: Springer; 2010

[5] Kim SB, Gillen DL. An alternative perspective on consensus priors with applications to Phase I clinical trials. Jacobs Journal of Biostatistics. 2016;**1**(1):006

[6] Shao K, Small MJ. Potential uncertainty reduction in model-averaged benchmark dose estimates informed by an additional dose study. Risk Analysis. 2011;**31**:1156-1175

[7] Bedrick EJ, Christensen R, Johnson W. A new perspective on priors for generalized linear models. Journal of the American Statistical Association. 1996;**91**(436):1450-1460

[8] International Agency for Research on Cancer. IARC Monographs on the Evaluation of Carcinogenic Risks to Humans. Vol. 69. Lyon: IARC; 1997. ISBN 92-832-1269-X

[9] Kociba RJ, Keyes DG, Beyer JE, Carreon RM, Wade CE, Dit-tenber DA. et al. Results of a two-year chronic toxicity andoncogenicity study of 2,3,7,8-tetrachlorodibenzo-p-dioxin in rats. Toxicology and Applied Pharmacology. 1978;**46**(2):279-303

[10] Hoeting JA, Madigan D, Raftery AE, Volinsky CT. Bayesian model averaging: a tutorial. Statistical Science. 1999;**14**(4):382-417

[11] Simmons SJ, Chen C, Li X, Wang Y, Piegorsch WW, Fang Q, Hu B, Dunn GE. Bayesian model averaging for benchmark dose estimation. Environmental and Ecological Statistics. 2015;**22**(1):5-16

[12] Crump KS. A new method for determining allowable daily intakes. Fundamental and Applied Toxicology. 1984;**4**:854-871

[13] EPA (US Environmental Protection Agency). Benchmark dose technical guidance, EPA/100/R-12/001, Risk Assessment Forum. Washington, DC: U.S. Environmental Protection Agency; 2012

[14] Whitehead J, Williamson D. Bayesian decision procedures based on logistic regression models for dose-finding studies. Journal of Biopharmaceutical Statistics. 1998;**8**:445-467

[15] O'Quigley J, Conaway M. Continual reassessment and related dose-finding designs. Statistical Science. 2010;**25**:202-216

[16] Bartroff J, Lai TL. Incorporating individual and collective ethics into Phase I cancer trial designs. Biometrics. 2011;**67**:596-603

[17] O'Quigley J, Pepe M, Fisher L. Continual reassessment method: A practical design for Phase 1 clinical trials in cancer. Biometrics. 1990;**46**:33-48

[18] Whitehead J, Brunier H. Continual reassessment method: Bayesian decision procedures for dose determining experiments. Statistics in Medicine. 1995;**14**:885-893

Chapter 9

Two Examples of Bayesian Evidence Synthesis with the Hierarchical Meta-Regression Approach

Pablo Emilio Verde

Additional information is available at the end of the chapter

http://dx.doi.org/10.5772/intechopen.70231

Abstract

This is the Information Age. We can expect that for a particular research question that is empirically testable, we should have a collection of evidence which indicates the best way to proceed. Unfortunately, this is not the case in several areas of empirical research and decision making. Instead, when researchers and policy makers ask a specific question, such as "What is the effectiveness of a new treatment?", the structure of the evidence available to answer this question may be complex and fragmented (e.g. published experiments may have different grades of quality, observational data, subjective judgments, etc.).

Meta-analysis is a branch of statistical techniques that helps researchers to combine evidence from a multiplicity of indirect sources. A main hurdle in meta-analysis is that we not only combine results from a diversity of sources but we also combine their multiplicity of biases. Therefore, commonly applied meta-analysis methods, e.g. random-effects models, could be misleading.

In this chapter we present a new method for meta-analysis that we have called: the "Hierarchical Meta-Regression" (HMR). The HMR is an integrated approach for evidence synthesis when a multiplicity of bias, coming from indirect and disparate evidence, has to be incorporated in a meta-analysis.

Keywords: Bayesian hierarchical models, meta-analysis, multi-parameters evidence synthesis, conflict of evidence, randomized control trials, retrospective studies

1. Introduction

In today's information age one can expect that the digital revolution can create a knowledge-based society surrounded by global communications that influence our world in an efficient and convenient way. It is recognized that never in human history we have accumulated such an astronomical amount of data, and we keep on generating data at in an alarming rate. A new term, "big data," was coined to indicate the existence of "oceans of data" where we may expect to extract useful information for any problem of interest.

In this technological society, one could expect that for a particular research question we should have a collection of high quality evidence which indicates the best way to proceed. Paradoxically, this is not the case in several areas of empirical research and decision making. Instead, when researchers and policy makers ask a specific and important question, such as "What is the effectiveness of a new treatment?", the structure of the evidence available to answer this question may be complex and fragmented (e.g., published experiments may have different grades of quality, observational data, subjective judgments, etc.). The way how researchers interpret this multiplicity of evidence will be the basis for their understanding of reality and it will determine their future decisions.

Bayesian meta-analysis, which has its roots in the work of Eddy et al. [1], is a branch of statistical techniques for interpreting and displaying results of different sources of evidence, exploring the effects of biases and assessing the propagation of uncertainty into a coherent statistical model. A gentle introduction of this area can be found in Chap. 8 of Spiegelhalter et al. [2] and a recent review in Verde and Ohmann [3].

In this chapter we present a new method for meta-analysis that we have called: the "Hierarchical Meta-Regression" (HMR). The aim of HMR is to have an integrated approach for bias modeling when disparate pieces of evidence are combined in meta-analysis, for instance randomized and non-randomized studies or studies with different qualities. This is a different application of Bayesian inference than those applications with which we could be familiar, for instance an intricate regression model, where the available data bear directly upon the question of interest.

We are going to discuss two recent meta-analyses in clinical research. The reason for highlighting these two cases is that they illustrate a main problem in evidence synthesis, which is the presence of a multiplicity of bias in systematic reviews.

1.1. An example of meta-analysis of therapeutic trials

The first example, is a meta-analysis of 31 randomized controlled trials (RCTs) of two treatment groups of heart disease patients, where the treatment group received bone marrow stem cells and the control group a placebo treatment, Nowbar et al. [4]. The data of this meta-analysis appear in the Appendix, see **Table 1**. **Figure 1** presents the forest plot of these 31 trials, where the treatment effect is measured as the difference of the ejection fraction between groups, which measures the improvement of left ventricular function in the heart.

At the bottom of **Figure 1** we see average summaries represented by two diamonds: the first one corresponds to the fixed effect meta-analysis model. This model is based under the assumption that studies are identical and the between study variability is zero. The widest diamond represents the results of a random effects meta-analysis model, which assume a substantial heterogeneity between studies. In this meta-analysis both models confirmed a positive treatment of effect of a mean difference 3.95 95% CI [3.43; 4.47] and 2.92 and a 95% CI of [1.47, 4.36], respectively.

http://dx.doi.org/10.5772/intechopen.70231

Study	TE	seTE	95%-CI	Weight (fixed)	Weight (random)
1	1.50	3.6700	1.50 [−5.69; 8.69]	0.5%	2.1%
2	1.10	2.0900	1.10 [−3.00; 5.20]	1.6%	3.3%
3	−1.70	2.9100	−1.70 [−7.40; 4.00]	0.8%	2.7%
4	0.80	2.7800	0.80 [−4.65; 6.25]	0.9%	2.8%
5	7.00	0.6300	7.00 [5.77; 8.23]	18.0%	4.4%
6	7.80	2.7600	7.80 [2.39; 13.21]	0.9%	2.8%
7	14.00	4.0500	14.00 [6.06; 21.94]	0.4%	1.9%
8	5.40	2.4400	5.40 [0.62; 10.18]	1.2%	3.0%
9	2.70	1.2000	2.70 [0.35; 5.05]	4.9%	4.1%
10	4.10	0.9800	4.10 [2.18; 6.02]	7.4%	4.2%
11	2.20	0.8800	2.20 [0.48; 3.92]	9.2%	4.3%
12	2.50	1.0100	2.50 [0.52; 4.48]	7.0%	4.2%
13	2.50	3.9600	2.50 [−5.26; 10.26]	0.5%	2.0%
14	−0.20	1.1700	−0.20 [−2.49; 2.09]	5.2%	4.1%
15	−2.00	2.0900	−2.00 [−6.10; 2.10]	1.6%	3.3%
16	−3.00	2.0300	−3.00 [−6.98; 0.98]	1.7%	3.4%
17	3.60	2.5200	3.60 [−1.34; 8.54]	1.1%	3.0%
18	4.00	1.2800	4.00 [1.49; 6.51]	4.4%	4.0%
19	−1.50	1.7500	−1.50 [−4.93; 1.93]	2.3%	3.6%
20	8.80	6.0700	8.80 [−3.10; 20.70]	0.2%	1.1%
21	4.00	3.8900	4.00 [−3.62; 11.62]	0.5%	2.0%
22	−0.20	1.5400	−0.20 [−3.22; 2.82]	3.0%	3.8%
23	−3.20	3.6300	−3.20 [−10.31; 3.91]	0.5%	2.2%
24	10.10	1.2100	10.10 [7.73; 12.47]	4.9%	4.0%
25	5.40	2.5400	5.40 [0.42; 10.38]	1.1%	3.0%
26	3.50	1.0400	3.50 [1.46; 5.54]	6.6%	4.2%
27	1.25	1.5600	1.25 [−1.81; 4.31]	2.9%	3.8%
28	6.70	3.1600	6.70 [0.51; 12.89]	0.7%	2.5%
29	−0.10	2.0300	−0.10 [−4.08; 3.88]	1.7%	3.4%
30	−3.90	2.6200	−3.90 [−9.04; 1.24]	1.0%	2.9%
31	10.40	1.0100	10.40 [8.42; 12.38]	7.0%	4.2%
Fixed effect model			**3.95 [3.43; 4.47]**	**100.0%**	**--**
Random effects model			**2.92 [1.47; 4.36]**	**--**	**100.0%**
Prediction interval			**[−4.33; 10.16]**		

Heterogeneity: $I^2 = 84\%$, $\tau^2 = 11.99$, $p < 0.01$

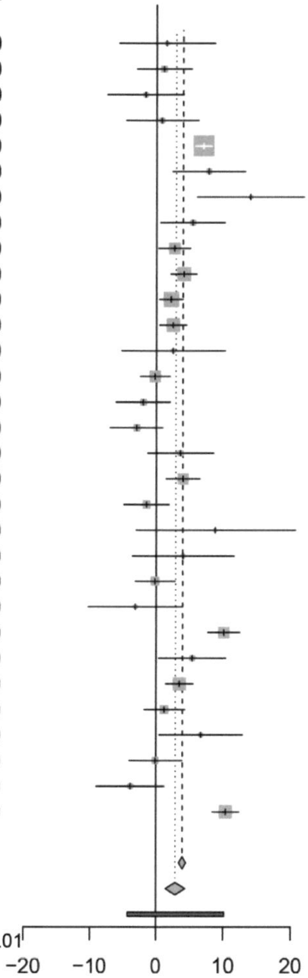

Figure 1. Meta-analysis results of studies applying treatments based on bone marrow stem cells to improve the left ventricular function.

Could we conclude that we have enough evidence to demonstrate the efficacy of the treatment? Unfortunately, these apparently confirming results are completely misleading. The problem is that these 31 studies are very heterogeneous, which resulted in a wide 95% prediction interval [−4.33; 10.16] covering the no treatment effect, and a large number of contradictory evidence displayed in **Figure 1**.

In order to explain the sources of heterogeneity in this area Nowbar et al. [4] investigated whether detected discrepancies in published trials, might account for the variation in reported effect sizes. They define a discrepancy in a trial as two or more reported facts that cannot both be true because they are logically or mathematically incompatible. In other words, the term discrepancies is a polite way to indicate that a published study suffers from poor reporting, could be implausible or its results have been manipulated. For example, as we see at the bottom of **Table 1** in the appendix, it would be difficult to believe in the results of a study with 55 discrepancies. In Section 2 we present a HMR model to analyze a possible link between the risk of bias results and the amount of discrepancies.

1.2. An example of meta-analysis of diagnostic trials

The topic of Section 3 is the meta-analysis of diagnostic trials. These trials play a central role in personalized medicine, policy making, healthcare and health economics. **Figure 2** presents our example in this area. The scatter plot shows the diagnostic summaries of a meta-analysis investigating the diagnostic accuracy of computer tomography scans in the diagnostic of appendicitis [5]. Each circle identifies the true positive rate vs. the false positive rate of each study, where the different circles' sizes indicate different sample sizes. One characteristic of this meta-analysis is the combination of disparate data. From 51 studies 22 were retrospective and 29 were prospective, which is indicated by the different grey scale of the circles.

The main problem in this area is the multiple sources of variability behind those diagnostic results. Diagnostic studies are usually performed under different diagnostic setups and patients' populations. For a particular diagnostic technique we may have a small number of studies which may differ in their statistical design, their quality, etc. Therefore, the main question in meta-analysis of diagnostic test is: How can we combine the multiplicity of diagnostic accuracy rates

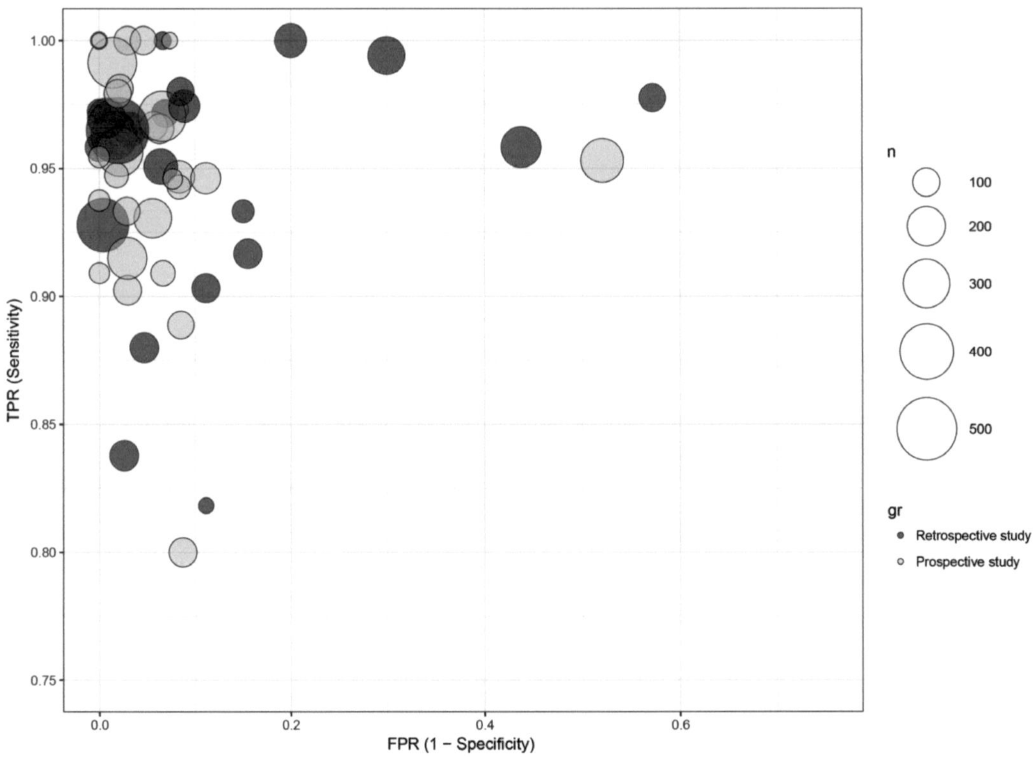

Figure 2. Display of the meta-analysis results of studies performing computer tomography scans in the diagnostic of appendicitis. Each circle identifies the true positive rate vs. the false positive rate of each study. Different colors are used for different study designs and different diameters for sample sizes.

in a single coherent model? A possible answer to this question is a HMR presented in Section 3. This model has been introduced by Verde [5] and it is available in the R's package **bamdit** [6].

2. A Hierarchical Meta-Regression model to assess reported bias

Figure 3 shows the reported effect size and the 95% confidence intervals of 31 trials from [4] against the number of discrepancies (in logarithmic scale). The authors reported a positive statistical significant correlation between the size effect and the number of discrepancies detected in the papers. However, a direct correlation analysis of aggregated results is threatened by ecological bias and it may lead to misleading conclusions. The amount of variability presented by the 95% confidence intervals is very big to accept a positive correlation at face value. In this section we are going to present a HMR model to link the risk of reporting bias with the amount of reported discrepancies. This model assumes that the connection between discrepancies and size effect could be much more subtle.

The starting point of any meta-analytic model is the description of a model for the pieces of evidence at face value. In statistical terms, this means the likelihood of the parameter of interest. Let $y_1, \ldots, y_N$ and $SE_1, \ldots, SE_N$ be the reported effect sizes and their corresponding standard errors, we assume a normal likelihood of θ_i the treatment effect of study i:

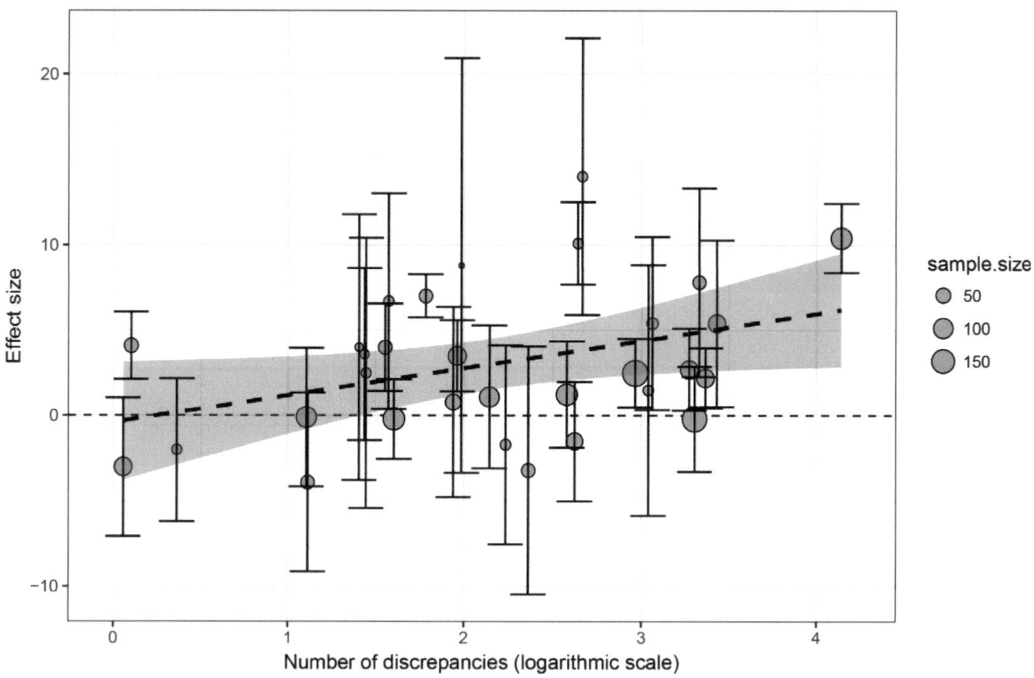

Figure 3. Relationship between effect size and number of discrepancies. The vertical axis corresponds to the effect size, the treatment group received a treatment based on bone marrow stem cells and the control group a placebo treatment. The horizontal axis corresponds to the number of discrepancies (in the logarithmic scale) found in the publication.

$$y_i|\theta_i \sim N(\theta_i, SE_i^2), \quad i = 1, \ldots, N. \tag{1}$$

If a prior assumption of exchangeability was considered reasonable, a random effects Bayesian model incorporates all the studies into a single model, where the $\theta_1, \ldots, \theta_N$ are assumed to be a random sample from a prior distribution with unknown parameters, which is known as a hierarchical model.

In this section we assume that exchangeability is unrealistic and we wish to learn how the unobserved treatment effects $\theta_1, \ldots, \theta_N$ are linked with some observed covariate x_i.

Let x_i be the number of observed discrepancies in the logarithmic scale. We propose to model the association between the treatment effect θ_i and the observed discrepancies x_i with the following HMR model:

$$\theta_i|I_i, x_i \sim I_i N(\mu_{\text{biased}}, \tau^2) + (1 - I_i) N(\mu, \tau^2), \tag{2}$$

where the non-observable variable I_i indicates if study i is at risk of bias:

$$I_i|x_i = \begin{cases} 1 & \text{if study } i \text{ is biased} \\ 0 & \text{otherwise.} \end{cases} \tag{3}$$

The parameter μ corresponds to the mean treatment effect of studies with low risk of bias. We assume that in our context of application biased studies could report higher effect sizes and the biased mean μ_{biased} can be expressed as:

$$\mu_{\text{biased}} = \mu + K, \text{ with } K > 0. \tag{4}$$

In this way, K measures the average amount of bias with respect to the mean effect μ. Eq. (4) also ensures that μ and μ_{biased} are identifiable parameters in this model. The parameter τ measures the between-studies variability in both components of the mixture distributions.

We model the probability that a study is biased as a function of x_i as follows:

$$\text{logit}(\Pr(I_i = 1|x_i)) = \alpha_0 + \alpha_1 x_i. \tag{5}$$

In Eq. (5) positive values of α_1 indicate that an increase in the number of discrepancies is associated with an increased risk of study bias.

In this HMR model the conditional mean is given by

$$\text{E}(\theta|x_i) = \Pr(I_i = 1|x_i)\mu_{\text{biased}} + (1 - \Pr(I_i = 1|x_i))\mu. \tag{6}$$

Eqs. (5) and (6) can be calculated as functional parameters for a grid of values of x. Their posteriors intervals are calculated at each value of x.

This HMR not only quantifies the average bias K and the relationship between bias and discrepancies in Eq. (5), but also allows to correct the treatment effect θ_i by its propensity of being biased:

$$\theta_i^{\text{corrected}} = (\theta_i - K)\Pr(I_i = 1|x_i) + \theta_i(1 - \Pr(I_i = 1|x_i)), \tag{7}$$

where the amount $(\theta_i - K)$ measures the bias of study i and $\Pr(I_i = 1 \mid x_i)$ its propensity of being biased.

The HMR model presented above is completed by the following vague hyper-priors: For the regression parameters α_0, $\alpha_1 \sim N(0, 100)$. We give to the mean $\mu_1 \sim N(0, 100)$ and for the bias parameter $K \sim \text{Uniform}(0, 50)$. Finally, for the variability between studies we use $\tau \sim \text{Uniform}(0, 100)$, which represent a vague prior within the range of possible study deviations.

The model presented in this section is mathematically non-tractable. We approximated the posterior distributions of the model parameters with Markov Chain Monte Carlo (MCMC) techniques implemented in *OpenBUGS*.

BUGS stands for *Bayesian Analysis Using Gibbs Sampling*, the *OpenBUGS* software constructs a Directed Acyclic Graph (DAG) representation of the posterior distribution of all model's parameters. This representation allows to automatically factorize the DAG as a product of each node (parameters or data) conditionally on its parents and children. The software scans each node and proposes a method of sampling. The kernel of the Gibbs sampling is built upon this algorithm.

Computations were performed with the statistical language *R* and MCMC computations were linked to *R* with the package **R2OpenBUGS**. We used two chains of 20,000 iterations and we discarded the first 5000 for the burn-in period. Convergence was assessed visually by using the R package **coda**.

The diagonal panels of **Figure 4** summarize the resulting posterior distributions for μ, K, τ, α_0 and α_1. The posterior of μ clearly covers the zero indicating that the stem cells treatment is not effective. The bias parameter K indicates a considerable over-estimation of treatment effects reported for some trials. The posterior of α_1 is concentrated in positive values, which indicates that an increase in discrepancies is associated with an increase of the risk of reporting bias. The posteriors of α_0 and α_1 also present a large variability, which is expected when a hidden effect is modeled.

Further results of the Hierarchical Meta-Regression model appears in **Figure 5**, where posteriors 95% intervals are plotted against the number of discrepancies. On the left panel, we can see the relationship between the number of discrepancies and the probability that a study is biased. We can observe an increase of probability with an increase of the number of discrepancies, but also a large amount of variability. On the right panel appears the conditional mean of effect size as a function of the number of discrepancies, which corresponds to Eq. (6). Our analysis shows that the 95% posterior intervals of the conditional mean covers the zero effect in most of the range of discrepancies. Only for studies with more than 33 (exp(3.5)) discrepancies

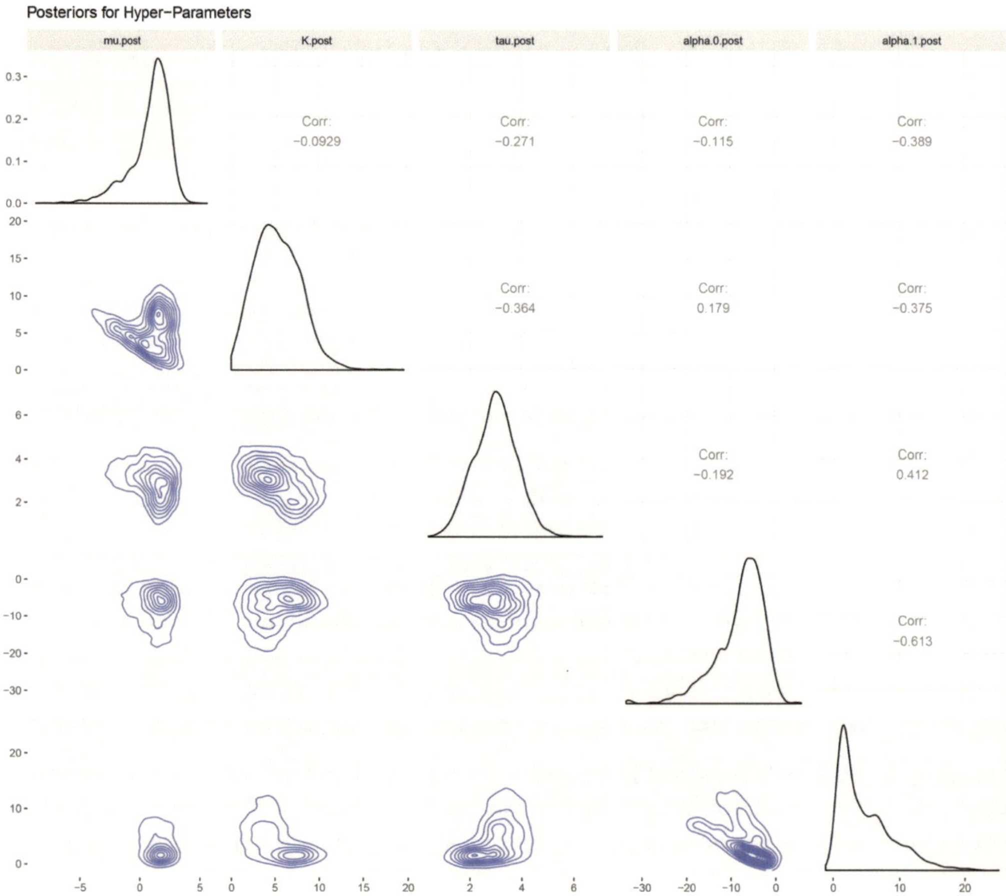

Figure 4. Posterior distributions for the hyper-parameters of the HRM model. The diagonal displays the posterior distributions, the upper panels the pairwise correlations and the lower panels the pairwise posterior densities.

the model predicts a positive effect. One interesting result of this analysis is, that a horizontal line which may represent a zero correlation is also predicted by the model. This means that the regression calculated directly from the aggregated data contains an ecological bias and it is misleading. We have added this regression line to the plot to highlight this issue.

The results presented so far indicate that increases in the amount of discrepancies increases the propensity of bias. The question is: How can we correct a particular study for its bias? Eq. (7) gives the bias correction of treatment effect in this HMR model.

In **Figure 6** we can see HMR bias correction in action. We display two studies which have 21 and 18 discrepancies respectively. The solid lines correspond to the likelihood functions of these studies. These likelihoods represent the information of the effect size at face value. The dashed lines correspond to the posterior treatment effects after bias correction. Clearly, we can see a strong bias correction with the conclusion of no treatment effect.

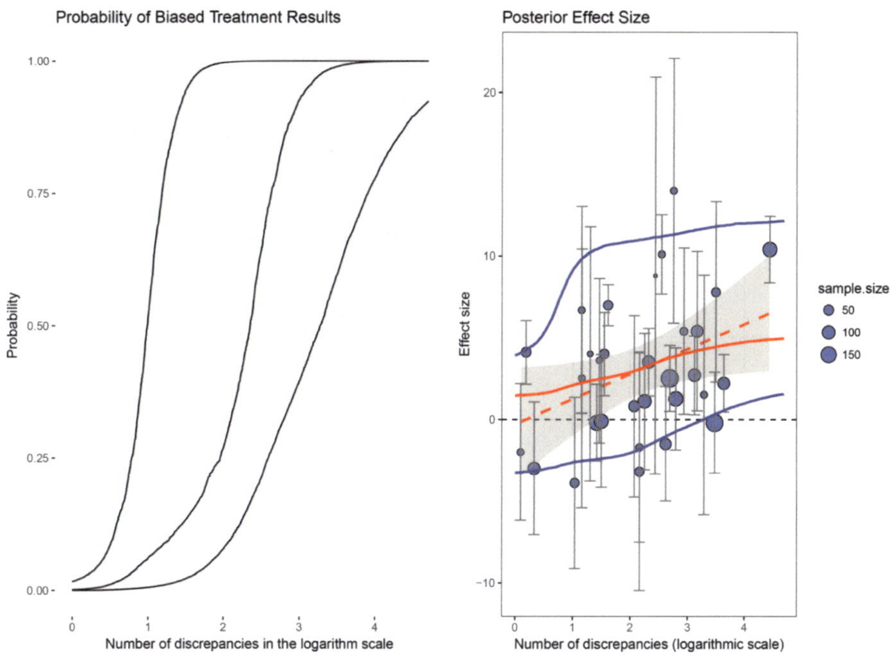

Figure 5. Results of the Hierarchical Meta-Regression model. The posterior median and 95% intervals are displayed as solid lines. Left panel: relationship between the number of discrepancies and probability that a study is biased. Right panel: conditional mean of effect size as a function of the number of discrepancies.

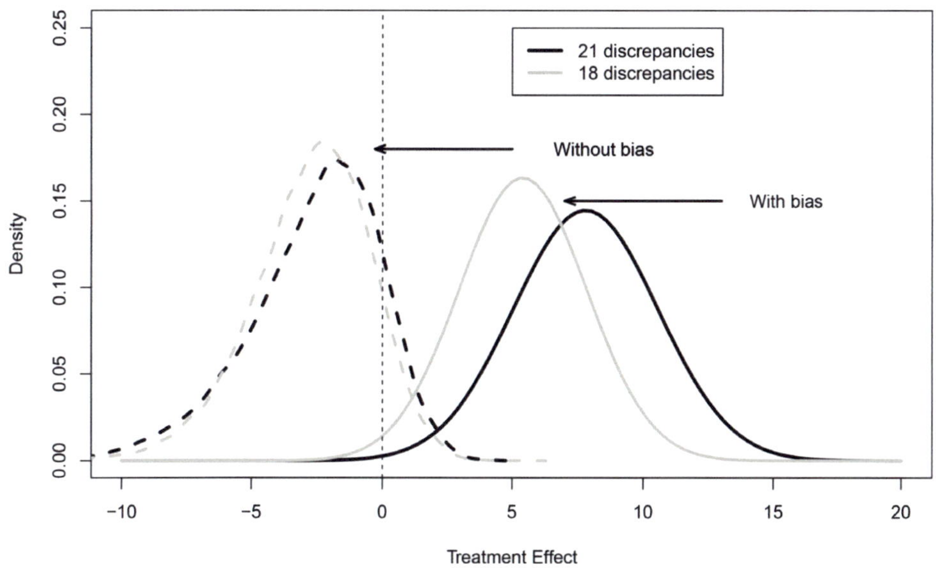

Figure 6. Bias correction for two studies with 21 and 18 discrepancies respectively. The solid lines correspond to the likelihood functions of effect sizes. The dashed lines represent the posteriors for treatment effect after bias correction.

3. Hierarchical Meta-Regression analysis for diagnostic test data

In meta-analysis of diagnostic test data, the pieces of evidence that we aim to combine are the results of N diagnostic studies, where results of the ith study ($i = 1, \ldots, N$) are summarized in a 2×2 table as follows:

		Patient status	
		With disease	**Without disease**
Test	+	tp_i	fp_i
Outcome	−	fn_i	tn_i
Sum:		$n_{i,1}$	$n_{i,2}$

where tp_i and fn_i are the number of patients with positive and negative diagnostic results from $n_{i,1}$ patients with disease, and fp_i and tn_i are the positive and negative diagnostic results from $n_{i,2}$ patients without disease.

Assuming that $n_{i,1}$ and $n_{i,2}$ have been fixed by design, we model the tp_i and fp_i outcomes with two independent Binomial distributions:

$$tp_i \sim B(\mathrm{TPR}_i, n_{i,1}) \quad \text{and} \quad fp_i \sim B(\mathrm{FPR}_i, n_{i,2}), \tag{8}$$

where TPR_i is the true positive rate or sensitivity, Se_i, of study i and FPR_i is the false positive rate or complementary specificity, i.e., $1 - \mathrm{Sp}_i$.

At face value, diagnostic performance of each study is summarized by the empirical true positive rate and true negative rate or specificity

$$\widehat{\mathrm{TPR}}_i = \frac{tp_i}{n_{i,1}} \quad \text{and} \quad \widehat{\mathrm{TNR}}_i = \frac{tn_i}{n_{i,2}} \tag{9}$$

and the complementary empirical rates of false positive rate and false negative diagnostic results,

$$\widehat{\mathrm{FPR}}_i = \frac{fp_i}{n_{i,2}} \quad \text{and} \quad \widehat{\mathrm{FNR}}_i = \frac{fn_i}{n_{i,1}}. \tag{10}$$

In this type of meta-analysis we could separately model TPR_i and FPR_i (or Sp_i), but this approach ignores that these rates could be correlated by design. Therefore, it is more sensible to handle TPR_i and FPR_i jointly.

We define the random effect D_i which represents the study effect associated with the diagnostic discriminatory power:

$$D_i = \log\left(\frac{\mathrm{TPR}_i}{1 - \mathrm{TPR}_i}\right) - \log\left(\frac{\mathrm{FPR}_i}{1 - \mathrm{FPR}_i}\right). \tag{11}$$

However, diagnostic results are sensitive to diagnostic settings (e.g., the use of different thresholds) and to populations where the diagnostic procedure under investigation is applied. These issues are associated with the *external validity* of diagnostic results. To model external validity bias we introduce the random effect S_i:

$$S_i = \log\left(\frac{\text{TPR}_i}{1 - \text{TPR}_i}\right) + \log\left(\frac{\text{FPR}_i}{1 - \text{FPR}_i}\right). \tag{12}$$

This random effect quantifies variability produced by patients' characteristics and diagnostic setup, that may produce a correlation between the observed $\widehat{\text{TPR}}$s and $\widehat{\text{FPR}}$s. In short, we called S_i **the threshold effect** of study i and it represents an adjustment of external validity in the meta-analysis.

We could assume exchangeability of pairs (D_i, S_i), but study's quality is known to be an issue in diagnostic studies. For this reason we model the *internal validity* of a study by introducing random weights $w_1, \ldots, w_N$. Conditionally to a study weight w_i, the study effects D_i and S_i are modeled as exchangeable between studies and they follow a *scale-mixture of bivariate Normal* distributions with the following mean and variance:

$$E\left[\begin{pmatrix} D_i \\ S_i \end{pmatrix} \middle| w_i\right] = \begin{pmatrix} \mu_D \\ \mu_S \end{pmatrix} \quad \text{and} \quad var\left[\begin{pmatrix} D_i \\ S_i \end{pmatrix} \middle| w_i\right] = \frac{1}{w_i}\begin{pmatrix} \sigma_D^2 & \rho\sigma_D\sigma_S \\ \rho\sigma_D\sigma_S & \sigma_S^2 \end{pmatrix} = \Sigma_i, \tag{13}$$

and scale mixing density

$$w_i \sim \text{Gamma}\left(\frac{\nu}{2}, \frac{\nu}{2}\right). \tag{14}$$

The inclusion of the random weights w_i into the model was proposed by [5]. This approach was generalized in [6] in two ways: firstly, by splitting w_i in two weights $w_{1,i}$ and $w_{2,i}$ corresponding to each component D_i and S_i respectively. Secondly, by putting a prior on the degrees of freedom parameter ν, which corresponds to an adaptive robust distribution of the random-effects.

The Hierarchical Meta-Regression representation of the model introduced above is the model based on the conditional distribution of $(D_i | S_i = x)$ and the marginal distribution of S_i. This HMR model was introduced by [7], who followed the stepping stones of the classical Summary Receiving Operating Characteristic (SROC) [8].

The conditional mean of $(D_i | S_i = x)$ is given by:

$$E(D_i | S_i = x) = A + Bx \tag{15}$$

where the functional parameters A and B are

$$A = \mu_D, \quad \text{and} \quad B = \rho\frac{\sigma_D}{\sigma_S}. \tag{16}$$

We define the *Bayesian SROC Curve* (BSROC) by transforming back results from (S, D) to (FPR, TPR) with

$$\mathrm{BSROC(FPR)} = \mathrm{g}^{-1}\left[\frac{A}{(1-B)} + \frac{B+1}{(1-B)}\,\mathrm{g(FPR)}\right], \tag{17}$$

where g(p) is the logit(p) transformation, i.e. logit(p) = log(p/(1 − p)).

The BSROC curve is obtained by calculating TPR in a grid of values of FPR which gives a posterior conditionally on each value of FPR. Therefore, it is straightforward to give credibility intervals for the BSROC for each value of FPR.

One important aspect of the BSROC is that it incorporates the variability of the model's parameters, which influences the width of its credibility intervals. In addition, given that FPR is modeled as a random variable, the curve is corrected by measurement error bias in FPR.

Finally, we can define a *Bayesian Area Under the SROC Curve* (BAUC) by numerically integrating the BSROC for a range of values of the FPR:

$$\mathrm{BAUC} = \int_0^1 \mathrm{BSROC}(x)\,dx. \tag{18}$$

In some applications it is recommend to use the limits of integration within the observed values of $\widehat{\mathrm{FPR}}$s.

In order to make this complex HMR model applicable in practice, we have implemented the model in the R's package **bamdit**, which uses the following set of hyper-priors:

$$\mu_D \sim \mathrm{Logistic}(m_1, v_1), \quad \mu_S \sim \mathrm{Logistic}(m_2, v_2) \tag{19}$$

and

$$\sigma_D \sim \mathrm{Uniform}(0, u_1), \quad \sigma_S \sim \mathrm{Uniform}(0, u_2). \tag{20}$$

The correlation parameter ρ is transformed by using the Fisher transformation,

$$z = \mathrm{logit}\left(\frac{\rho+1}{2}\right) \tag{21}$$

and a Normal prior is used for z:

$$z \sim \mathrm{N}(m_r, v_r). \tag{22}$$

Modeling priors in this way guarantees that in each MCMC iteration the variance-covariance matrix of the random effects θ_1 and θ_2 is positive definite. The values of the constants m_1, v_1, m_2, v_2, u_1, u_2, m_r and v_r have to be given. They can be used to include valid prior information which might be empirically available or they could be the result of expert elicitation. If such information is not available, we recommend setting these parameters to values that represent weakly informative priors. In this work, we use $m_1 = m_2 = m_r = 0$, $v_1 = v_2 = 1$, $u_1 = u_2 = 5$ and $v_r = \sqrt{1.7}$ as weakly informative prior setup.

These values are fairly conservative, in the sense that they induce prior uniform distributions for TPR_i and FPR_i. They give locally uniform distributions for μ_1 and μ_2; uniforms for σ_1 and σ_2; and a symmetric distribution for ρ centered at 0.

Figure 7 summarizes the meta-analysis results of fitting the bivariate random-effect model to the computer tomography diagnostic data. The Bayesian Predictive Surface are presented by contours at different credibility levels and compare these curves with the observed data represented by the circles with varying diameters according to the sample size of each study. The scattered points are samples from the predictive posteriors and the histograms correspond to the posterior predictive marginals. This result was generated by using the functions `metadiag()` and `plot` in the R package **bamdit**.

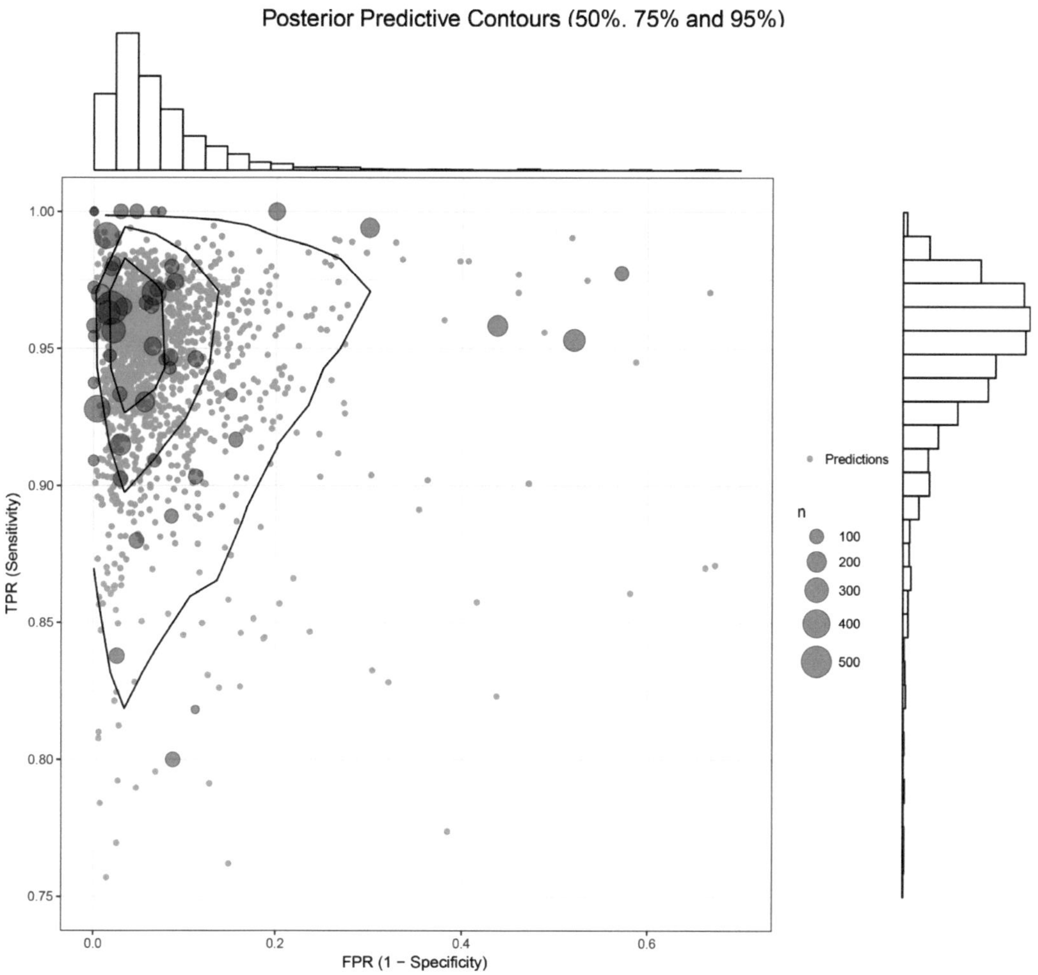

Figure 7. Results of the meta-analysis: Bayesian Predictive Surface by contours at different credibility levels.

Figure 8 displays the posteriors of each components' weights. The left panel shows that prospective studies number 25 and 33 deviate with respect to the prior mean of 1, while on the right panel we see that a prospective study (number 47) and five retrospective studies (number 1, 3, 4, 8 and 29) have substantial variability.

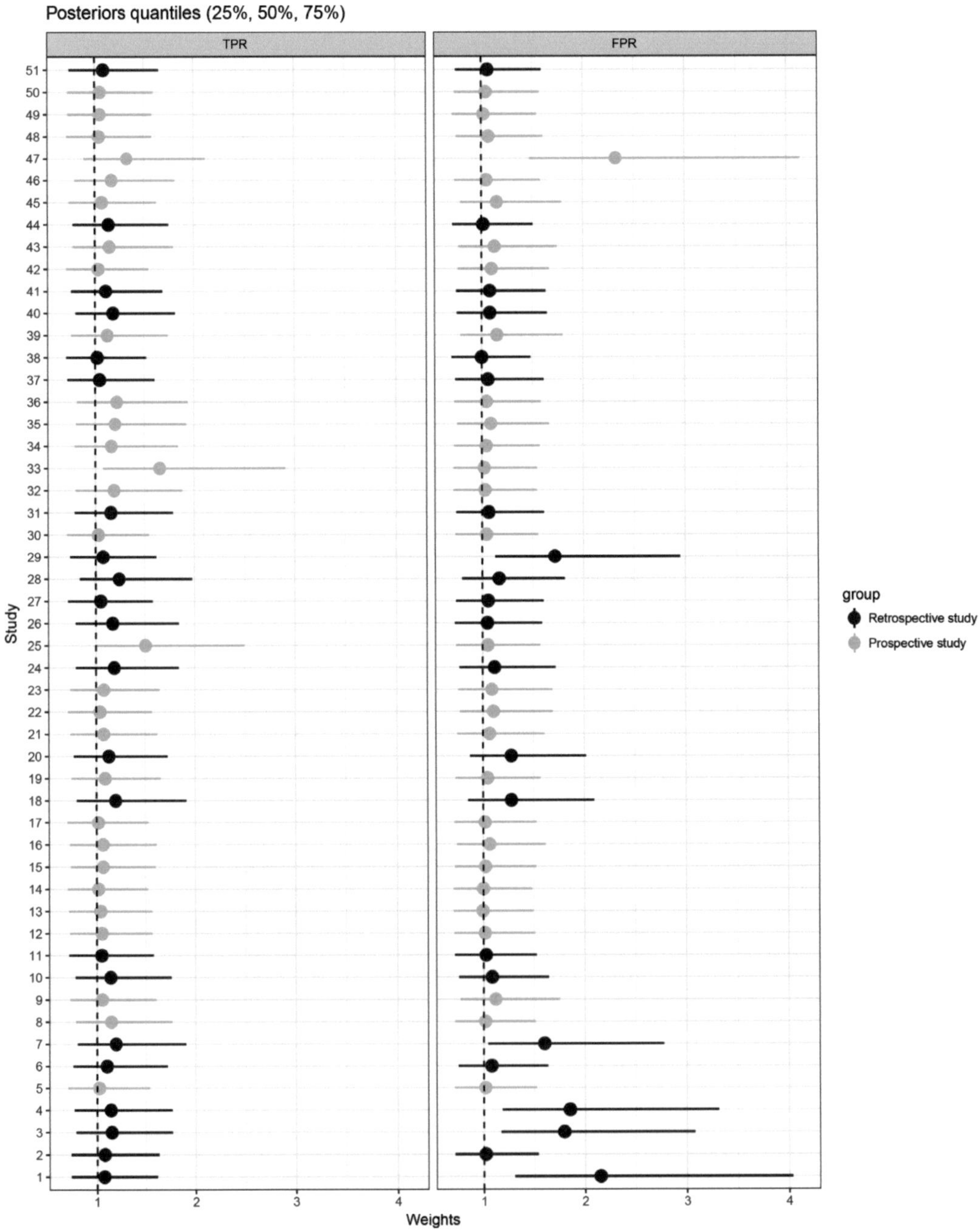

Figure 8. Posterior distributions of the component weights: it is expected that the posterior is centered at 1. Studies with retrospective design tend to present deviations in FPR.

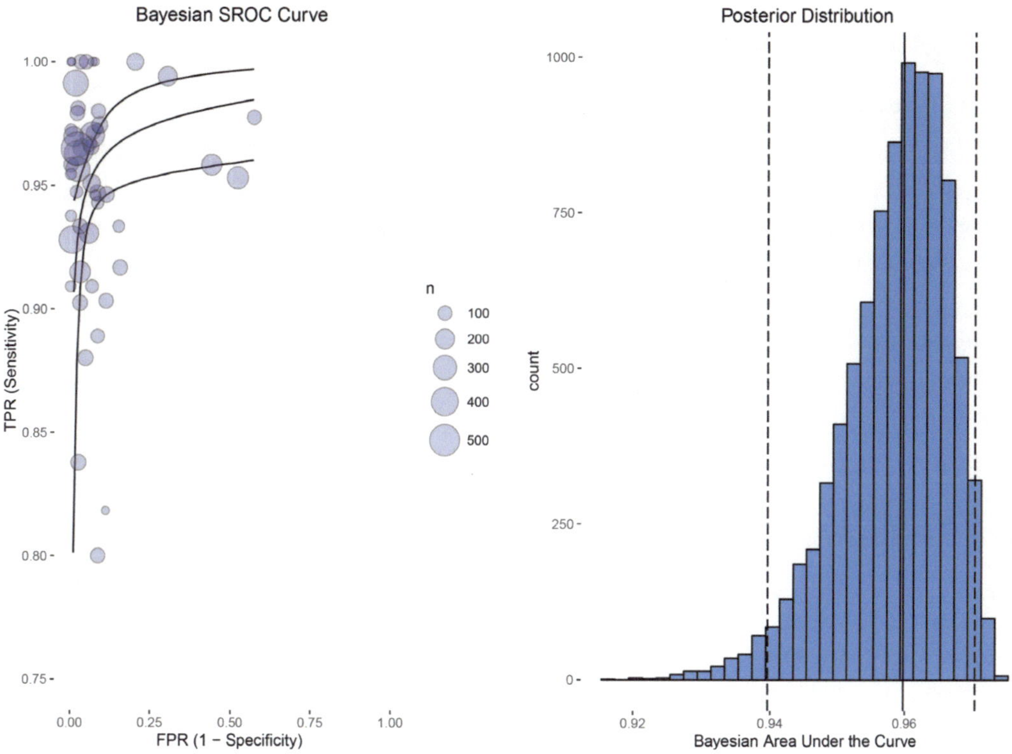

Figure 9. Hierarchical Meta-Regression model: left panel shows the BSROC curve, the central line corresponds to the posterior median and the upper and lower curves correspond to the quantiles of the 2.5 and 97.5%, respectively. The right panel displays the posterior distribution of the area under the BSROC curve.

An important aspect of w_i is its interpretation as **estimated bias correction**. *A priori* all studies included in the review have a mean of $E(w_i) = 1$. We can expect that studies which are unusually heterogeneous will have posteriors substantially greater than 1. Unusual studies' results could be produced by factors that may affect the quality of the study, such as errors in recording diagnostic results, confounding factors, loss to follow-up, etc. For that reason, the studies' weights w_i can be interpreted as an adjustment of studies' **internal validity bias**.

The BSROC curve and its area under the curve are presented in **Figure 9**. The left panel shows this HMR as a meta-analytic summary for this data. On the right panel the posterior distribution of the BAUC show quite a high diagnostic ability for computer tomography scans as diagnostic of appendicitis.

4. Conclusions

In this work we have seen the HMR in action. This approach of meta-analysis is based on a simple strategy: two sub-models are defined in the meta-analysis, one which models the problem of interest, for instance the treatment effect, and one which handles the multiplicity of bias. The meta-analysis is summarized by understanding how these components interact with each other.

The examples presented in this work have shown that we could have misleading conclusions from indirect evidence, if it were analyzed as directly contributing to the problem of interest.

For instance, in the first example, Section 2, we have seen in **Figure 1** that pooling studies gave a wrong conclusion about the effect of stem cells treatment. The positive correlation between the aggregated effect size and the number of discrepancies exaggerates its relationship.

Actually, in **Figure 5** the HMR has shown that it is possible to simultaneously have a zero correlation between effect size and discrepancies while still having a risk of reporting bias. In addition, the HMR allows to extract the amount of bias in the meta-analysis and to correct the treatment effect at the level of the study (**Figure 6**).

In the second example, Section 3, biases come from the external validity of diagnostic studies and the internal validity due to their quality. In this example the HMR showed that it was possible to simultaneously model these two types of subtle biases.

To account for internal validity bias, the application of a scale mixture of normal distributions allows us to detect conflictive studies, which can be considered as outliers. The Bayesian Summary Receiving Operative Curve accounts for the external validity bias due to changes in factors that affected the diagnostic results. In addition, the posterior for its Area Under the Curve (AUC) summarizes the results of the meta-analysis.

Acknowledgements

This work was supported by the German Research Foundation project DFG VE 896 1/1.

Appendix: Source Data for Sections 1.1 and Section 2

Trial ID	Effect size	SE (effect size)	Sample size	Number of discrepancies	Author or principal investigator	Year	Country
t01	1.5	3.67	21	17	Quyyumi	2011	USA
t02	1.1	2.09	100	7	Lunde	2007	Norway
t03	−1.7	2.91	23	7	Srimahachota	2011	Thailand
t05	0.8	2.78	60	4	Meyer	2006	Germany
t06	7	0.63	40	4	Meluzín	2006	Czech Republic
t09	7.8	2.76	38	21	Piepoli	2010	Italy
t11	14	4.05	20	13	Suárez de Lezo	2007	Spain
t12	5.4	2.44	77	18	Huikuri HV	2008	Finland
t13	2.7	1.2	82	16	Perin	2012	USA
t15	4.1	0.98	46	0	Assmus	2006	Germany

Trial ID	Effect size	SE (effect size)	Sample size	Number of discrepancies	Author or principal investigator	Year	Country
t16	2.2	0.88	79	27	Assmus	2013	Germany
t17	2.5	1.01	187	11	Assmus	2010	Germany
t18	2.5	3.96	20	2	Hendrikx	2006	Belgium
t19	−0.2	1.17	127	3	Hirch	2011	The Netherlands
t20	−2	2.09	20	0	Perin	2012	USA
t23	−3	2.03	81	0	Traverse	2011	USA
t24	3.6	2.52	17	2	Ang	2008	UK
t25	4	1.28	40	3	Rodrigo	2012	The Netherlands
t26	−1.5	1.75	66	8	Herbolts	2009	Belgium
t29	8.8	6.07	10	6	Castellani	2010	Italy
t30	4	3.89	14	2	Maureira	2012	France
t31	−0.2	1.54	183	19	Ribero dos Santos	2012	Brazil
t32	−3.2	3.63	40	7	Traverse	2010	USA
t35	10.1	1.21	20	11	Patel	2004	USA
t38	5.4	2.54	27	15	Tse	2007	Hong Kong
t42	3.5	1.04	86	6	Cao	2009	China
t45	1.25	1.56	118	9	Sürder	2013	Switzerland
t46	6.7	3.16	20	2	Ge	2006	China
t47	−0.1	2.03	112	2	Traverse	2012	USA
t48	−3.9	2.62	40	1	Wöhrle	2010	Germany
t49	10.4	1.01	116	55	Yousef (Strauer)	2009	Germany

Table 1. Results from 31 randomized controlled trials of heart disease patients, where the treatment group received bone marrow stem cells and the control group a placebo treatment. The source of this table is Nowbar et al. [4].

Author details

Pablo Emilio Verde

Address all correspondence to: pabloemilio.verde@hhu.de

Coordination Center for Clinical Trials, University of Duesseldorf, Duesseldorf, Germany

References

[1] Eddy DM, Hasselblad V, Shachter R. Meta-Analysis by the Confidence Profile Method: The Statistical Synthesis of Evidence. San Diego, CA: Academic Press; 1992

[2] Spiegelhalter DJ, Abrams KR, Myles JP. Bayesian Approaches to Clinical Trials and Health-Care Evaluation. The Atrium, Southern Gate, Chichester, West Sussex PO19 8SQ, England: John Wiley & Sons, Ltd.; 2004

[3] Verde PE, Ohmann C. Combining randomized and non-randomized evidence in clinical research: A review of methods and applications. Research Synthesis Methods. Vol. 6. 2014. DOI: 10.1002/jrsm.1122

[4] Nowbar AN, Mielewczik M, Karavassilis M, Dehbi HM, Shun-Shin MJ, Jones S, Howard JP, Cole GD, Francis DP. Discrepancies in autologous bone marrow stem cell trials and enhancement of ejection fraction (damascene): Weighted regression and meta-analysis. BMJ. 2014;**348**:1-9

[5] Verde PE. Meta-analysis of diagnostic test data: A bivariate Bayesian modeling approach. Statistics in Medicine. 2010;**29**(30):3088-3102

[6] Verde PE. bamdit: An R package for Bayesian meta-analysis of diagnostic test data. Journal of Statistical Software. 2017, in press

[7] Verde PE. Meta-analysis of diagnostic test data: Modern statistical approaches. PhD Thesis, University of Düsseldorf. Deutsche Nationalbibliothek. July, 2008. Available from: http://docserv.uni-duesseldorf.de/servlets/DocumentServlet?id=8494

[8] Moses LE, Shapiro D, Littenberg B. Combining independent studies of a diagnostic test into a summary roc curve: Data-analytic approaches and some additional considerations. Statistics in Medicine. 1993;**12**:1293-1316

Chapter 10

Bayesian Modeling in Genetics and Genomics

Hafedh Ben Zaabza, Abderrahmen Ben Gara and
Boulbaba Rekik

Additional information is available at the end of the chapter

http://dx.doi.org/10.5772/intechopen.70167

Abstract

This chapter provides a critical review of statistical methods applied in animal and plant breeding programs, especially Bayesian methods. Classical and Bayesian procedures are presented in pedigree-based and marker-based models. The flexibility of the Bayesian approaches and their high accuracy of prediction of the breeding values are illustrated. We show a tendency of the superiority of Bayesian methods over best linear unbiased prediction (BLUP) in accuracy of selection, but some difficulties on elicitation of some complex prior distributions are investigated. Genetic models including marker and pedigree information are more accurate than statistical models based on markers or pedigree alone.

Keywords: accuracy of prediction, breeding value, Bayesian methods, BLUP, pedigree, markers

1. Introduction

Quantitative genetics result from the (connection) combination of statistics and the principles of animal and plant breeding. In quantitative genetics, selection for economically important traits refers to use of phenotypic values of the individual and pedigree information. Genomic is based on the use of dense markers through the whole genome to predict the breeding value of the individuals [1]. Linear models (univariate and multivariate) are of fundamental importance in applied and theoretical quantitative genetics [2]. In animal breeding, two major methods were particularly applied, restricted maximum likelihood (REML) and Bayesian methods. REML has emerged as the method of choice in animal breeding for variance component estimation [3]. Bayesian analysis is gaining popularity because of its more comprehensive assumptions than those of classical approaches and its flexibility in resolving a wide range of biological problems [4, 5]. In the Bayesian approach, the idea is to combine what is known about the statistical ensemble before the data are observed (prior probability distributions) with the information coming from the data, to obtain a posterior distribution from which inferences are made using the standard probability calculus techniques [2, 6]. In recent years, Bayesian methods were broadly used to solve many of the difficulties faced by conventional statistical methods and extend the applicability of statistics on animal and plant breeding data [7]. Furthermore, Markov chain Monte Carlo (MCMC) has an important impact in applied statistics, especially from Bayesian perspective for the estimation of genetic parameters in the linear mixed effect model [2, 5]. The specific objective of this chapter was to illustrate applications of Bayesian inference in quantitative genetics and genomics. First, Bayesian models in the quantitative genetics theory are examined. Second, and in the context of the genomic selection, we presented the details of statistical modeling, using BLUP and Bayesian analyses. Third, a critical review with a focus on the prior distributions is illustrated. Finally, genomic predictions from several methods used in many countries are discussed.

2. A brief introduction to Bayesian analyses

In Bayesian inference, the idea is to combine what is known about the statistical ensemble before the data are observed (prior probability distributions) with the information coming from the data, to obtain a posterior distribution from which inferences are made using the standard probability calculus techniques.

$$P(\theta/y)\alpha\ \mathrm{P}(\mathrm{y}/\theta)\mathrm{P}(\theta) \tag{1}$$

$P(\theta)$ is the prior distribution, which reflects the relative uncertainty about the possible values of θ before the data are seen. $P(y/\theta)$ is the likelihood function of observing the data given the parameter which represents the contribution of y to knowledge about the parameter θ. $P(\theta/y)$ is the posterior distribution of the parameter θ given the previous information on the data.

3. Bayesian analyses of linear models

3.1. The mixed linear model

The mixed linear model is of great importance in genetics and is one of the most used statistical models. Arguably, variance components and genetic parameters are important because they give an indication of the ability of species to respond to selection and thus the potential of that species to evolve. Mixed linear model is the simplest method for estimating the variance components for quantitative traits in population. In the "frequentist" view, mixed linear model is one included linearly the fixed and random effects. In the Bayesian context, there is no distinction between fixed and random effects. Detailed Bayesian analyses of models with two or more component variances will be discussed.

3.1.1. The univariate linear additive genetic model

The mixed linear model is one that includes fixed and random effects.

Consider the linear model:

$$\mathbf{y} = \mathbf{X}\boldsymbol{\beta} + \mathbf{Za} + \mathbf{e} \tag{2}$$

$\mathbf{y}$ is a $\mathbf{n}\times\mathbf{1}$ vector of records on a trait; $\boldsymbol{\beta}$ is the vector of fixed effects affecting records; a is the vector of additive genetic effects; $\mathbf{e}$ is a vector of residual effects. $\mathbf{X}$ and $\mathbf{Z}$ are incidence matrices relating records to fixed effects and additive genetic effects, respectively. Data are assumed to be generated from the following distribution:

$$\mathbf{y}|\boldsymbol{\beta}, a, \sigma_e^2 \sim N(\mathbf{Xb} + \mathbf{Za}, \mathbf{I}\sigma_e^2)$$
$$\mathbf{e} \sim \mathrm{N}(0, \mathbf{I}\sigma_e^2)$$

where, $\mathbf{I}$ is an identity matrix of order $\mathbf{n}\times\mathbf{n}$ and σ_e^2 is the residual variance. Independence of various effects was assumed for the sake of simplicity in implementation. We assume a genetic model in which genes act additively within and between loci, and there are effectively an infinite number of loci. Under this infinitesimal model, and assuming further initial Hardy-Weinberg and linkage equilibrium, the distribution of additive genetic values conditional on the additive genetic covariance is multivariate normal.

$$\mathbf{a}|\mathbf{A}, \sigma_a^2 \sim N(0, \mathbf{A}\sigma_a^2)$$

where $\mathbf{A}$ is the numerator relationship matrix of order $\mathbf{q}\times\mathbf{q}$; $\boldsymbol{\beta}$ is assumed to have a uniform distribution with bounds $\boldsymbol{\beta}_{\mathbf{min}}$ and $\boldsymbol{\beta}_{\mathbf{max}}$.

$$\mathbf{P}(\sigma_i^2|v_i, S_i^2) \sim (\sigma_i^2)^{((\frac{v_i}{2}+1))}\exp\left(-\frac{v_i S_i^2}{2\sigma_i^2}\right), \quad (i = a, e)$$

where v_e, S_e^2 and v_a, S_a^2 are interpreted as degrees of belief and a priori values for residual and additive genetic covariances. Posterior conditional distributions derived from the likelihood and the prior distributions for these parameters are,

$\mathbf{b_i}|\ \mathbf{b_{-i}}, \mathbf{a}, \sigma_a^2, \sigma_e^2, \mathbf{y} \sim \mathrm{N}(\hat{b}_i, (x_i'x_i)^{-1}\sigma_e^2)$, with $(\boldsymbol{x}'_i\boldsymbol{x}_i)$ is the $\boldsymbol{i}$th element of the diagonal of $X'X$

3.1.2. The univariate linear additive genetic model with permanent and genetic group effects

The model equation [8] used to estimate genetic parameters and genetic breeding value for milk yield was as follows:

$$\mathbf{y} = \mathbf{X}\boldsymbol{\beta} + \mathbf{Za} + \mathbf{ZQg} + \mathbf{Wp} + \mathbf{e} \tag{3}$$

where $\mathbf{y}$ is the vector of milk yield, $\mathbf{b}$ is the vector of fixed effects, $\mathbf{a}$ is the vector of additive genetic effects, $\mathbf{g}$ is the vector of genetic group effects, $\mathbf{p}$ is the vector of random permanent environmental effects, and **e** is the vector of residual effects. **X, Z, W**, and **ZQ** are incidence matrices relating a record to fixed environmental effects in **b**, to a random animal effects in a, to a random permanent environment effects in **p**, and to genetic groups in **g**, respectively. g^* is the vector of genetic group effects, $\hat{\mathbf{a}}$ is a vector of breeding values. **A** is the numerator relationship matrix. where $\hat{\mathbf{a}}^* = Q\,\hat{\mathbf{g}} + \hat{\mathbf{a}}$.

The conditional distribution of observed yield is defined by:

$$\mathbf{y}|\mathbf{b},\ \mathbf{p},\ \mathbf{a}*, \sigma_e^2 \sim N(\mathbf{Xb}\ +\ \mathbf{Za}* + \mathbf{Wp}, \mathbf{I}\sigma_e^2)$$

with the assumption of P(**b**) being a constant; $\boldsymbol{a}^* | \boldsymbol{A}^*, \sigma_a^2 \sim \mathrm{N}(\mathbf{Qg}, \mathbf{A}^*\sigma_{\mathbf{a}}^2)$;

$$\mathbf{p}|\sigma_{\mathbf{p}}^2 \sim \mathrm{N}(0, \mathbf{I}\sigma_{\mathbf{p}}^2);\ \text{and}\ \mathbf{P}(\sigma_{\mathbf{i}}^2|\nu_{\mathrm{i}}, \mathbf{S}_{\mathbf{i}}^2) \sim (\sigma_{\mathbf{i}}^2)^{((\frac{\nu_i}{2}+1))} \exp\left(-\frac{\nu_{\mathrm{i}} S_{\mathrm{i}}^2}{2\sigma_{\mathrm{i}}^2}\right)$$

where S_i^2 are prior values for the variances, $\chi_{\nu_i}^{-2}$ are inverted chi-square distributions, and ν_i are degrees of freedom of parameters.

3.1.2.1. Management and environmental effects

The distribution of a fixed effect is:

$$\mathbf{b_i}|\mathbf{b}_{-\mathrm{i}}, \mathbf{a}^*, \sigma_a^2, \sigma_p^2, \sigma_e^2, \mathbf{y} \sim \mathrm{N}(\hat{\boldsymbol{b}}_i, (\mathbf{x}'_i\mathbf{x}_i)^{-1}\sigma_{\mathbf{e}}^2)$$

with $(\boldsymbol{x}'_i\, \boldsymbol{x}_i)\, \hat{b}_i = x' y - x'_i x'_{-i} b_{-i} - x'_i w_p - x'_i z a^*$,

where $(\boldsymbol{x}'_i\, \boldsymbol{x}_i)$ is the $\boldsymbol{i}$th element of the diagonal of X′X

3.1.2.2. Permanent environmental effects

The distribution of a permanent effect is:

$$p_i|b_i, p_{-i}, a^*, \sigma_a^2, \sigma_p^2, \sigma_e^2, y \sim \mathrm{N}(\hat{p}_i, (w'_i w_i + \delta)^{-1}\sigma_e^2)$$

with $(w'_i w_i + \delta)\hat{p}_i = w'_i\, y - w'_i\, Xb - (w_i W_{-i} + \delta)p_{-i} - w_i z_a{}^*$,

where $w'_i\, \boldsymbol{w_i}$ is the ith element of the diagonal of $W'W$.

3.1.2.3. Breeding values

The distribution of a breeding value is:

$$\mathbf{a}_{\mathbf{i}}^*|\mathbf{b}, \mathbf{p}_{-\mathrm{i}}, \mathbf{a}^*_{-\mathrm{i}}, \sigma_{\mathbf{a}}^2, \sigma_{\mathbf{p}}^2, \sigma_{\mathbf{e}}^2, \mathbf{y} \sim \mathrm{N}(\mathbf{a}_{\mathbf{i}}^*(\mathbf{z}'_{\mathrm{i}}\mathbf{z}_{\mathrm{i}} + \mathbf{A}_{\mathbf{i},\mathbf{i}}^{*-1}\alpha^{-1})\sigma_{\mathbf{e}}^2)$$

with $(z'_i z_i + A_{i,i}^{*-1}\alpha)\hat{a}_i = z'_i y - z'_i Xb - z'_i W_P - A_{i,i}^{*-1}\alpha a^*_{-i}$,

where $z'_i\, \boldsymbol{z_i}$ is the $\boldsymbol{i}$ th element of the diagonal of $Z'Z$.

3.1.2.4. Variance components

The additive genetic variance is defined by

$$\sigma_a^2|b, p, a^*, \sigma_p^2, \sigma_e^2, \mathbf{y} \sim \tilde{V}_a \tilde{S}_a^2 \chi_{\tilde{v}_a}^{-2}$$

with $\tilde{V}_a = n_a + V_a$, $\tilde{S}_a^2 = (a*' A^{*-1} a^* + V_a S_a^2)/\tilde{V}_a$, and $\boldsymbol{n_p}$ is the number of animals being evaluated.

The variance of permanent environmental effects is given by:

$$\sigma_p^2, |b, p, a^*, \sigma_a^2, \sigma_e^2, \mathbf{y} \sim \tilde{V}_p \tilde{S}_p^2 \chi_{\tilde{v}_p}^{-2}$$

with $\tilde{V}_p = n_p + V_p$, $\tilde{S}_p^2 = (p'p + V_p S_p^2)/\tilde{V}_p$, and n_p is the number of animals being evaluated.

Residual variance:

$$\sigma_e^2|b, p, a^*, \sigma_a^2, \sigma_p^2, \mathbf{y} \sim \tilde{V}_e \tilde{S}_e^2 \chi_{\tilde{v}_e}^{-2}$$

with $\tilde{V}_e = n_e + V_e$,

$$\tilde{S}_e^2 = [(y - Xb - Wp - Za*)' (y - Xb - Wp - Za*) + V_e S_e^2]/\tilde{V}_e,$$

and n_e is the total number of records.

Comparing genetic value predictions based on polygenic model in Tunisian Holstein Population using BLUP and Bayesian analyses, Ref. [8] reported that the rankings of animals with Bayesian methods are similar to those obtained by BLUP method. Spearman's rank correlation between genetic values estimated from Bayesian procedures and genetic values estimated from BLUP methods were high (0.99). Again, Bayesian and best linear unbiased estimator (BLUE) solutions of fixed effects (month of calving, herd-year, and age-parity) showed the same patterns. The same result is reported by Ref. [9]. However, Ref. [8] illustrated different correlation estimates between two methods (Bayesian and BLUP) for cow's and bull's breeding value.

4. Genomic selection

A massive quantity of genomic data is now available in animal and plant breeding with the revolutionary development in sequencing and genotyping. The cost of genotyping is dramatically reduced. Consequently, practices of genomic selection are nowadays possible with the high number of single nucleotide polymorphism (SNP) markers available. Therefore, it is feasible to perform analysis of the genome at a level that was not possible before [10–13]. The concept of genomic selection was introduced by Ref. [1]. The latter suggested that a set of markers covering the whole genome explain the all genetic variances and each marker is likely to be associated with a quantitative trait locus (QTL), and each QTL is in linkage disequilibrium with the markers. The number of effects per QTL to be estimated is very small. The estimated effects of all markers are summed in order to obtain the genetic value of the individual. Using simulation, Ref. [1] showed in simulation that with a high-density SNP marker, it is possible to predict the breeding value with an accuracy of 0.85 (where accuracy is the correlation between the estimated breeding value and true breeding value). The challenge in genomic evaluation is to find the best prediction method to obtain accurate genetic values of candidates. Many genomic evaluation methods have been proposed [14, 15]. The main objective of this section is to compare Bayesian methods to other methods used in genomic selection based on their predictive abilities. The study reported by Ref. [1] was considered an influential paper on dairy cattle breeding programs. First, the methods suggested correspond well to the data structures where the number of SNPs substantially exceeds the number of observations. Second, the methods of Ref. [1] constitute a logical evolution of the BLUP methodology, which is the reference method in animal genetics by considering specific variances of SNPs in the different loci. Third, the Bayesian approaches used in Ref. [1] that take into account unknown effects (measuring prior uncertainty) in a model, and combined with the ability of the Monte Carlo Markov chain, can be used in the majority of parametric statistical models.

4.1. Genomic BLUP (GBLUP)

The GBLUP method assumes that effects of all SNPs are sampled from the same normal distribution; the effects of all markers are assumed to be small with equal variance. Genomic BLUP was defined by the model:

$$y = 1\mu + Zg + e \tag{4}$$

where **y** is the data vector; μ is the overall mean; 1 is a vector of **n** ones; **Z** is a matrix of incidence, allocating records to the markers' effects; **g** is a vector of SNP effects assumed to be normally distributed $g \sim \mathrm{N}(0, G\sigma_g^2)$, where σ_g^2 is the additive genetic variance and **G** is the genomic relationship matrix; **e** is the vector of normal error, $e \sim \mathrm{N}(0, \sigma_e^2)$ where σ_e^2 is the error variance. The genomic relationship matrix was defined as $G = \frac{X'X}{\sum_{i=1}^{m} p_i(1-p_i)}$, where X is matrix for specified SNP genotype coefficient at each locus, p_i is the rare allele frequency for $\mathrm{SNP_i}$.

4.2. Bayesian approaches

In Bayesian estimation, the information from the data is combined with the information from the prior distribution of the variances of the markers. Several Bayesian statistical analyses have been used in genomic evaluation, which differ in the hypotheses of distributions of marker effects. At the level of the modeling of the variances of the effects of the markers, Meuwissen et al. [1] proposed different distributions a priori between the Bayes A and Bayes B methods.

4.2.1. Bayes A

Bayes A method assumes that variance of marker effects differ among loci (e.g., $\sigma_{g_j}^2$ is different across the **j**) [16]. The variances are modeled according to the scaled inverted chi-square distribution: The a priori distribution of the variances of the SNP effects is written:

http://dx.doi.org/10.5772/intechopen.70167

$P(\sigma^2_{g_j}) \sim \chi^{-2}(\nu, S)$, where S is the scale parameter and ν is the number of degrees of freedom. This has the advantage, if we consider a normal distribution of the data, to lead to an a posteriori conditional distribution of χ^{-2}.

$$P(\sigma^2_{g_j}|g_j) \sim \chi^{-2}(\nu + n_j, S + g'_j g_j),$$

where, n_j is the number of marker effects at segment j. The posterior distribution combines both the information provided by the data and the a priori distribution.

4.2.2. Bayes B

In a genomic evaluation context, Bayes B method [1, 17] assumes different variances of SNP effects, with many SNP contribute per zero effects, and a few contribute per a large effects on the trait. Meuwissen et al. [1] propose a model in which a proportion π (arbitrarily fixed at 0.95) of the markers having zero effect. The a priori distribution of the variances of the effects to the markers is then written:

$\sigma^2_g = 0$ with a probability π, $P(\sigma^2_{g_j}) \sim \chi^{-2}(\nu, S)$ with a probability $(1 - \pi)$, Gibbs sampling cannot be used to estimate the effects and variances of the Bayes B model because of the high probability on some markers of being of zero variance. We therefore use a Metropolis-Hastings algorithm which allows the simultaneous estimation of $\sigma^2_{g_j}$ and g_j. On the basis of the results of Ref. [1] and many subsequent works, the Bayes B method is often considered the "benchmark" in terms of genomic prediction efficiency, but it is extremely costly in computational time. However, Meuwissen [18] propose an alternative to the Bayes B method which relies on a fast algorithm.

4.2.3. Bayesian lasso

Legarra et al. [19] proposed a model of Bayesian lasso (BL) with different variances for residual and SNP effects which they termed BL2Var. It is therefore assumed that a large number of SNPs have an effect practically zero and that very few have large effects. Tibshirani [20] showed that the distribution of the lasso estimators can be written:

$$P(\sigma^2_{g\,j}|\sigma^2, \lambda) \sim \frac{\lambda}{2} \exp\left(-\lambda|g_j|\right)$$

He suggests that the lasso estimators can be interpreted as an a posteriori mode of a model in which the regression parameters would be independent and identically distributed according to a prior double exponential distribution. Park and Casella [21] propose to use a complete Bayesian approach by assuming an a priori distribution of regression coefficients such as:

$$P(\sigma^2_{g\,j}|\sigma^2, \lambda) \sim \frac{\lambda}{2\sqrt{\sigma^2}} \exp\left(-\frac{\lambda}{\sqrt{\sigma^2}}|g_j|\right)$$

where σ^2 represents the variance of residual effects of the model and the variance of the SNP effects. Applications of the Bayesian lasso to the genomic selection proposed by Refs. [22, 23] use the same variance σ^2 to model both the distribution of effects of SNPs and residuals. De los

Campos et al. [22] showed that the Bayesian lasso is close in terms of precision of prediction to the Bayes B method but with a significant reduction in the complexity of the calculations. In addition, these authors suggested using Bayesian lasso against the large number of markers included in regression models, which is typically larger than the number of records.

4.2.4. The Bayes C method

Bayesian methods such as Bayes A and Bayes B [1] have been widely used for genomic evaluation. Similar methods exist, with similar performances, developed in order to reduce computation times and to simplify statistical modeling. The Bayes C method [24] differs from Bayes B by assuming the variance associated with SNPs common to all markers. In Bayes C, as in Bayes B, the probability π that an SNP has a nonzero effect is assumed to be known. The model is similar to the Bayes B model but for a homogeneous variance of effects on all loci: $\sigma_g^2 = 0$ with a probability $1 - \pi;(\sigma_g^2) \sim \chi^{-2}(v,\ S)$. The main problem with the Bayes C method is that SNPs with a nonzero effect is assumed to be known. With the Bayes A method, the parameter π is equal to 1, which implies that all the markers have an effect. For the Bayes B method, π is strictly less than 1 in order to take into account the hypothesis that some SNPs may have a zero effect but is fixed arbitrarily while the intensity of the selection of variables is controlled by this parameter. Habier et al. [25] propose to modify the Bayes C method by estimating the parameter π: the parameter π is assumed to be unknown. Thus, the a priori distribution of π becomes uniform over [0, 1]. SNP modeling is the same as with Bayes C. $P(g_j|\pi, \sigma_g^2) = 0$ with a probability $1 - \pi$; $P(g_j|\pi, \sigma_g^2) \sim N(0, \sigma_g^2)$ where $P(\sigma_g^2) \sim \chi^{-2}(v, S)$ with a probability π. The various parameters of this model are estimated by MCMC methods, Markov Chain Monte Carlo [6, 26] as proposed by Ref. [25]. It is written as a function of the additive genetic variance σ_a^2. $\sigma_g^2 = \frac{\sigma_a^2}{(1-\pi)\sum_{j=1}^{p} 2p_j(1 - p_j)}$, where p_j is the allelic frequency of SNP j.

4.3. A critique

The extreme speed with which events are running handicaps the process of linking new development to extant theory, and the understanding of statistical models suggested up until now [27]. The latter authors criticize the theoretical and statistical concepts followed by Ref. [1] in three levels. The first is the connection between parameters (additive genetic variances with Bayesian view) from infinitesimal models with those from marker-based models. The second is the relationship between molecular marker genotypes and similarity between relatives. The third is the connection between infinitesimal genetic models and marker-based regression models. Gianola et al. [27] argued that the methods Bayes A and Bayes B proposed by Ref. [18] require specifying parameters. The latter used formulas for obtaining the variance of SNP effects, based on some knowledge of the additive genetic variance in the population. Their development begins on the assumption that the effects of the markers are fixed and in other development, they consider them as random without a clear demonstration. Meuwissen et al. [1] explained that affecting a priori a value $\sigma_g^2 = 0$ with a probability π means that the specific SNP does not have an effect on the trait. By contrast, Ref. [27] illustrated that a parameter having zero variance does not obligatory imply that the parameter takes zero value. The parameter could have any value, but with certainty. Gianola et al. [27] suggested the use of a nonparametric method as developed by Refs. [22, 28] because these methods do not impose hypotheses about mode of inheritance as Bayesian A and Bayesian B methods.

5. Applications in genomics

Major dairy breeding countries are now using genomic evaluation [27]. Several results have been reported around the world. Several authors reported that the reliabilities of genomic estimated breeding values (GEBV) were substantially greater than breeding values from estimated breeding values (EBV) based on pedigree information [29]. The accuracy of selection was different between countries [12]. The accuracy was dependent on the size of reference population, the heritability of the trait studied, the statistical models and approaches used for prediction of genetic values for quantitative traits, and the method achieved to estimate the accuracy [12, 27, 29]. Ref. [14] found the reliability of GEBV bulls of the Canadian and American Holstein population. A genotyping of 39,416 molecular markers of 3576 Holstein bulls was used to establish the prediction equations.

The prediction methods contained a linear model, in which marker effects are assumed to be normal, and a nonlinear model with a heavier tailed prior distribution to account for major genes as described by [1]. VanRaden et al. [14] reported that the combination of the polygenic effects based on pedigree information with the genomic predictions can improve the reliability to 23% greater than the reliability of polygenic effects only. The same study showed that the nonlinear model had a little advantage in reliability over the linear model for all traits except for fat and protein percentages. Genomic breeding values of 25 traits in New Zealand dairy cattle were estimated by Ref. [30]. The reference population consisted of 4500 bulls genotyped using the BovineSNP50Beadchip, containing 44,146 SNPs. Harris and Johnson [31] reported an increase in accuracy was found by using Bayesian approaches compared to BLUP methods. In Ref. [31], genomic breeding values (GBVs) for young bulls with no daughter information had accuracies ranging from 50 to 67% for milk traits, live weight, fertility, somatic cell, and longevity, versus an average 34% for progeny test. Meuwissen et al. [1] compared least squares method with BLUP and two Bayesian methods (Bayesian A and Bayesian B). The latter authors estimated the effects of 50,000 marker haplotypes from a limited number of observations (2200). Using least squares method, it is not possible to estimate all effects simultaneously. For this reason, different steps have been adopted to incorporate the effects of markers. First, they performed regression on markers for every segment of 1 cm each. Second, they calculated a Log-likelihood, which assumed to be normal at every segment of chromosome. Third, they summed all segments corresponding to a likelihood peak into multiple regression models. Using BLUP analyses, Ref. [1] considered that all SNP effects were independent and identically distributed with a known variance. Bayes A method was as BLUP at the level of the data, but differs in the variance of the chromosome segments, which assumed to have an inverted chi-square distribution. A mixture prior distribution of genetic variances was used in Bayes B method. **Table 1** shows the accuracy of selection obtained by Ref. [1] from the GBLUP methods, the least squares regression and the

Methods	ρ	b
Least squares	0.318	0.285
GBLUP	0.732	0.896
Bayes A	0.798	0.827
Bayes B	0.848	0.946

Table 1. Comparing estimated versus the breeding value [1].

Bayes A and Bayes B approaches. The predictive abilities of the different methods are estimated by calculating the correlation (ρ) between true and estimated breeding values and the regression (b) of true on estimated breeding value.

The least squares method is the least efficient because it overestimates effects on QTL [32]. The Bayes B approach is the most accurate both in terms of correlation and regression. However, the regression coefficient obtained by the Bayesian methods was still less than 1, and probably due to the hypothesis of a priori distribution χ^{-2} for Bayes A and Bayes B being different from the simulated distribution of the variances. Goddard and Hayes [11] compared the correlation of 0.85 as reported by Ref. [1] to results obtained on real data by Refs. [14, 33, 34]. VanRaden et al. [14] produced a mean correlation over several characters of 0.71 from a reference population of more than 3500 bulls. Studies have shown the superiority of genomic evaluation [35] or marker-assisted selection in France [36] on classical infinitesimal model of quantitative genetics. Several authors have applied the first genomic evaluation methods described by Ref. [1] or their derived methods on real data. The Bayes A and Bayes B approaches have found results that are often similar or slightly superior to GBLUP in terms of accuracy of genetic value prediction for the Australian Holstein-Friesian cattle breed (+0.02 to +0.07 of correlation gain between predicted and observed values), for example [12] and New Zealand (+2% correlation gain, [31]). However, the GBLUP method required less computing time than the Bayes A method [32, 37]. Gredler et al. [38] demonstrated the superiority of the Bayes B method, in terms of the accuracy of genomic estimates, on a modified Bayes A method for integrating a polygenic effect [39]. Thus, although the Bayes B method seems slightly more efficient than the Bayes A method, numerous studies showed that the Bayes B method is not so much better in terms of accuracy of the genomic estimates than a GBLUP model [40]. Again, all researches indicate that the Bayesian approaches, which assume an a priori distribution of SNPs, increase the reliability of breeding values over traditional BLUP methods [1, 12, 14]. A common conclusion is that for most quantitative traits, the hypothesis of the traditional BLUP method, that all markers are associated with equal variances, is far from reality. By comparing the results obtained in the various populations around the world, clearly, the accuracies of GEBVs were greater than breeding values estimated from progeny test based on pedigree information. Several researches suggested combining the progeny test based on pedigree information with the breeding value from genomic to calculate the final GEBV [5, 25]. Accuracy based on modeling molecular marker and pedigree information was generally superior to that of the model including only genomic or pedigree information. Hayes et al. [12] reported that a main advantage of using the both sources of information coming from polygenic breeding values and genomic information is that any QTL not detected by the marker effects may be detected by the progeny test based on pedigree information. A significant reduction in posterior mean of residual variance component was reported by Ref. [22] when pedigree and markers were considered jointly compared to pedigree-based model. In the same study, Spearman's rank correlation of estimated breeding value between model including marker information and pedigree-based model was close to 1.

6. Conclusion

Standard quantitative genetic model based on phenotypic and pedigree information has been very successful in term of genetic value prediction. Also, the availability of genome-wide dense markers leads researchers to be able to perform advanced genetic evaluation of quantitative traits with a high accuracy of prediction of genetic value. However, a main problem is how this information should be included into statistical genetic models. Bayesian MCMC methods appear to be convenient for genetic value prediction with a focus on the precision of the choice of prior distribution for the different parameters.

Author details

Hafedh Ben Zaabza[1]*, Abderrahmen Ben Gara[2] and Boulbaba Rekik[2]

*Address all correspondence to: hafedhbenzaabza@gmail.com

1 Institut National Agronomique, Tunis-Mahrajène, Tunisie

2 Département des productions animales, Ecole supérieure d'Agriculture de Mateur, Mateur, Tunisie

References

[1] Meuwissen THE, Hayes BJ, Goddard ME. Prediction of total genetic value using genome-wide dense marker maps. Genetics. 2001;**157**:1819-1829

[2] Sorenson D, Gianola D. Likelihood, Bayesian, and MCMC Methods in Quantitative Genetics. 1st ed. New York: Springer-Verlag; 2002. p. 740

[3] Neumaier A, Groeneveld E. Restricted maximum likelihood estimation of covariances in sparse linear models. Genetics Selection Evolution. 1997;**30**(1):3-26

[4] Waldmann P. Easy and flexible Bayesian inference of quantitative genetic parameters. Evolution. 2009;**63**(6):1640-1643. DOI: 10.1111/j.1558-5646.2009

[5] Hallander J, Waldmann P, Chunkao W, Sillanpaa MJ. Bayesian inference of genetic parameters based on conditional decompositions of multivariate normal distributions. Genetics. 2010;**185**:645-654. DOI: 10.1534/genetics.110.114249

[6] Robert CP. Le choix bayésien Principes et pratique. 1st ed. Paris: Springer-Verlag France; 2006. p. 638

[7] Ben Zaabza H, Ben Gara A, Hammami H, Ferchichi MA, Rekik B. Estimation of variance components of milk, fat, and protein yields of Tunisian Holstein dairy cattle using Bayesian and REML methods. Archives Animal Breeding. 2016;**59**:243-248. DOI: 10.5194/aab-59-243-2016

[8] Ben Gara A, Rekik B, Bouallègue M. Genetic parameters and evaluation of the Tunisian dairy cattle population for milk yield by Bayesian and BLUP analyses. Livestock Science. 2006;**100**:142-149. DOI: 10.1016/j.livsci.2005.08.012

[9] Schenkel FS, Schaeffer LR, Boettcher PJ. Comparison between estimation of breeding values and fixed effects using Bayesian and empirical BLUP estimation under selection on parents and missing pedigree information. Genetic Selection Evolution. 2002;**34**:41-59. DOI: 10.1051/gse:2001003

[10] Gianola D, Fernando RL, Stella A. Genomic-assisted prediction of genetic value with semi-parametric procedures. Genetics. 2006;**173**(3):1761-1776. DOI: 10.1534/genetics.105.049510

[11] Goddard ME, Hayes BJ. Genomic selection. Journal of Animal Breeding and Genetics. 2007;**124**:323-330. DOI: 10.1111/j.1439-0388.2007

[12] Hayes BJ, Bowman PJ, Chamberlain AJ, Goddard ME. Genomic selection in dairy cattle: Progress and challenges. Journal of Dairy Science. 2009;**92**:433-443. DOI: 10.3168/jds.2008-1646

[13] Wittenburg D, Melzer N, Reinsch N. Including non-additive genetic effects in Bayesian methods for the prediction of genetic values based on genome-wide markers. BMC Genetics. 2011;**12**(74):14

[14] VanRaden PM, Van Tassell CP, Wiggans GR, Sonstegard TS, Schnabel RD, Taylor JF, et al. Reliability of genomic predictions for North American Holstein bulls. Journal of Dairy Science. 2009;**92**:16-24. DOI: 10.3168/jds.2008-1514

[15] Colombani C, Croiseau P, Fritz S, Guillaume F, Legarra A, Ducrocq V, Robert-Granié C. A comparison of partial least squares (PLS) and sparse PLS regressions in genomic selection in French dairy cattle. Journal of Dairy Science. 2012;**95**:2120-2131. DOI: 10.3168/jds.2011-4647

[16] Su G, Guldbrandtsen B, Gregersen VR, Lund MS. Preliminary investigation on reliability of genomic estimated breeding values in the Danish Holstein population. Journal of Dairy Science. 2010;**93**(3):1175-1183. DOI: 10.3168/jds.2009-2192

[17] Villumsen TM, Janss L, Lund MS. The importance of haplotype length and heritability using genomic selection in dairy cattle. Journal of Animal Breeding and Genetics. 2009;**126**(1):3-13. DOI: 10.1111/j.1439-0388.2008

[18] Meuwissen THE. Accuracy of breeding values of "unrelated" individuals predicted by dense SNP genotyping. Genetics Selection Evolution. 2009;**41**:35. DOI: 10.1186/1297-9686-41-35

[19] Legarra A, Robert-Granié C, Croiseau P, Guillaume F, Fritz S. Improved lasso for genomic selection. Genetics Research. 2011;**93**(1):77-87. DOI: 10.1017/S0016672310000534

[20] Tibshirani R. Regression shrinkage selection via the LASSO. Journal of the Royal Statistical Society Series B. 1996;**73**(3):273-282. DOI: 10.2307/41262671

[21] Park T, Casella G. The Bayesian lasso. Journal of the American Statistical Association. 2008;**103**(482)681-686. DOI: 10.1198/016214508000000337

[22] De los Campos G, Naya H, Gianola D, Crossa J, Legarra A, Manfredi E, et al. Predicting quantitative traits with regression models for dense molecular markers and pedigree. Genetics. 2009;**182**:375-385. DOI: 10.1534/genetics.109.101501

[23] Weigel KA, De los Campos G, González-Recio O, Naya H, Wu XL, Long N, et al. Predictive ability of direct genomic values for lifetime net merit of Holstein sires using selected subsets of single nucleotide polymorphism markers. Journal of Dairy Science. 2009;**92**(10): 5248-5257. DOI: 10.3168/jds.2009-2092

[24] Kizilkaya K, Fernando RL, Garrick DJ. Genomic prediction of simulated multi-breed and purebred performance using observed fifty thousand single nucleotide polymorphism genotypes. Journal of Animal Science. 2010;**88**(2):544-551. DOI: 10.2527/jas.2009-2064

[25] Habier D, Fernando RL, Kizilkaya K, Garrick DJ. Extension of the Bayesian alphabet for genomic selection. BMC Bioinformatics. 2011;**12**:12

[26] Metropolis N, Rosenbluth AW, Rosenbluth MN, Teller AH, Teller E. Equations of state calculations by fast computing machines. Journal of Chemical Physics. 1953;**21**:1087-1092

[27] Gianola D, Manfredi E, Fernando RL. Additive genetic variability and the Bayesian alphabet. Genetics. 2009;**183**:347-363. DOI: 10.1534/genetics.109.103952

[28] Gianola D, Van Kam JBCHM. Reproducing kernel hilbert spaces regression methods for genomic assisted prediction of quantitative traits. Genetics. 2008;**178**(4):2289-2303. DOI: 10.1534/genetics.107.084285

[29] Su G, Madsen P, Nielsen US, Mäntysaari EA, Aamand GP, Christensen OF, et al. Genomic prediction for Nordic Red Cattle using one-step and selection index blending. Journal of Dairy Science. 2012;**95**:909-917. DOI: 10.3168/jds.2011-4804

[30] Harris BL, Johnson DL, Spelman RJ. Genomic selection in New Zealand and the implications for national genetic evaluation. In: Proceeding Interbull Meeting; 2008; Canada. The 36th International Committee for Animal Recording (ICAR) Session, held June 16-20, in Niagara Falls; 2008

[31] Harris BL, Johnson DL. Genomic predictions for New Zealand dairy bulls and integration with national genetic evaluation. Journal of Dairy Science. 2009;**93**(3):1243-1252. DOI: 10.3168/jds.2009-2619

[32] Moser G, Tier B, Crump RE, Khatkar MS, Raadsma HM. A comparison of five methods to predict genomic breeding values of dairy bulls from genome-wide SNP markers. Genetics Selection Evolution. 2009;**41**(56). DOI: 10.1186/1297-9686-41-56

[33] Legarra A, Misztal I. Technical note: Computing strategies in genome-wide selection. Journal of Dairy Science. 2008;**91**(1):360-366. DOI: 10.3168/jds.2007-0403

[34] González-Recio O, Gianola G, Rosa GJM, Weigel KA, Kranis A. Genome-assisted prediction of a quantitative trait measured in parents and progeny: Application to food conversion rate in chickens. Genetics Selection Evolution. 2009;**41**(3):10. DOI: 10.1186/1297-9686-41-3

[35] VanRaden P. Efficient methods to compute genomic predictions. Journal of Dairy Science. 2008;**91**(11):4414-4423. DOI: 10.3168/jds.2007-0980

[36] Boichard D, Fritz S, Rossignol MN, Bosher MY, Malafosse A, Colleau JJ. Implementation of marker-assisted selection in French dairy cattle. In: 7th World Congress on Genetics Applied to Livestock Production; 19-23 August 2002; Montpellier, France. 2002. Session 22. Exploitation of molecular information in animal breeding. Electronic communication 22-03. p. 4

[37] Solberg TR, Sonesson AK, Woolliams JA, Meuwissen THE. Reducing dimensionality for prediction of genome-wide breeding values. Genetics Selection Evolution. 2009;**41**(29):8. DOI: 10.1186/1297-9686-41-29

[38] Gredler B, Nirea KG, Solberg TR, Egger-Danner C, Meuwissen THE, Solkner J. Genomic selection in Fleckvieh/Simmental—First results. In: Proceedings of the Interbull Meeting; 21-24 August 2009; Interbull Bulletin, Barcelone, Espagne; 2009;**40**:209-213

[39] Hayes BJ. Genomic selection in the era of the $1000 genome sequence. In: Symposium Statistical Genetics of Livestock for the Post-Genomic Era; USA: Wisconsin-Madison, USA; 2009

[40] Habier DJ, Tetens J, Seefried FR, Lichtner P, Thaller G. The impact of genetic relationship information on genomic breeding values in German Holstein cattle. Genetics Selection Evolution. 2010;**42**(5). DOI: 10.1186/1297-9686-42-5

Chapter 11

Bayesian Two-Stage Robust Causal Modeling with Instrumental Variables using Student's t Distributions

Dingjing Shi and Xin Tong

Additional information is available at the end of the chapter

http://dx.doi.org/10.5772/intechopen.70393

Abstract

In causal inference research, the issue of the treatment endogeneity is commonly addressed using the two-stage least squares (2SLS) modeling with instrumental variables (IVs), where the local average treatment effect (LATE) is the causal effect of interest. Because practical data are usually heavy tailed or contain outliers, using traditional 2SLS modeling based on normality assumptions may result in inefficient or even biased LATE estimate. This study proposes four types of Bayesian two-stage robust causal models with IVs to model normal and nonnormal data, and evaluates the performance of the four types of models with IVs. The Monte Carlo simulation results show that the Bayesian two-stage robust causal modeling produces reliable parameter estimates and model fits. Particularly, in different types of the two-stage robust models with IVs, the models that take outliers into consideration and use Student's t distributions in the second stage to model heavy-tailed data or data containing outliers provide more accurate and efficient LATE estimates and better model fits than other distribution models when data are contaminated. The preferred models are recommended to be adopted in general in the two-stage causal modeling with IVs.

Keywords: Bayesian methods, two-stage causal modeling with instrumental variables, nonnormal data, robust method using Student's t distributions

1. Introduction

Causal inference and experimental researchers are often interested in the average treatment effect (ATE), measured by the outcome difference between participants who are assigned to the treatment and those being assigned to the control. The estimation of ATE for the whole population is neither reliable nor feasible when certain conditions are not achieved or assumptions are violated [6, 9]. The treatment effects for only a subset of participants is instead estimated, which is called the local average treatment effect (LATE) [2, 13]. Different studies may have different

LATEs, depending on the subgroup of interest. Often the subgroup of interest is those who have been assigned to the treatment and have actually received the treatment [3]. One way to estimate the LATE is to incorporate instrumental variables (IVs), which are correlated with both the endogenous regressors and error terms when the linearity assumption of the traditional linear models is violated and the endogenous regressors are correlated with the errors. Instrumental variables are incorporated in the analysis to estimate the LATE, or a part of the treatment effect whose estimation is not contaminated by the violation of the linearity assumption.

Two-stage least squares (2SLS) modeling [1] is widely used to estimate the LATE with IVs. In the first stage, IVs are used to predict the partial treatment effect that can be explained by the variations of IVs, and in the second stage, the fitted treatment values are used to predict the experimental outcome, and to estimate the LATE. In estimating the LATE in traditional 2SLS modeling with IVs, it is typically assumed that the measurement errors at both stages are normally distributed. However, practical data in social and behavioral research usually violate the normality assumption and often have heavy tails or contain outliers [25]. Failure to take the nonnormal data into consideration but instead treating the heavy-tailed data or data containing outliers as if they were normally distributed may result in unreliable parameter estimates and inflated type I error rates [35, 38–40], which will eventually lead to misleading statistical inference.

Routine methods to accommodate heavy-tailed data or data with outliers include data transformation and data truncation. However, transformed data are often difficult to interpret especially when the raw scores have meaningful scales [17], and the exclusion of outliers may lead to underestimated standard errors and reduced efficiency [14, 32]. Alternatively, different robust procedures have been developed to provide reliable parameter estimates, the associated standard errors, and statistical tests. The rationale of most robust procedures is to weigh each observation according to its distance from the center of the majority of the data, so that outliers that are far from the center of the data are downweighted [10, 11, 37]. In recent research, more and more robust methods have been used to estimate complex models, such as linear and generalized linear mixed-effects models [19, 26], structural equation models [15, 31], and hierarchical linear and nonlinear models [20, 29].

Over the past decades, robust procedures based on Student's t distributions have been developed and advanced to model heavy-tailed data or data containing outliers [14, 33]. For example, Student's t distributions have been applied under the structural equation modeling framework and were found to produce reliable parameter estimates and inferences [15, 16]; in robust mixture models, Wang et al. [30] used the multivariate t distribution to fit heavy-tailed data and data with missing information, Shoham [24] implemented a robust clustering algorithm in mixture models by modeling data that are contaminated by outliers using multivariate t distributions, Seltzer et al. [21] and Seltzer and Choi [22] conducted sensitivity analysis employing Student's t distributions in robust multilevel models and downweighted outliers in level two (the between-subject level), and Tong and Zhang [28] and Zhang et al. [36] advanced the Student's t distributions to robust growth curve models and provided online software to carry out the analysis. Although robust methods based on Student's t distributions have been used in different modeling frameworks, few have been adopted in the causal modeling, where heavy-tailed data or data containing outliers are not uncommon [18].

Recently, Shi and Tong [23] implemented a robust Bayesian estimation method using Student's t distributions to the two-stage causal modeling with IVs to fit data that contain outliers or are normally distributed concurrently at both stages. However, in the two-stage causal models with IVs, data at either stage are equally likely having outliers or are nonnormally distributed. Previous studies have noticed such a situation. For example, Pinheiro et al. [19] used a robust estimation to the linear mixed-effects model and applied the multivariate t distribution to both the random effects and intraindividual errors simultaneously. Tong and Zhang [28] conducted a robust estimation to growth curve modeling and modeled the measurement errors and random effects separately with t distributions or normal distributions rather than the same distribution for the two effects. Therefore, this article extends the study of Shi and Tong [23] and proposes four possible types of two-stage causal models with IVs to the data. The study evaluates the performance of the robust method in four types of models. In the following section, the robust method based on Student's t distributions is reviewed. Then, the two-stage causal models with IVs, the associated LATE, and the corresponding four types of models are introduced. Next, a Monte Carlo simulation study is conducted to evaluate the performance of the robust method in four possible types of two-stage causal models with IVs. In the end, conclusions are summarized and discussions are provided.

2. Robust methods based on Student's t distributions

As a robust procedure, the fundamental idea of using Student's t distributions to model heavy-tailed data or data containing outliers is to assign a weight to each case and properly downweight cases that are far from the center of the majority of the data [10, 11, 37]. Suppose a population of k random variables, $\mathbf{y}$, follow a multivariate t distribution, with mean vector $\boldsymbol{\mu}$, scale matrix $\boldsymbol{\Psi}$, and degrees of freedom ν, denoted by t ($\boldsymbol{\mu}$, $\boldsymbol{\Psi}$, ν). The probability density function of $\mathbf{y}$ can be expressed as:

$$p(\boldsymbol{y}|\boldsymbol{\mu}, \boldsymbol{\Psi}, \nu) = \frac{|\boldsymbol{\Psi}|^{-\frac{1}{2}}\Gamma\left(\frac{\nu+k}{2}\right)}{\left(\Gamma\left(\frac{1}{2}\right)^{k}\Gamma\left(\frac{\nu}{2}\right)\nu^{\frac{k}{2}}\right)} \times \left(1+\frac{(\boldsymbol{y}-\boldsymbol{\mu})^{T}\boldsymbol{\Psi}^{-1}(\boldsymbol{y}-\boldsymbol{\mu})}{\nu}\right)^{-\frac{(\nu+k)}{2}}. \tag{1}$$

The maximum likelihood estimates of model parameters under the model with t distribution assumptions satisfy

$$\Sigma_{i=1}^{n} w_i \boldsymbol{A}_i \boldsymbol{\Psi}_i^{-1}\left(\boldsymbol{y}_i - \boldsymbol{\mu}\right) = 0, \tag{2}$$

where n is the total sample size, $\boldsymbol{y}_i$ is a sample from $\boldsymbol{y}$, $\boldsymbol{A}_i$ is the partial derivatives of $\boldsymbol{\mu}$, and

$$w_i = \frac{\nu + \tau_i}{\nu + \sigma_i^2} \tag{3}$$

is the weight assigned to case i. In the equation for w_i, τ_i is the dimension of the parameter for each i and σ_i^2 is the squared Mahalanobis distances $\sigma_i^2 = (\boldsymbol{y}_i - \boldsymbol{\mu})^T \boldsymbol{\Psi}^{-1} (\boldsymbol{y}_i - \boldsymbol{\mu})$. Note that $(\boldsymbol{y}_i - \boldsymbol{\mu})$ is the distance between each observation and the population mean, and a large $(\boldsymbol{y}_i - \boldsymbol{\mu})$ indicates a potential outlier as well as a large squared Mahalanobis distance σ_i^2. The outliers are downweighted in the analysis because the weight w_i decreases with increasing squared Mahalanobis distances σ_i^2, given fixed degrees of freedom ν, and dimensions τ_i [14].

The shape of a t distribution is controlled by its degrees of freedom ν, and ν can be set a priori or estimated in the analysis. Under certain conditions, the degrees of freedom have been recommended setting a priori. Lange et al. [14] and Zhang et al. [36] suggested fixing the value for the degrees of freedom of Student's t distributions when sample size is small, as small sample sizes could lead to biased degrees of freedom estimate. Moreover, Tong and Zhang [28] argued that by fixing the degrees of freedom, more accurate parameter estimates and credible intervals can be obtained when model specification is built on solid substantive theories. In contrast, estimating the degrees of freedom can make the model more flexible. When the degrees of freedom ν are freely estimated, Student's t distributions have an additional parameter ν, compared with normal distributions. As the degrees of freedom ν increase, the Student's t distribution approaches a normal distribution.

There are several advantages in using Student's t distributions for robust data analysis [28]. First, unlike the nonparametric robust analysis, Student's t distributions have parametric forms, and inferences based on them can be carried out relatively easily through maximum likelihood estimation or Bayesian estimation methods. Second, the degrees of freedom of Student's t distributions control the weight of outliers and can flexibly set a priori or be estimated. Third, when data have heavy tails or contain outliers, considering Student's t distribution as a natural extension of the normal distribution is rather intuitive.

3. Bayesian two-stage robust causal modeling with IVs

In causal Ordinary Least Squares (OLS) regression, when the error terms are related to some regressors, the estimated ATE is biased due to the violation of the linearity assumption. Variables that are related to both endogenous regressors and errors are used as instruments to differentiate the correlations between endogenous regressors and errors, leaving only a part of the treatment effects that have not been contaminated by the violation of the linearity assumption to be estimated, and such variables are called instrumental variables (IVs). The ATE of interest becomes the LATE of interest. For example, Currie and Yelowitz [8] studied the effect of public housing voucher program of having a larger housing unit on housing quality and educational attainment. Based on the fact that some families in voucher program tradeoff physical housing amenities and reductions in rental payments that are bad and have negative effects for the housing quality and their children, some regressors are correlated with errors and become endogenous. Previous theory supports that a household having an extra number of kids is

http://dx.doi.org/10.5772/intechopen.70393

entitled to a larger housing unit, whether there are extra kids in the household and the sex decomposition of the extra kids are chosen as the IVs, to study the voucher program effect to participants who have one girl and one boy (i.e., having sex decomposition) in the household. It was found that the voucher program participants who have the sex decomposition in the household are more likely to have better housing quality and educational attainment. The example shows that when IVs are introduced, the external validity is traded for the improvement of the internal validity, and the ATE (i.e., all of the voucher program participants) becomes the LATE (i.e., program participants who have extra kids and who have the sex decomposition).

One commonly used framework to estimate LATE is the 2SLS modeling with IVs. Let d_i and y_i be the treatment and the outcome for individual i, respectively, and $\mathbf{Z}_i = (z_{i1}, \ldots, z_{iJ})'$ be a vector of instrumental variables for individual i ($i = 1, \ldots, N$). Here, N is the sample size and J is the total number of instrumental variables. In the first stage of the 2SLS model, the IVs $\mathbf{Z}$ are used to predict the treatment $\mathbf{d}$. In other words, the portion of variations in the treatment $\mathbf{d}$ is identified and estimated by the IVs $\mathbf{Z}$; and then the second stage relies on the estimated exogenous portion of treatment variations in the form of the predicted treatment values to estimate the treatment effect on the outcome $\mathbf{y}$. A typical form of the 2SLS model with IVs can be expressed as:

$$d_i = \pi_{10} + \boldsymbol{\pi}_{11}\mathbf{Z}_i + e_{1i}, \tag{4}$$

$$y_i = \pi_{20} + \pi_{21}\widehat{d}_i + e_{2i}, \tag{5}$$

where π_{10} and $\boldsymbol{\pi}_{11} = (\pi_{11}, \ldots, \pi_{1J})'$ are the intercept and regression coefficients for the linear model where the treatment $\mathbf{d}$ is regressed on the IVs $\mathbf{Z}$, respectively; and π_{20} and π_{21} are the intercept and slope for the linear model where the outcome y is regressed on the predicted treatment values of $\widehat{\mathbf{d}}$, respectively. The IVs help estimate the treatment effects in which the causal effect of IVs on the treatment is first estimated in Eq. (4), and the causal effect of this estimated partial treatment effect on the outcome is then estimated in Eq. (5). From the model, $\boldsymbol{\pi}_{11}$ is the causal effect of the IVs $\mathbf{Z}$ on the treatment Z, and $\boldsymbol{\pi}_{21}$ is the treatment effect on the outcome $\mathbf{y}$ for a subset of participants whose treatment effect has been partialled out and explained by the IVs $\mathbf{Z}$. π_{21} is the causal effect of interest and is called LATE. There are several advantages in using 2SLS modeling to estimate LATE. First, unlike method of point estimate such as Wald estimator [4], 2SLS modeling also provides standard error estimate and confidence intervals of the LATE, making statistical inferences more efficient. Second, when 2SLS models are used, covariates could be controlled simultaneously at both stages of the 2SLS model when the effect of $\mathbf{Z}$ on $\mathbf{d}$ and the effect of $\widehat{d}$ on $\mathbf{y}$ are estimated. Mathematically, the estimated LATE $\widehat{\pi}_{21}$ in 2SLS can be derived as:

$$\widehat{\pi}_{21} = \frac{cov\left(y_i, \widehat{d}_i\right)}{var\left(\widehat{d}_i\right)} = \frac{cov\left(y_i, \widehat{\pi}_{10} + \widehat{\pi}_{11}z_{i1} + \cdots + \widehat{\pi}_{1J}z_{iJ}\right)}{var\left(\widehat{\pi}_{10} + \widehat{\pi}_{11}z_{i1} + \cdots + \widehat{\pi}_{1J}z_{iJ}\right)} = \frac{\widehat{\pi}_{11}cov\left(y_i, z_{i1}\right) + \cdots + \widehat{\pi}_{1J}cov\left(y_i, z_{iJ}\right)}{\widehat{\pi}^2{}_{11}var(z_{i1}) + \cdots + \widehat{\pi}^2{}_{11}var(z_{i1})}. \tag{6}$$

Traditional causal 2SLS models with IVs are commonly estimated using OLS methods or maximum likelihood estimation from the frequentist approach. The measurement errors at both stages, e_{1i} and e_{2i}, are assumed to be normally distributed as $e_{1i} \sim N(0, \sigma^2_{e_1})$ and

$e_{2i} \sim N(0, \sigma_{e_2}^2)$. Because practical data usually violate the normality assumption, it was proposed from a Bayesian approach that the normal distributions can be replaced by Student's t distributions for heavy-tailed data or data containing outliers [23, 28, 36]. In the two-stage causal model with IVs, data at either stage are equally likely to be nonnormal or containing outliers. Therefore, we propose four possible types of Bayesian two-stage causal models to data with (a) normal measurement errors at both stages, denoted as *Bayesian normal model*, (b) t measurement errors in the first stage and normal measurement errors in the second stage, denoted as *Bayesian nonnormal-s1 model*, (c) normal measurement errors in the first stage and t measurement errors in the second stage, denoted as *Bayesian nonnormal-s2 model*, and (d) t measurement errors at both stages, denoted as *Bayesian nonnormal-both model*. The four types of Bayesian two-stage causal models have the same mathematical model expressions as those from the frequentist approach. Namely, for the Bayesian normal model, measurement errors are assumed to be distributed as $e_{1i} \sim N(0, \sigma_{e_1}^2)$ and $e_{2i} \sim N(0, \sigma_{e_2}^2)$; for the Bayesian nonnormal-s1 model, the measurement errors are assumed to be distributed as $e_{1i} \sim t(0, \sigma_{e_1}^2, \nu_1)$ and $e_{2i} \sim N(0, \sigma_{e_2}^2)$; for the Bayesian nonnormal-s2 model, the measuremenet errors are assumed to be distributed as $e_{1i} \sim N(0, \sigma_{e_1}^2)$ and $e_{2i} \sim t(0, \sigma_{e_2}^2, \nu_2)$; finally, for the Bayesian nonnormal-both model, the measurement errors are assumed to be distributed as $e_{1i} \sim t(0, \sigma_{e_1}^2, \nu_1)$ and $e_{2i} \sim t(0, \sigma_{e_2}^2, \nu_2)$. All four types of models are estimated using Bayesian methods.

In the Bayesian approach, we obtain the joint posterior distributions of the parameters based on the prior distributions of the parameters and the likelihood of the data information. Making statistical inferences directly from the joint posterior distributions is usually difficult. Gibbs sampling, a Markov chain Monte Carlo (MCMC) method is a widely used algorithm to draw a sequence of samples from the joint posterior distribution of two or more random variables, given that the conditional posterior distributions of the model parameters can be obtained [7]. In specific, Gibbs sampling alternately samples parameters one at a time from their conditional posterior distribution on the current values of other parameters, which are treated as known. After a sufficient number of iterations, the sequence of samples constitutes a Markov chain that converges to a stationary distribution. This stationary distribution is the sought-after joint posterior distribution of the parameters [12].

The Gibbs sampling algorithm is used to obtain the LATE estimate for the two-stage causal model with IVs. Because the t distribution can be viewed as a normal distribution with variance weighted by a Gamma distribution, the data augmentation method is used here to simplify the posterior distribution. Specifically, a Gamma random variable ω is augmented with a normal random variable because if $\omega_i \sim G(\frac{\nu}{2}, \frac{\nu}{2})$, and $\boldsymbol{y}_i | \omega_i \sim N(\boldsymbol{\mu}, \boldsymbol{\Psi}/\omega_i)$, then $\boldsymbol{y}_i \sim t(\boldsymbol{\mu}, \boldsymbol{\Psi}, \nu)$. The detailed steps of the Gibbs sampling algorithm for the Bayesian nonnormal-s2 model are given below. The Gibbs sampling procedures for the other models are similar.

1. Start with initial values $\boldsymbol{\pi}_1^{(0)}, \boldsymbol{\pi}_2^{(0)}, \sigma_{e1}^{2(0)}, \sigma_{e2}^{2(0)}, \nu^{(0)}, \omega_i^{(0)}$, where $\boldsymbol{\pi}_1^{(0)} = \left(\pi_{10}^{(0)}, \pi_{10}^{(0)}\right)'$ and $\boldsymbol{\pi}_2^{(0)} = \left(\pi_{20}^{(0)}, \pi_{21}^{(0)}\right)'$.

2. Assume at the jth iteration, we have $\boldsymbol{\pi}_1^{(j)}, \boldsymbol{\pi}_2^{(j)}, \sigma_{e1}^{2\,(j)}, \sigma_{e2}^{2\,(j)}, \nu^{(j)}, \omega_i^{(j)}$, where $\boldsymbol{\pi}_1^{(j)} = \left(\pi_{10}^{(j)}, \pi_{10}^{(j)}\right)'$ and $\boldsymbol{\pi}_2^{(j)} = \left(\pi_{20}^{(j)}, \pi_{21}^{(j)}\right)'$.
   At the (j+1)th iteration,
3. Step 3
   3.1 Sample $\boldsymbol{\pi}_1^{(j+1)}$ from $p\left(\boldsymbol{\pi}_1|\sigma_{e1}^{2\,(j)}, d_i, \mathbf{Z}_i, i = 1, \ldots, N\right)$;

   3.2 Sample $\sigma_{e1}^{2\,(j+1)}$ from $p\left(\sigma_{e1}^2|\boldsymbol{\pi}_1^{(j+1)}, d_i, \mathbf{Z}_i, i = 1, \ldots, N\right)$;

   3.3 Sample $\sigma_{e2}^{2\,(j+1)}$ from $p\left(\sigma_{e2}^2|\boldsymbol{\pi}_2^{(j)}, \widehat{d}_i, y_i, \omega_i^{(j)}, i = 1, \ldots, N\right)$;

   3.4 Sample $\nu^{(j\,+\,1)}$ from $p\left(\nu|\omega_i^{(j)}, i = 1, \ldots, N\right)$;

   3.5 Sample $\omega_i^{(j+1)}, i = 1, \ldots N$, from $p\left(\omega_i|\nu^{(j+1)}, \sigma_{e2}^{2\,(j+1)}, \widehat{d}_i, y_i, \boldsymbol{\pi}_2^{(j)}, i = 1, \ldots, N\right)$;

   3.6 Sample $\boldsymbol{\pi}_2^{(j+1)}$ from $p\left(\boldsymbol{\pi}_2|\omega_i^{(j+1)}, \sigma_{e2}^{2\,(j)}, \widehat{d}_i, y_i, i = 1, \ldots, N\right)$.
4. Repeat Step 3.

4. Evaluation of four types of distributional 2SLS models

In this section, the performance of the four types of two-stage robust causal models is evaluated through a Monte Carlo simulation study. Data are generated from a general causal inference model as presented in Eq. (7). Full Bayesian methods are used for the estimation of all four types of two-stage causal models. In specific, noninformative priors are applied to all model parameters, conditional posterior distributions of all model parameters are obtained and Markov chains are generated through Gibbs sampling algorithm, convergence tests are conducted and finally statistical inferences for the model parameters are made. Free software (R Development Core Team, 2011) R [41] and OpenBUGS [42] (Thomas, O'Hara, Ligges, & Sturtz, 2006) were used for the implementation of MCMC algorithms and model estimation. A total of 20,000 iterations was conducted for each simulation condition, with the first 10,000 iterations as the burn-in period.

4.1. Study design

Data are generated from a general causal inference model

$$y_i = 3 + 0.5x_i + e_i, \tag{7}$$

where y_i is the causal outcome, x_i is the causal treatment, and e_i is the measurement error. Three potential influential factors are considered. First, sample size (N) is either 200 or 600. Second, correlation between $\mathbf{x}$ and $\mathbf{e}(\Phi)$ is manipulated to be either 0.3 or 0.7, reflecting relatively weak or strong linear relationship between the treatment and the measurement error. Third, a proportion of observations that contains outliers is manipulated. The proportion of outliers (OP) is considered to be 0, 5, or 10%. When the OP is 0%, data contain no outliers and measurement errors e_i are normally distributed. When the OP is above zero, data contain outliers. For outliers, the measurement errors are generated from a different normal distribution with the same standard deviation, but a larger mean (eight times of the standard deviation). An IV is also generated from a normal distribution and correlated with **x** with the correlation coefficient being 0.6.

If we fit a linear regression to the generated data, we will immediately notice that the residuals and the regressors are not independent. Therefore, we adopt the two-stage causal model with IVs. The four types of two-stage models (normal model, nonnormal-s1 model, nonnormal-s2 model, and nonnormal-both model) are used to fit the data. In the first stage, the IV is used to predict the endogenous treatment, and the estimated treatment is then used in the second stage to estimate the LATE. Based on Eq. (6), the theoretical LATE is 5/6.

As discussed previously, Bayesian methods using Gibbs sampling algorithm are used to obtain the LATE estimates in four types of two-stage causal models. The bias and standard error (SE) of the LATE estimate for each of the four distributional models are assessed. In addition, the deviance information criterion (DIC) [27] for each condition is examined to study the model fit. A lower value of DIC indicates a better model fit.

4.2. Results

The bias and SEs of the LATE estimates from four types of models when $\varphi = 0.3$ are presented in **Table 1**.

In almost all cases, models that use normal distributions to model the normal data and that use Student's t distributions to model the data with outliers provide the best estimates with smaller bias and SEs among other types of two-stage causal models. For example, when N = 200, the normal model provides smaller bias and SE for normal data; similarly, nonnormal-s2 and nonnormal-both models lead to the smaller bias and SEs when they are used to fit data containing outliers. This shows that using Student's t distributions to model data containing

			Normal model		Nonnormal-s1 model		Nonnormal-s2 model		Nonnormal-both model	
N	**Data**	**OP**	**Bias**	**SE**	**Bias**	**SE**	**Bias**	**SE**	**Bias**	**SE**
200	Normal	0%	0.001	0.154	−0.004	0.154	−0.004	0.155	−0.003	0.155
	Nonnormal	5%	0.210	0.283	0.200	0.281	−0.022	0.177	−0.020	0.171
		10%	0.342	0.379	0.341	0.378	−0.060	0.157	−0.050	0.155
600	Normal	0%	−0.021	0.076	−0.023	0.076	−0.023	0.077	−0.023	0.076
	Nonnormal	5%	0.180	0.168	0.170	0.167	−0.064	0.099	−0.060	0.096
		10%	0.390	0.230	0.380	0.210	−0.077	0.099	0.070	0.098

Table 1. Bias and SEs of the LATE estimates for all the conditions when $\Phi = 0.3$.

outliers is an effective way to accommodate heavy-tailed data or data containing outliers, and this finding is consistent with the previous research [34, 36]. In causal inference study, because practical data at either stage are equally likely having outliers or are normally distributed in the two-stage causal model with IVs, we fit all four types of distributional models and try to decide which one is the best-fitted model. From the results, modeling heavy-tailed data or data containing outliers with nonnormal-both model provides more reliable parameter estimates than traditional methods that ignore the data distributions and model all data exclusively with normal distributions.

Although it is always a good choice to model normal data with normal distributions and heavy-tailed data or data containing outliers with Student's t distributions, in practice, researchers may not know whether the first stage or the second stage of the model should account for the nonnormality. The simulation results show that when data contain outliers, the nonnormal-s2 model and nonnormal-both model that use t distributions in the second stage produce the smallest bias and SEs of the LATE estimates. This is probably because the causal effect of interest, LATE, is housed in the second stage, and using Student's t distribution to model outliers in that stage is effective in capturing the LATE. On the contrary, in the normal model or the nonnormal-s1 model, the normal distribution is being used to model the second stage data that are heavy tailed or contain outliers. For example, for all the nonnormal data that contain outliers (i.e., OP = 5 or 10%), the nonnormal-s2 model and the nonnormal-both model, both of which use t distributions to model data in the second stage, outperform other models, providing smaller bias and SEs of the LATE estimates regardless of sample size (N) and proportion of outliers (OP). Comparing between nonnormal-s2 and nonnormal-both models, the nonnormal-both models perform slightly better than the nonnormal-s2 model does. Take N = 600 and OP = 10% as an example, the bias and SEs for the nonnormal-s2 model are −0.077 and 0.099, whereas those for the nonnormal-both model are slightly smaller to be 0.070 and 0.098, showing that fitting the nonnormal data with Student's t distributions at both stages has the best performance in terms of accuracy and efficiency of the LATE estimate.

Table 2 presents the results for DICs for the four types of two-stage causal models when $\Phi = 0.3$.

In practice, DIC can be used as a model selection criteria. To select the best-fitted parsimonious model, we first fit all four types of models to the data, and then select the model with the

N	Data	OP	Normal model	Nonnormal-s1 model	Nonnormal-s2 model	Nonnormal-both model
200	Normal	0%	1145.09	1145.83	1145.82	1146.54
	Nonnormal	5%	1380.18	1380.82	1241.48	1242.04
		10%	1488.71	1489.43	1315.18	1315.87
600	Normal	0%	3418.20	3419.25	3419.23	3420.53
	Nonnormal	5%	4126.86	4128.00	3705.62	3706.93
		10%	4448.88	4450.07	3922.32	3923.73

Table 2. DICs of all the distributional models when $\Phi = 0.3$.

smallest DIC. Notice that for normal data, all four types of models have similar DIC values. When data contain outliers, nonnormal-s2 and nonnormal-both models provides the smallest DIC, indicating that these types of models fit the data better. In all data conditions in the study, the DICs of the nonnormal-s2 model and the nonnormal-both model are very similar, and either model can be adopted.

The proportions of outliers contained in the data have effect on the performance of the nonnormal-s2 model and the nonnormal-both model. Specifically, the larger the proportions of outliers, the more salient the advantages of the nonnormal-s2 and nonnormal-both models. For example, for the nonnormal data with N = 200 and OP = 5%, the bias from the normal model, the nonnormal-s2 model and the nonnormal-both model is 0.210, −0.022, and −0.020, respectively; when OP becomes 10%, the bias from the normal model jumps to 0.342, whereas the bias from the nonnormal-s2 model changes slightly to −0.060 and that from the nonnormal-both model is −0.050. Similarly, the preferred models provide less biased LATE estimates when sample size is small, and the advantage of the preferred models is more apparent under small sample conditions (e.g., [23]).

When $\Phi = 0.7$, consistent with the results from previous conditions when $\Phi = 0.3$, when data have outliers, using Student's t distributions to model the data provides more accurate and efficient LATE estimates and better model fits than using normal distribution to model the data. The advantage of using t distributions is more obvious when sample size is small and the proportion of outliers is large.

5. Discussion

In causal inference research, the issue of the treatment endogeneity is commonly addressed in the 2SLS model with IVs, where the LATE is the causal effect of interest. Because practical data usually violate the normality assumption, using normal distributions to model heavy-tailed data or data containing outliers may result in inefficient or even biased LATE estimate. In the 2SLS model with IVs, data at either stage are equally likely having outliers or are normally distributed. To address this problem, this study proposes four possible types of Bayesian two-stage robust causal models with IVs to the data, and evaluates the performance of the robust method using Student's t distributions in the causal modeling. The Monte Carlo simulation results show that modeling normal data with normal distributions and normal or heavy-tailed data or data containing outliers with Student's t distributions gives good performance in terms of accuracy, efficiency, and model fit. When data are normally distributed, the methods that either use normal distributions or the Student's t distributions perform equally well as they provide similar bias, SEs and DICs. In the presence of outliers, the nonnormal-s2 and the nonnormal-both models that take outliers into consideration and use Student's t distributions in the second stage to model heavy-tailed data or data containing outliers outperform other distribution models that use normal distributions to model either exclusively all the data or the second stage data in two-stage causal models with IVs with smaller bias and higher efficiency. In addition, the nonnormal-s2 model and the nonnormal-both model have smaller DICs than the other two models, suggesting evidence of better model fit. The nonnormal-s2 and nonnormal-both models are especially preferred when sample size is small and the proportion of outliers is large as they produce more accurate and efficient LATE estimates.

Note that fitting the nonnormal-both model to data may require longer Markov chains as degrees of freedom for t distributions at both stages need to be estimated. We also want to be cautious to simply use Student's t distributions to model all the data as this method is numerically not optimal all the time and computationally time consuming [28]. Additionally, Student's t distributions are sensitive to the skewness, so some nonnormally distributed data may not be modeled by them. If data are highly skewed, alternative robust method, such as robust methods based on skewed-t distributions may be considered [5].

Author details

Dingjing Shi and Xin Tong*

*Address all correspondence to: xtong@virginia.edu

Department of Psychology, University of Virginia, Charlottesville, Virginia, USA

References

[1] Angrist JD, Imbens G. Two stage least squares estimation of average causal effects in models with variable treatment intensity. Journal of the American Statistical Association. 1995;**90**:431-442

[2] Angrist JD, Imbens G, Rubin D. Identification of causal effects using instrumental variables. Journal of the American Statistical Association. 1996;**91**:444-455

[3] Angrist JD, Pischke J. Mastering Metrics: The Path from Cause to Effect. Princeton, NJ: Princeton University Press; 2014

[4] Angrist JD, Pischke J. Mostly Harmless Econometrics: An Empiricist's Companion. Princeton, NJ: Princeton University Press; 2008

[5] Azzalini A, Genton MG. Robust likelihood methods based on the skew-t and related distributions. International Statistical Review. 2008;**76**:106-129

[6] Baiocchi M, Cheng J, Small D. Instrumental variable methods for causal inference. Statistics in Medicine. 2014;**33**:2297-2340

[7] Casella G, George EI. Explaining the Gibbs sampler. The American Stastician. 1992;**46**: 167-174

[8] Currie J, Yelowitz A. Are public housing projects good for kids? Journal of Public Economics. 2000;**75**:99-124

[9] Gerber AS, Green DP. Field Experiments: Design, Analysis and Interpretation. New York, NY: W.W.Norton & Company; 2011

[10] Hampel FR, Ronchetti EM, Rousseeuw PJ, Stahel WA. Robust Statistics: The Approach Based on Influence Functions. New York: John Wiley & Sons, Inc; 1986

[11] Huber PJ. Robust Statistics. New York: John Wiley & Sons, Inc; 1981

[12] Geman S, Geman D. Stochastic relaxation, Gibbs distributions, and the Bayesian restoration of images. IEEE Transactions on Pattern Analysis and Machine Intelligence. 1984;**6**:721-741

[13] Imbens G, Angrist JD. Identification and estimation of local average treatment effects. Econometrica. 1994;**62**:467-475

[14] Lange KL, Little RJ, Taylor JM. Robust statistical modeling using the t distribution. Journal of the American Statistical Association. 1989;**84**:881-896

[15] Lee SY, Xia YM. Maximum likelihood methods in treating outliers and symmetrically heavy-tailed distributions for nonlinear structural equations. Psychometrika. 2006;**71**:565-585

[16] Lee SY, Xia YM. A robust Bayesian approach for structural equation models with missing data. Psychometrika. 2008;**73**:343-364

[17] Osbourne, J. W. (2002). Notes on the Use of Data Transformation. Practical Assessment, Research & Evaluation, **8**(6), n6.

[18] Osborne, J. W., & Overbay, A. (2004). The power of outliers (and why researchers should always check for them). Practical assessment, research & evaluation, **9**(6), 1-12.

[19] Pinheiro JC, Liu C, Wu Y. Efficient algorithms for robust estimation in linear mixed-effects models using the multivariate t distribution. Journal of Computational and Graphical Statistics. 2001;**10**:249-276

[20] Rachman-Moore D, Wolfe RG. Robust analysis of a nonlinear model for multilevel educational survey data. Journal of Educational Statistics. 1984;**9**:277-293

[21] Seltzer M, Novak J, Choi K, Lim N. Sensitivity analysis for hierarchical models employing t level-1 assumptions. Journal of Educational and Behavioral Statistics. 2002;**27**:181-222

[22] Seltzer M, Choi K. Sensitivity analysis for hierarchical models: Downweighting and identifying extreme cases using the t distribution. Multilevel Modeling: Methodological Advances, Issues, and Applications. 2003;**1**:25-52

[23] Shi D, Tong X. Robust Bayesian estimation in causal two-stage least squares modeling with instrumental variables. In: van der Ark LA, Culpepper S, Douglas JA, Wang W-C, Wiberg M, editors. Quantitative Psychology Research. Springer: New York; 2017

[24] Shoham S. Robust clustering by deterministic agglomeration EM of mixtures of multivariate t-distributions. Pattern Recognition. 2002;**35**:1127-1142

[25] Simmons J, Nelson L, Simon S. False-positive psychology: Undisclosed flexibility in data collection and analysis allows presenting anything as significant. Psychological Science. 2011;**22**:1359-1366

[26] Song P, Zhang P, Qu A. Maximum likelihood inference in robust linear mixed-effects models using multivariate t distribution. Statistica Sinica. 2007;**17**:929-943

[27] Spiegelhalter D, Best N, Carlin B, van der Linder A. Bayesian measures of model complexity and fit (with discussion). Journal of the Royal Statistical Society, Series B. 2002;**64**:583-639

[28] Tong X, Zhang Z. Diagnostics of robust growth curve Modeling using Student's t distribution. Multivariate Behavioral Research. 2012;**47**:493-518

[29] Wang J, Lu Z, Cohen AS. The sensitivity analysis of two-level hierarchical linear models to outliers. Quantitative Psychology Research. New York: Springer; 2015. 307-320

[30] Wang H, Zhang Q, Luo B, Wei S. Robust mixture modelling using multivariate t-distribution with missing information. Pattern Recognition Letters. 2004;**25**:701-710

[31] Yuan K-H, Bentler PM. Structural equation modeling with robust covariances. Sociological Methodology. 1998;**28**:363-396

[32] Yuan K-H, Bentler PM. On normal theory based inference for multilevel models with distributional violations. Pschometrika. 2002;**67**:539-561

[33] Yuan K-H, Zhang Z. Structural equation modeling diagnostics using R package semdiag and EQS. Structural Equation Modeling. 2012;**19**:683-702

[34] Yuan K-H, Bentler PM, Chan W. Structural equation modeling with heavy tailed distributions. Psychometrika. 2004;**69**:421-436

[35] Yuan K-H, Lambert PL, Fouladi RT. Mardia's multivariate kurtosis with missing data. Multivariate Behavioral Research. 2004;**39**:413-437

[36] Zhang Z, Lai K, Lu Z, Tong X. Bayesian inference and application of robust growth curve models using Student's t distribution. Structural Equation Modeling. 2013;**20**:47-78

[37] Zhong X, Yuan K-H. Weights. In: Salkind NJ, editors. Encyclopedia of Research Design. Thousand Oaks, CA: Sage; 2010. pp. 1617-1620

[38] Zimmerman D. A note on the influence of outliers on parametric and nonparametric tests. Journal of General Psychology. 1994;**121**:391-401

[39] Zimmerman D. Invalidation of parametric and nonparametric statistical tests by concurrent violation of two assumptions. Journal of Experimental Education. 1998;**67**:55-68

[40] Zu J, Yuan K-H. Local influence and robust procedures for mediation analysis. Multivariate Behavioral Research. 2010;**45**:1-44

[41] R Development Core Team. R: A Language and Environment for Statistical Computing. Vienna, Austria: R Foundation for Statistical Computing; 2011

[42] Thomas A, O'Hara B, Ligges U, Sturtz S. Making BUGS open. R News. 2006:**6**;12-17

Chapter 12

Bayesian Hypothesis Testing: An Alternative to Null Hypothesis Significance Testing (NHST) in Psychology and Social Sciences

Alonso Ortega and Gorka Navarrete

Additional information is available at the end of the chapter

http://dx.doi.org/10.5772/intechopen.70230

Abstract

Since the mid-1950s, there has been a clear predominance of the Frequentist approach to hypothesis testing, both in psychology and in social sciences. Despite its popularity in the field of statistics, Bayesian inference is barely known and used in psychology. Frequentist inference, and its null hypothesis significance testing (NHST), has been hegemonic through most of the history of scientific psychology. However, the NHST has not been exempt of criticisms. Therefore, the aim of this chapter is to introduce a Bayesian approach to hypothesis testing that may represent a useful complement, or even an alternative, to the current NHST. The advantages of this Bayesian approach over Frequentist NHST will be presented, providing examples that support its use in psychology and social sciences. Conclusions are outlined.

Keywords: Bayesian inference, Bayes factor, NHST, quantitative research

1. Introduction

> *"Scientific honesty then requires less than had been thought: it consists in uttering only highly probable theories: or even in merely specifying, for each scientific theory, the evidence, and the probability of the theory in the light of this evidence".* Lakatos [1, p. 208].

The nature and role of experimentation in science found its origins in the rise of natural sciences during the sixteenth and seventeenth centuries [2]. Since then, knowledge meant that theories have to be corroborated either by the power of the intellect or by the evidence of the senses [1]. However, until the mid-late 1800s, "psychological experiments had been performed, but the science was not yet experimental" [3, p. 158]. It was not until 1875 that—either at

Wundt laboratory in Leipzig or at James' laboratory in Harvard—experimental procedures were introduced and contributed to the development of psychology as an independent science [3]. From almost one and a half centuries, scientific research mostly relies on empirical findings to provide support to their hypotheses, models, or theories. From this point of view, psychology and social sciences must take distance from rhetorical speculations, desist from unproven statements and build its knowledge on the basis of empirical evidence [1, 4]. Almost a decade ago, Curran reemphasized that the aim of any empirical science is to pursue the construction of a cumulative base of knowledge [5]. However, it has also been emphasized that such a cumulative knowledge—for a true psychological science—is not possible through the current and widespread paradigm of hypothesis testing [5–9]. Since approximately two decades ago, some explicit claims have appeared in peer review articles, such as *"Psychology will be a much better science when we change the way we analyze data"*[7], *"We need statistical thinking, not statistical rituals"* [10], *"Why most research findings are false"* [11] or *"Yes, psychologists must change the way they analyze their data…"* [12]. Most critiques have been directed toward the current—and still predominant—approach to hypothesis testing (i.e., NHST) and its overreliance on *p-values* and *significance levels* [6, 11, 13], emphasizing its pervasive consequences against the construction of a cumulative base of knowledge in psychological science [8]. Despite all warnings, they seem not to have generated a noteworthy echo in the scientific community, even though "it is evident that the current practice of focusing exclusively on a … decision strategy of null hypothesis testing can actually impede scientific progress" [14, p. 100]. Therefore, it seems reasonable to suggest that there is a need to make considerable changes to how we usually carry out research, especially if the goal is to ensure research integrity [6]. Regarding this matter, a frequently proposed alternative has been moving from the exclusive focus on *p-values* to incorporate other existing techniques such as "power analysis" [15] and "meta-analysis" [16], or to report and interpret "effect sizes" and "confidence intervals" [7]. However, in our view, a sounder alternative would be to move from a Frequentist paradigm to a Bayesian approach, which allows us not only to provide evidence against the null hypothesis but also in favor of it [17]. Furthermore, Bayesian analysis allows us to compare two (or more) competing models in light of the existent data and not only based in "theoretical probability distributions," as in the Frequentist approach to hypothesis testing [18].

A Bayesian approach would offer some interesting possibilities for both individual psychology researchers and the research endeavor in general. First, Bayesian analysis allows us to move from a dichotomous way of reasoning about results (e.g., either an effect exists of it does not) to a less artificial view that interprets results in terms of magnitude of evidence (e.g., the data are more likely under H_0 than H_a), and therefore, allows us to better depict to which extent a phenomenon may occur. Second, a Bayesian approach naturally allows us to directly test the plausibility of both the null and the alternative hypothesis, but the current NHST paradigm does not. In fact, when a researcher does not reach a desired *p-value* oftentimes it is—falsely—assumed that the effect "does not exist." As a consequence, the researcher's chances of getting his or her results published decrease dramatically, which moves us to our third argument. As broadly known, the most scientific peer-reviewed journals do not show much interest in results, which are "non-statistically significant." This common practice—or scientific standard—sadly reinforces the idea of thinking in terms of relevant or irrelevant findings. In our view, such standards do not promote scientific advance and quickly lead us to ignore some promising but "non-significant" findings that may be further explored, fed into meta-analysis, of just be considered by other researchers in the field. Of course, systematically ignoring a portion of the research undermines the primary goal of scientific inquiry that is to collect evidence and not only to reject hypothesis. The facts and ideas exposed in this introductory section set forth the necessity to reanalyze the way in which scientific evidence has been conceived during the NHST era.

The following sections will: (a) concisely address the NHST procedure, (b) introduce a Bayesian framework to hypothesis testing, (c) provide an example that highlights the advantages of a Bayesian approach over the current NHST in terms of the way in which scientific evidence is quantified, and (d) briefly summarize and discuss the benefits of a Bayesian approach to hypothesis testing.

2. Null hypothesis significance testing (NHST)

> "*Never use the unfortunate expression: accept the null hypothesis.*" Wilkinson and the Task Force on Statistical Inference APA Board of Scientific Affairs[19, p. 602].

The most influential methods to modern null hypothesis significance testing (NHST) were developed by Fisher, and by Neyman and Pearson in the early and mid-1900s [20]. Since then, the NHST has been broadly used to provide an association between empirical evidence and models or theories [21]. In the traditional NHST procedure, two hypotheses are postulated: a null hypothesis (i.e., H_0) and a research hypothesis, also called alternative (i.e., H_a), which describe two contrasting conceptions about some phenomenon [22]. When conducting a NHST, researchers usually pursue to reject the null hypothesis (H_0) on the basis of a *p-value*. When the observed *p-value* is lower than a predetermined significance level (i.e., alpha, usually corresponding to $\alpha = 0.05$), the conclusion is that such *p-value* constitutes supporting evidence that favors the plausibility of the alternative hypothesis [23]. However, a more important feature of this procedure that remains unknown for most scientists, including psychology researchers, is that the NHST constitutes an amalgamation of two irreconcilable schools of thought in modern statistics: the Fisher test of significance, and the Neyman and Pearson hypothesis test [24, 25]. To this respect, Goodman stated that "it is not generally appreciated that the *p-value*, as conceived by Fisher, is not compatible with the Neyman and Pearson hypothesis in which it has become embedded" [25, p. 485]. In this synthesized NHST, the Fisherian approach includes a test of significance of *p-values* obtained from the data, whereas the Neyman and Pearson method incorporates the notion of error probabilities from the test (i.e., Type I and Type II).

2.1. Origins and rationale of NHST

First, in the early 1900s, Fisher [26, 27] developed a method that tested a single hypothesis (i.e., null or H_0), which has been mainly referred to as a hypothesis of "no effect" between variables (e.g., relationship, difference). The null hypothesis, as conceived by Fisher, has a known distribution of the test statistic t. Thus, as the test statistic moves away from its expected value, then the null hypothesis becomes progressively less plausible. In other words, it appears less likely to occur by chance. Then, if H_0 achieves a probability of occurrence sufficiently lower than the significance level (i.e., a small *p-value*) then it should be rejected. Otherwise, no conclusion can be reached. Subsequently, the question that logically arises is: what *p-value* is sufficiently small to reject H_0? The answer to this question was clearly addressed by Fisher when he stated that this threshold should be determined by the context of the problem, and it was not until the 1950s that Fisher presented the first significance tables to establishing rejection thresholds [22]. However, Fisher [28] refused the idea of establishing a conventional significance level and, in its place, recommended reporting the exact *p-value* instead of a significance level (e.g., $p = 0.019$, but not $p < 0.05$; see [10]). Similarly, May et al. indicated that the choice of a significance level should depend on the consequences of rejecting or failing to reject the null hypothesis [29]. Despite these recommendations about threshold determination, most scientists from different research fields adopted standard significance levels (i.e., $\alpha = 0.05$ or $\alpha = 0.01$), which have been used—or misused—regardless of the hypotheses being tested.

Later, in 1933, Neyman and Pearson proposed a procedure in which two explicitly stated rival hypotheses were contrasted, being one of them still considered as the "null" hypothesis, as in the Fisher test [30]. Neyman and Pearson rejected Fisher's idea of only testing the null hypothesis. In this scenario, there are now two hypotheses (i.e., the null and the alternative), and based on the observed *p-value*, the researcher has to decide whether to reject or not to reject the null hypothesis. This decision rule faces the researcher with the probability of committing two kinds of errors: Type I and Type II. As defined by Neyman and Pearson, the Type I error is the probability of falsely rejecting H_0 (i.e., null) when H_0 is true [30]. Conversely, the probability of failing to reject H_0 when H_0 is false is the Type II error. For the sake of simplicity, an analogy of both kinds of errors can be found in the classic fairy tale "The boy who cried wolf!" When the young shepherd, called Peter, shouted out: "Help! the wolf is coming!" The village's people believed the young boy warning and quickly came to help him. However, when they found out that all was a joke, they got angry. To believe in the boy's false, alarm can be considered as a Type I error. Peter repeated the same joke a couple of times and, when the wolf actually appeared, the villagers did not believe the young shepherd's desperate calls. This situation is analogous to be engaged in a Type II error [31].

Within this NHST framework, the Fisher's *p-value* is then used to dichotomize effects into two categories: significant and non-significant results [21]. Consequently, on one hand, obtaining significant results led us to assume that the phenomenon under investigation can be considered as "existing" and, therefore, can be used as supporting evidence for a particular model or theory. On the other hand, non-significant results are usually (and erroneously) considered as "noise," implicating the nonexistence of an effect [21]. In this last case, there are no findings that could be reported. From this view, the evidence in favor of a research finding is then solely judged on the ability to reject H_0 when a sufficiently low *p-value* is observed. This simple and appealing decision rule may constitute a very seductive way of thinking about results, that is: A phenomenon either exists or it does not. However, thinking in this fashion is fallacious, led to misinterpretations of results and findings, and more importantly "it can distract us from a higher goal of scientific inquiry. That is, to determine if the results of a test have any practical value or not" [32, p. 7].

2.2. NHST: Common misconceptions and criticisms

As previously stated, most problems and criticisms to the current NHST paradigm appear as a result of the mismatch of these essentially incompatible statistical approaches [10, 33, 34]. In this line, Nickerson stated that "A major concern expressed by critics is that such testing is misunderstood by many of those who use it" [35, p. 241]. Some of these misconceptions are common among researchers and are interpretative in nature. As a matter of fact, Badenes-Ribera et al. recently reported the results of a survey conducted to 164 academic psychologists who were questioned about the meaning of *p-values* [36]. Results confirmed previous findings regarding the occurrence of wrongful interpretations of *p-values*. For instance, the false belief that the *p-value* indicates the conditional probability of the null hypothesis given certain data (i.e., $p\ (H_0 | D)$), instead of the probability of witnessing a given result, assuming that the null hypothesis is true [37]. This wrong interpretation of a *p-value* is known as "the inverse probability" fallacy. Another common misconception regarding *p-values* is that they provide direct information about the magnitude of an effect, that is, a *p-value* of 0.00001 represents evidence of a bigger effect than a *p-value* of 0.01. This conclusion is wrong because the only way to estimate the magnitude of an effect is to calculate the value of the effect size with the appropriate statistic and its confidence interval (e.g., Cohen's *d*; see [38]). This erroneous interpretation of a *p-value* is known as "the effect size" fallacy. A comprehensive review of these and other common misconceptions is out of the scope of this chapter, but several resources on these topics are available for the interested readers (see [14, 35, 37–40]).

Likewise, the rationale under the NHST has been largely criticized. Most criticisms against NHST are focused on the way in which data are (unsoundly) analyzed and interpreted, for example:

a. NHST only provides evidence against the plausibility of H_0, but does not provide probabilistic evidence in favor of the plausibility of H_a.

b. NHST uses inference procedures based on hypothetical data distributions, instead of being based on actual data.

c. NHST does not provide clear rules for stopping data collection; therefore, as long as sample size increases any H_0 can be rejected (see [9, 18]).

However, an issue that is of particular interest for this chapter is related to the use of *p-values* as a way to quantify statistical evidence [13, 41]. As previously stated in this chapter, rejecting H_0 does not provide evidence in favor of the plausibility of H_a, and all that can be concluded is that H_0 is unlikely [9]. Conversely, failing to reject H_0 simply allows us to state that—given the evidence at hand—one cannot make an assertion about the existence of some effect or phenomenon [42]. Hence, rejecting H_0 is not a valid indicator of the magnitude of evidence of a result [43]. In Schmidt's words: "… reliance on statistical significance testing in psychology and the other social sciences has led to frequent serious errors in interpreting the meaning of data, errors that have systematically retarded the growth of cumulative knowledge" [16, p. 120]. Despite the existence of scientific literature that highlights the weaknesses of NHST [9, 16, 21, 22, 39, 43–46], it is still considered as the: "*sine qua non* of the scientific method" [10, p. 199]. Moreover, NHST is arguably the most widely used method of data analysis in psychology since the mid-1950s and still governs the interpretation of quantitative data in social science research [35, 47]. In Krueger's words: "NHST is the researcher's workhorse for making inductive inferences" [45, p. 16]. An immediate matter of concern is that most of scientific discoveries, in a wide range of research fields, are based on a procedure that still generates controversy (see [12, 48–50]). Since the focus of research should be on what data tell us about the magnitude of effects, it seems necessary to shift from our reliance on NHST to more robust alternatives [14]. Some recommended practices include estimates based on effect sizes, confidence intervals, and meta-analysis [6]. However, a sounder alternative comes from the Bayesian paradigm through the use of a simple estimate of the magnitude of evidence called Bayes factor (BF) [17]. This approach to hypothesis testing has shown several benefits. First, it is not oriented to pursue the rejection of H_0; on the contrary, it provides a way to obtain evidence for and against H_0. Second, it does not use arbitrary thresholds (i.e., significance levels) to reach dichotomous decisions about the plausibility or implausibility of H_0; on the contrary, it directly contrasts the magnitude of evidence for and against both H_0 and H_a. Third, it permits the continuous update of evidence as long as new data are available, which is in line with the nature of scientific inquiry. Bayesian methods have been largely suggested as a practical alternative to NHST [9, 17, 23, 51], but—until now—they have not received enough attention from researchers in psychology and social sciences.

3. Bayesian hypothesis testing: An alternative to NHST

> "(…) *prior and posterior are relative terms, referring to the data. Today's posterior is tomorrow's prior.*" Lindley [52, p. 301].

In the field of statistics, probabilities can be interpreted under two predominant paradigms: Frequentist inference and Bayesian inference. The former makes predictions about experiments whose outcomes depend basically upon random processes [53]. The latter assigns probabilities to any statement, even when a random process is not involved [54]. In a Bayesian framework, a probability is a way to embody an individual's degree of belief in a statement. Since the mid-1950s, there has been a clear predominance of the Frequentist approach to hypothesis testing, both in psychology and social sciences. The hegemony of Frequentist inference and its null hypothesis significance testing (NHST) might be partially attributed to the massive incorporation of such approaches in psychology undergraduate programs [9] and also to the fact that the Neyman and Pearson approach had the most well-developed computational software to conduct statistical inference [18]. However, the current scenario has drastically changed, and the development of sampling techniques like Markov-Chain Monte Carlo (MCMC; see [55, 56]) along with the availability and improvement of specifically developed software (e.g., WinBUGS, see [57, 58]; JAGS, see [59, 60]; JASP, see [61]) makes exact Bayesian inferences possible even in very complex models. As a result, "Bayesian applications have found their way into most social science fields" [22, p. 665], and psychologists can now easily implement Bayesian analysis for many common experimental situations (see for example JASP Statistics: https://jasp-stats.org/).

http://dx.doi.org/10.5772/intechopen.70230

3.1. Bayes in a nutshell

In Bayesian inference, our degrees of belief about a set of hypotheses are quantified by probability distributions over those hypotheses [47, 62], which makes the Bayesian approach fundamentally different from the Frequentist approach, which relies on sampling distributions of data [47]. A Bayesian analysis usually implicates the updating of prior knowledge or information in light of newly available experimental data [63]. The latter clearly reflects the aim of any empirical science, which is to strive for the elaboration of a cumulative base of knowledge. Any Bayesian analysis implies the combination of three sources of information as follows:

a. a model that specifies how latent parameters (e.g., θ) generate data (e.g., D);

b. prior information about those parameters (i.e., prior distribution); and

c. the observed data (i.e., likelihood).

This prior information, represented by $p(\theta)$, represents our degree of uncertainty about the parameters included in the model. Conversely, this prior distribution may also represent our degree of knowledge about the same parameters. Then, the more informative is our prior distribution, the less will be our degree of uncertainty about the parameters. The likelihood is the conditional probability of observing the data under some latent parameter (i.e., $p(D \mid \theta)$). Following the Bayes theorem [64], the combination of these three elements produces an updated knowledge about the model parameters after the data have been observed, which is also known as the posterior distribution. The change from the prior to the posterior distribution reflects what has been learned from the data (see **Figure 1**). Thus, within a Bayesian framework, a researcher can invest more effort in the specification of prior distributions by translating existing knowledge about the phenomenon under study into prior distributions [65]. As suggested by Lee and Wagenmakers "such knowledge may be obtained by eliciting prior beliefs from experts, or by consulting the literature for earlier work on similar problems" [65, p. 110].

As shown in **Figure 1**, the strength of each source of information is indicated by the narrowness of its curve. A narrower curve is more informative about the value of parameters, whereas a wider one is less informative.

Bayes' rule specifies how the prior information $p(\theta)$ and the likelihood $p(D \mid \theta)$ are combined to arrive at the posterior distribution denoted by $p(\theta \mid D)$, in Eq. (1):

$$p(\theta|D) = \frac{p(D|\theta)\,p(\theta)}{p(D)} \tag{1}$$

Eq. (1) is usually paraphrased as:

$$p(\theta|D) \propto p(D|\theta)p(\theta) \tag{2}$$

which means, "the posterior is proportional (i.e.,$\propto$) to the likelihood times the prior." In other words, the observed data (i.e., likelihood) increases our previous degree of knowledge (i.e.,

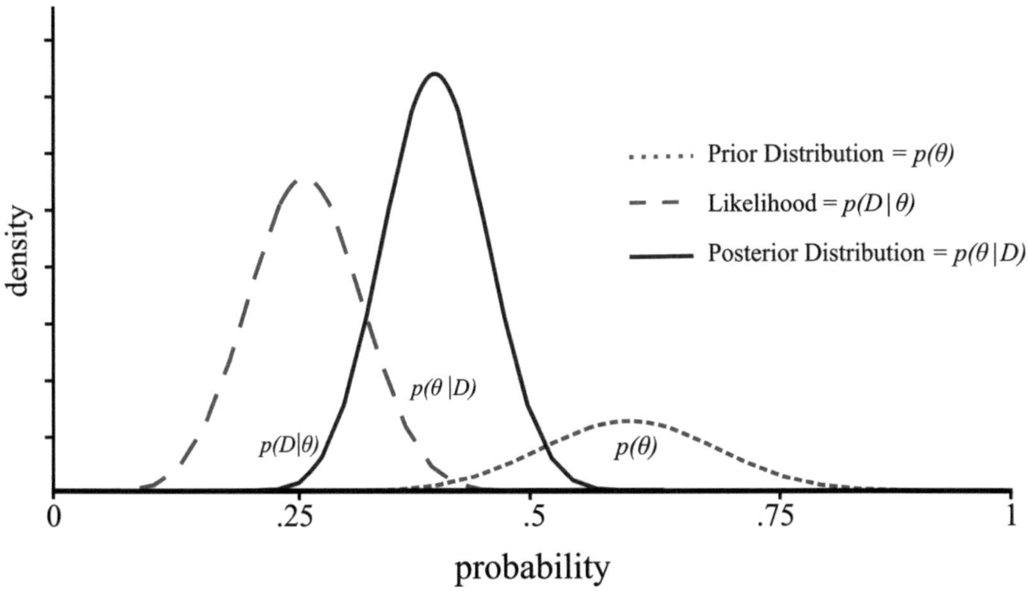

Figure 1. Prior, likelihood and posterior probability distributions.

prior) in a proportional way to its informative strength, producing a new state of knowledge about the parameters of the model (i.e., posterior). One of the benefits of the Bayesian approach is that the prior (i.e., $p(\theta)$; our present knowledge about the model parameters moderates the influence provided by the data (i.e., $p(D|\theta)$). This compromise leads to less pessimism when data are unexpectedly bad and less optimism when it is unexpectedly good [66]. Both influences are beneficial and help us to make more realistic inferences and take better decisions. For more detailed information on Bayesian inference, see, for instance, O'Hagan and Forster [54], Kruschke [59], and Jackman [67].

3.2. Bayes factor

Bayesian approaches for hypothesis testing are comparative in nature. Different models often represent competing theories or hypotheses, and the focus of interest is on which one is more plausible and better supported by the data [65]. Therefore, the Bayesian approach allows to quantify the plausibility of a given model or hypothesis (i.e., H_0) against that of an alternative model (i.e., H_a). For any comparison of two competing models or hypotheses (e.g., H_a vs. H_0), we can rely on an estimate of evidence known as the Bayes factor [52]. One of the attractive features of the Bayes factor is that it follows the principle of parsimony: When two models fit the data equally well, the Bayes factor prefers the simple model over the more complex one [68]. Nonetheless, in contrast to the NHST approach, "Bayesian statistics assigns no special status to the null hypothesis, which means that *Bayes factors* can be used to quantify evidence for the null hypothesis just as for any other hypothesis" [65, p. 108].

Before observing the data, the *prior odds* of H_a over, e.g., H_0, are $p(H_a)/p(H_0)$, and after having observed the data we have the *posterior odds* $p(H_a|D)/p(H_0|D)$. Therefore, the ratio of the posterior odds and the prior odds is defined as the Bayes factor:

$$BF_{H_aH_0} = \frac{(D|H_a)}{(D|H_0)} = \frac{\frac{\{p(H_a|D)\}}{\{p(H_0|D)\}}}{\frac{\{p(H_a)\}}{\{p(H_0)\}}} = \frac{posterior\, odds}{prior\, odds} \tag{3}$$

Eq. (3) shows the Bayes factor for given data D and two competing hypotheses (i.e., H_0 vs. H_a), which is a measure of the evidence for H_a against H_0 provided by the data. In other words, the Bayes factor is the probability of the data under one hypothesis relative to the other. For instance, a $BF_{H_aH_0} = 3$ indicates that H_a is three times more plausible relative to H_0 than it was a priori. From this view, the Bayes factor may be considered as analogous to the Frequentist likelihood ratio. Nevertheless, in the Bayesian context there is no reference at all to theoretical probability distributions as it is customary in a Frequentist approach. In a Bayesian framework, all inferences are made conditional on the observed data, and therefore, the Bayes factor has to be interpreted as a summary measure of the information provided by the data about the relative plausibility of two models or hypotheses (e.g., H_a vs. H_0). Jeffreys [52] suggests the following scale for interpreting the Bayes factor (**Table 1**), although some people argue against the use of thresholds, least we fall in a different version of the old $p < 0.05$ ritual (see, for instance, [69]).

Bayes factor			**Interpretation**
	>	100	Extreme evidence for H_a
30	–	100	Very strong evidence for H_a
10	–	30	Strong evidence for H_a
3	–	10	Moderate evidence for H_a
1	–	3	Anecdotal evidence for H_a
1			No evidence
1/3	–	1	Anecdotal evidence for H_0
1/10	–	1/3	Moderate evidence for H_0
1/30	–	1/10	Strong evidence for H_0
1/100	–	1/30	Very strong evidence for H_0
	<	1/100	Extreme evidence for H_0

Adapted from Jeffreys [52, p. 433], and Lee and Wagenmakers [65, p. 105].

Table 1. Evidence categories for the Bayes factor.[1]

4. Bayesian vs. Frequentist approaches to hypothesis testing: An example

Bayes factors to evaluate the amount of evidence in favor or against H_0 and H_a are one of the big selling points of the Bayesian framework.[1] As stated in the previous section, the core idea is that the magnitude of evidence in favor of the null hypothesis compared to that of the alternative hypothesis can be estimated (or vice-versa). As we have seen, this approach has multiple advantages, such as departing from a *hit-or-miss* approach to results reporting, or being able to show evidence in favor of the null. The possibility of providing evidence in favor of both the null and the alternative hypotheses has some important advantages. One of them is that it helps to overcome one of the most common issues behind the well-known file-drawer effect, in that results do not suddenly become meaningless when the *p-value* is over certain threshold. Another advantage is that it gives us more freedom when establishing hypothesis, particularly in topics where hypothesizing the absence of differences may be necessary for theoretical advance.

In this section, an example from a field known as Bayesian reasoning will be presented, which deals with how people update their beliefs when new evidence is available (e.g., when receiving a positive result in a medical test, how likely it is that I have a disease?). There is a long standing debate in the field about why people are unable to solve medical screening problems such as the one shown in **Table 2** when the information is shown in a standard probability format (i.e., single-event probabilities; for instance, 1% have cancer), but have a comparatively better time when the same information is shown in a standard frequency format (i.e., natural frequencies; for instance, 10 in 1000 have cancer). As it is often the case, the debate about these issues is very complex (for a review, see [71]), and the present example will focus on a single unnuanced aspect with the goal of showing the usefulness of the Bayesian statistics paradigm.

Standard probability format

The probability of breast cancer is 1% for women at age 40 who participate in routine screening. If a woman has breast cancer, the probability is 80% that she will get a positive mammography. If a woman does not have breast cancer, the probability is 9.6% that she will also get a positive mammography.
A woman in this age group had a positive mammography in a routine screening. What is the probability that she actually has breast cancer? _____%

Standard frequency format

Ten out of every 1000 women at age 40 who participate in routine screening have breast cancer. Eight of every 10 women with breast cancer will get a positive mammography. Ninety-five out of every 990 women without breast cancer will also get a positive mammography.
Here is a new representative sample of women at age 40 who got a positive mammography in routine screening. How many of these women do you expect to actually have breast cancer? ____out of____

Table 2. Standard probability and standard frequency format problems, as shown by Gigerenzer and Hoffrage [72].

[1]However, we recommend the interested reader to revise a recent paper by Lakens [70], which describes an approach to test for equivalence within a Frequentist framework.

Some authors [73, 74] argue that the crucial factor explaining the differences between the two versions is not the representation format (i.e., probabilities or natural frequencies), but the reference class or more specifically the computational complexity is caused by the reference class of the problems [75]. In brief, as the probability version has a relative reference class, and all the numbers refer to the group above them (e.g., 80% from the 1% who have breast cancer will get a positive mammography). To solve the problem, we need to use the base-rates (in this example, percentage of women with and without breast cancer; 1 and 99%), and the percentage of women who got a positive mammography amongst those two groups (e.g., 80 and 9.6%; see Eq. (4)). In the frequency version, as the reference class is absolute, and all numbers can be seen as referring to the 1000 women, we can ignore the base-rates and directly use the positive mammographies for women with and without cancer (8 and 95; see Eq. (5)). The above-mentioned authors hypothesized that when reference class and computational complexity are taken into account, there is no difference between probabilities and natural frequencies. In other words, they expect the null hypothesis to be true (**Figure 2**).

$$p(H|D) = \frac{1\% \times 80\%}{1\% \times 80\% + 99\% \times 9.6\%} = 0.077 \quad (4)$$

$$p(H|D) = \frac{8}{8 + 95} = 0.077 \quad (5)$$

Now, imagine two PhD students, a Frequentist (i.e., Student 1) and a Bayesian (i.e., Student 2). After reading a critical but often ignored Fiedler's paper [73], they had the idea that computational complexity class (and not representation format) is the key issue when trying to understand how people solve Bayesian reasoning problems. They devise a very simple experiment where two different groups of people will be asked to solve one Bayesian reasoning problem that will be shown either in single-event probabilities or in natural frequencies. In both cases, the arithmetic complexity (i.e., number of arithmetic steps required to solve the problem) will be exactly 2. That is, to solve the problems, participants would need to do two arithmetic operations, a sum and a division. They used a test with a 100% sensitivity and 0% specificity, which could not have any clinical application, but it is useful to get a few arithmetic steps out of the probability format and check if computational complexity underlies Bayesian reasoning. With this manipulation, the algorithms to solve the probability and frequency versions become

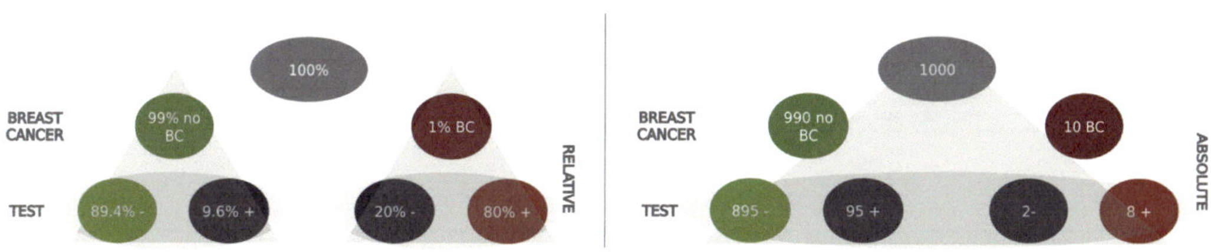

Figure 2. Relative and absolute reference classes represented by the reference of the last row (test results). In the Relative reference class, the information about the test, for example, 80% positive (+) and 20% negative results (−) refers to the 1% women with BC, but not to the 100% of the women (it is not an 80% of the 100%!). However, in the absolute reference class, the same information, 8+ and 2−, refers to the women with BC, but also to the 1000 women directly. This translates in the need to use Eq. (4) for relative probabilities and Eq. (5) for absolute frequencies.

Eqs. (6) and (7), respectively. It is easy to see how both have become roughly equivalent now in terms of arithmetic complexity.

$$p(H|D) = \frac{10\% \times 100\%}{10\% \times 100\% + 90\% \times 100\%} = \frac{10\%}{10\% + 90\%} = 0.1 \quad (6)$$

$$p(H|D) = \frac{10}{10 + 90} = 0.1 \quad (7)$$

As it can be deduced, Student 1 would have a Fisherian approach to statistics and Student 2 a Bayesian approach. Both run an experiment with a total of 62 participants (31 per group),[2] and have the following results:

Contingency tables

	Representation format		
Accuracy	**Natural frequencies**	**Probabilities**	**Total**
0	23	24	47
1	8	7	15
Total	31	31	62

4.1. PhD Student 1—Frequentist

Student 1, as the most good NHST practitioners would do, conducts a Chi-square test and reports that he did not obtain a significant effect of representation format when arithmetic steps were equal (χ^2 = 0.088, p = 0.767). He is happy, because this is congruent with his hypothesis. He then writes a brief report detailing his idea and experimental results and sends the manuscript draft to his advisor. A few days later, he receives his advisor feedback, telling him that his non-significant results could be caused by a number of reasons, and as a consequence, the non-significant results are hard to interpret.

Chi-square tests

	Value	*df*	*p*
χ^2	0.088	1	0.767
N	62		

[2]Of course, the sample size and manipulation for this experiment is more congruent with a pilot experiment than a real one that could be sent to a journal on its own. As a side note, take into account that one of the advantages of the Bayesian framework some authors propose is a sequential sampling rule, where sampling stops when the evidence (BF) is over a predetermined threshold (e.g., $BF_{10} > 10 \mid < 0.1$), see Lindley [76].

http://dx.doi.org/10.5772/intechopen.70230

His advisor suggests carrying out a few more experiments using variations of the task and decent sample-sizes, to be able to perform a meta-analysis that could convince the editorial board of a journal that their endeavor is noteworthy, as they would probably have a hard time publishing those non-significant results by themselves.

4.2. PhD Student 2—Bayesian

Student 2, instead of performing a Chi-square test, prefers to use a well-known analysis among Bayesian statisticians called Bayes factor (BF; see [17, 65]). He uses a very simple to use software called JASP [61], that incorporates Bayesian contingency tables, and outputs BF results in ready to use APA formatted tables. He finds that when arithmetic steps are equal, there is a BF_{01} of 4.656, that is, there is 4.6 times more evidence in favor of the null-hypothesis than the alternative-hypothesis. Along his advisor, they send the manuscript to a journal, pushing for the relative importance of arithmetic complexity over representation format. In practical terms, it is more likely that the editor will be willing to publish this interesting result, although the amount of evidence in favor of the null would be considered moderate by some standards (see [53]).

Bayesian contingency tables tests

	Value
BF_{0+} independent multinomial	4.656
N	62

Note: For all tests, the alternative hypothesis specifies that group *Natural-Frequencies* is greater than group *Probabilities.*

As the evidence for the null effect is not very strong, they would need to run a few more studies with variations to replicate the finding and show, using BF, how much more evidence there is for the null hypothesis compared to the alternative hypothesis. Alternatively, they could increase the sample size in their experiment until the stopping rule threshold (e.g., $BF_{10} < 0.1$) is reached.

This example was aimed to describe (in a very simplified manner) one of the practical advantages of the Bayesian framework, that is, being able to present the amount of evidence for and against both the null and alternative-hypotheses. This, combined with the incremental nature of the Bayesian inference process, allows us to move further from the *hit-or-miss* approach generally reinforced by the NHST framework, in which significant results are seen as more valuable than non-significant ones.

5. Conclusion

During the past 70 years, the NHST has dominated the way in which knowledge is produced and interpreted and still governs the way in which researchers analyze their data, reach conclusions, and report results [10, 45]. This approach has been largely criticized [9, 16, 21, 22, 39, 43–46], and "a major concern expressed by critics is that such testing is misunderstood by many of those who use it" [35, p. 241]. Some authors [9, 13] emphasized that one of the most pervasive influences of the NHST approach has been its over reliance on *p-values,* and in particular, in the way that *p-values* have been interpreted (see, for instance [35, 36, 77]). One of the most common misinterpretations of *p-values* it has been to consider a *p-value* as a valid indicator of the magnitude of evidence of a result (i.e., effect size fallacy). Regarding this point, Cohen emphasized that the only way to estimate the magnitude of an effect is to calculate the value of the effect size with the appropriate statistic and its confidence interval [38]. The correct way to interpret *p-values* is two-fold. On one hand, to reject H_0 only allows us to conclude that H_0 is unlikely. On the other hand, failing to reject H_0 simply allows us to state that—given the evidence at hand—one cannot make an assertion about the existence of some effect or phenomenon [42]. An immediate consequence of the wrong way in which a big number of researchers interpret *p-values* is that null results have been usually considered as the absence of evidence of the existence of an effect. This perspective regarding the decisions made when a given *p-value* threshold is not reached (i.e., $p < 0.05$) do not promote scientific advance and quickly leads us to a systematic bias toward ignoring promising but "non-significant" findings that may be further explored, fed into meta-analysis, of just be considered by other researchers in the field. This fact is against the pursue of any empirical science and may be harmful to the construction of a cumulative base of knowledge [5].

As a way to provide a complementary (or alternative) method to deal with the current NHST practice, we described here a Bayesian approach to hypothesis testing. A Bayesian approach allows us to think about phenomena in terms of the magnitude of evidence that supports the existence of an effect, instead of a dichotomous and artificial way of thinking in which an effect either exists or does not exist [21]. As described in previous sections, a Bayesian approach provides us a measure of evidence for and against both the null and the alternative hypotheses (i.e., Bayes factor, BF; see [17]). The use of Bayes factors helps to overcome one of the most common issues behind the well-known file-drawer effect, reducing the existent bias through which results suddenly become meaningless when the *p-value* is over certain threshold (e.g., $p > 0.05$). A straightforward feature of this approach is that "Bayesian statistics assigns no special status to the null hypothesis, which means that *Bayes factors* can be used to quantify evidence for the null hypothesis just as for any other hypothesis" [65, p. 108]. Therefore, a Bayesian approach gives us more freedom when establishing hypothesis, for example in topics where hypothesizing the absence of differences may be necessary for theoretical advance.

However, a major problem with Bayesian statistics has historically been that they require complex and intricate mathematical calculations that were analytically intractable, at least without the required techniques and specialized software. However, this scenario changed dramatically during the 1990s with the development of sampling techniques like Markov-Chain Monte Carlo (MCMC; see [55]) along with the availability and improvement of specifically developed software (e.g., WinBUGS, see [57, 58]; JAGS, see [59, 60]) that makes exact Bayesian inferences possible even in very complex models. Nowadays, the relatively recent implementation and availability of Bayesian analysis in "easy-to-use" and open software such as JASP [61], R toolboxes such as Bayes factor [78], or more specialized ones like WinBUGS,

JAGS, or Stan (http://mc-stan.org/) makes Bayesian statistics more accessible to all researchers, academics and students. This widespread availability, paired with the advantages of the Bayesian approach described in this chapter, and several times elsewhere [79–82], should help establish the Bayesian paradigm as a viable and popular alternative to NHST.

Despite all the important Bayesian paradigm advantages, as always, there is potential for misuse. As pointed out by Morey, Bayes factor interpretation is very natural (i.e., as the amount of evidence in favor of one hypothesis in comparison to another), and does not need specific decision thresholds, as it is the case of *p-values* [83]. However, some standards that could help to communicate BF results have been proposed (see [53]) and may be helpful to people that are not familiar with them. Nonetheless, the introduction of these labels also creates an opportunity for misuse, as they could be misinterpreted as decision boundaries. It is very important to be aware of this fact, and be careful when using them, to avoid making "$BF > 3$" the new "$p < 0.05$."

To sum up, the main goal of this chapter has been to increase the degree of awareness regarding the limitations of the NHST approach and highlight the advantages of the Bayesian approach. We expect that the inclusion of an easy-to-understand example of a specific case where a Bayesian paradigm shows its practical utility may offer the newborn readers on this matter a glimpse to the usefulness of this alternative to the way in which they can analyze and interpret their data. As a final remark, we would like to point an often-heard recommendation for people interested in starting to use BF, which is to introduce them alongside *p-values* and effect size measures, to ease the transition to the new paradigm, and make them comprehensible to people not yet familiarized with them.

Author details

Alonso Ortega[1]* and Gorka Navarrete[2]

*Address all correspondence to: alonso.ortega@uai.cl

1 School of Psychology, Universidad Adolfo Ibáñez, Chile

2 Center for Social and Cognitive Neuroscience (CSCN), School of Psychology, Universidad Adolfo Ibáñez, Chile

References

[1] Lakatos I. Falsification and the methodology of scientific research programmes. In: Harding S, editor. Can Theories be Refuted? Dordrecht: Holland: D. Reidel Publishing Company; 1976. pp. 205-259

[2] Radder H. Toward a more developed philosophy of scientific experimentation. In: Radder H, editor. The Philosophy of Scientific Experimentation. Pittsburgh: University of Pittsburgh Press; 2003. pp. 1-18

[3] Harper RS. The first psychological laboratory. Isis. 1950;**41**(2):158-161

[4] Popper KR. Degree of confirmation. The British Journal for the Philosophy of Science. 1954;**5**(18):143-149

[5] Curran PJ. The seemingly quixotic pursuit of a cumulative psychological science: Introduction to the special issue. Psychological Methods. 2009;**14**(2):77-80

[6] Cumming G. The new statistics why and how. Psychological Science. 2013;**25**(1):7-29

[7] Loftus GR. Psychology will be a much better science when we change the way we analyze data. Current Directions in Psychological Science. 1996;**5**(6):161-171

[8] Rossi JS. A case study in the failure of psychology as a cumulative science: The spontaneous recovery of verbal learning. In: Harlow L, Mulaik S, Steiger J, editors. What If There Were No Significance Tests. Mahwah, NJ: Erlbaum Associates Publishers; 1997. pp. 175-197

[9] Wagenmakers E-J. A practical solution to the pervasive problems of *p* values. Psychonomic Bulletin & Review. 2007;**14**(5):779-804

[10] Gigerenzer G. We need statistical thinking, not statistical rituals. Behavioral and Brain Sciences. 1998;**21**(2):199-200

[11] Ioannidis JP. Why most published research findings are false. PLOS Medicine. 2005;**2**(8): e124

[12] Wagenmakers EJ, Wetzels R, Borsboom D, Van Der Maas HL. Why psychologists must change the way they analyze their data: The case of psi: Comment on Bem (2011). Journal of Personality and Social Psychology. 2011;**100**(3):426-432

[13] Llobell JP, Dolores M, Navarro F, et al. Usos y abusos de la significación estadística: propuestas de futuro ("Necesidad de nuevas normativas editoriales"). Metodologia de las Ciencias del Comportamiento, 2004; Volumen Especial: 465-469

[14] Kirk RE. The importance of effect magnitude. In: Davis SF, editor. Handbook of Research Methods in Experimental Psychology. Malden, MA: Blackwell Publishing; 2003. pp. 83-105

[15] Cohen J. A power primer. Psychological Bulletin. 1992;**112**(1):155-159

[16] Schmidt FL. Statistical significance testing and cumulative knowledge in psychology: Implications for training of researchers. American Psychological Association. 1996;**1**(2): 115-129

[17] Kass RE, Raftery AE. Bayes factors. Journal of the American Statistical Association. 1995;**90**(430):773-795

[18] Dienes Z. Bayesian versus Orthodox statistics: Which side are you on? Perspectives on Psychological Science. 2011;**6**(3):274-290

[19] Wilkinson L, Task Force on Statistical Inference APA Board of Scientific Affairs. Statistical methods in psychology journals: Guidelines and explanations. American Psychologist. 1999;**54**:594-604

[20] Levine TR, Weber R, Hullett C, Park HS, Lindsey LLM. A critical assessment of null hypothesis significance testing in quantitative communication research. Human Communication Research. 2008;**34**(2):71-187

[21] Dixon P. The p-value fallacy and how to avoid it. Canadian Journal of Experimental Psychology/Revue Canadienne de Psychologie Experimentale. 2003;**57**(3):189-202

[22] Gill J. The insignificance of null hypothesis significance testing. Political Research Quarterly. 1999;**52**(3):647-674

[23] Rouder JN, Speckman PL, Sun D, Morey RD, Iverson G. Bayesian t tests for accepting and rejecting the null hypothesis. Psychonomic Bulletin & Review. 2009;**16**(2):225-237

[24] Christensen R. Testing Fisher, Neyman, Pearson, and Bayes. The American Statistician. 2005;**59**(2):121-126

[25] Goodman SN. P values, hypothesis tests, and likelihood: Implications for epidemiology of a neglected historical debate. American Journal of Epidemiology. 1993;**137**(5):485-496

[26] Fisher RA. Two new properties of mathematical likelihood. Proceedings of the Royal Society of London. Series A, Containing Papers of a Mathematical and Physical Character. 1934;**144**(852):285-307

[27] Fisher RA. Statistical Methods for Research Workers. Edinburgh: Genesis Publishing Pvt Ltd; 1925

[28] Fisher RA. Statistical methods and scientific induction. Journal of the Royal Statistical Society. Series B (Methodological). 1955;**17**:69-78

[29] May RB, Masson MJ, Hunter MA. Application of Statistics in Behavioral Research. NY: Harper & Row; 1990

[30] Neyman J, Pearson ES. On the problem of the most efficient tests of statistical hypotheses. Philosophical Transactions of the Royal Society of London. 1933;**A231**:289-337

[31] Singh VB. Don't Confuse Type I and Type II errors. 2015. Available from: https://www.linkedin.com/pulse/dont-confuse-type-i-ii-errors-bhaskar-vijay-singh-frm?articleId=6077308381431951360 [Accessed: June 21, 2017]

[32] Nix TW, Barnette JJ. The data analysis dilemma: Ban or abandon. A review of null hypothesis significance testing. Research in the Schools. 1998;**5**(2):3-14

[33] Gigerenzer G. The superego, the ego, and the id in statistical reasoning. In: A Handbook for Data Analysis in the Behavioral Sciences: Methodological Issues. Hillsdale, NJ: L. Erlbaum Associates; 1993. pp. 311-339

[34] Sedlmeier P, Gigerenzer G. Teaching Bayesian reasoning in less than two hours. Journal of Experimental Psychology General. 2001;**130**(3):380-400

[35] Nickerson RS. Null hypothesis significance testing: A review of an old and continuing controversy. Psychological Methods. 2000;**5**(2):241-301

[36] Badenes-Ribera L, Frias-Navarro D, Iotti B, Bonilla-Campos A, Longobardi C. Misconceptions of the p-value among Chilean and Italian Academic Psychologists. Frontiers in Psychology. 2016;**7**:1247

[37] Kline RB. Beyond Significance Testing, Statistics Reform in the Behavioral Sciences. 2nd ed. Washington, DC: American Psychological Association; 2013

[38] Cohen J. The earth is round ($p < .05$). American Psychologist. 1994;**49**:997-1003

[39] Carver R. The case against statistical significance testing. Harvard Educational Review. 1978;**48**(3):378-399

[40] Rozeboom WW. The fallacy of the null-hypothesis significance test. Psychological Bulletin. 1960;**57**(5):416

[41] Wetzels R, Raaijmakers JG, Jakab E, Wagenmakers E-J. How to quantify support for and against the null hypothesis: A flexible WinBUGS implementation of a default Bayesian t test. Psychonomic Bulletin & Review. 2009;**16**(4):752-760

[42] Cohen J. The statistical power of abnormal-social psychological research: A review. The Journal of Abnormal and Social Psychology. 1962;**65**(3):145

[43] Shaver JP. What statistical significance testing is, and what it is not. The Journal of Experimental Education. 1993;**61**(4):293-316

[44] Carver RP. The case against statistical significance testing, revisited. The Journal of Experimental Education. 1993;**61**(4):287-292

[45] Krueger J. Null hypothesis significance testing: On the survival of a flawed method. American Psychologist. 2001;**56**(1):16

[46] Meehl PE. Theoretical risks and tabular asterisks: Sir Karl, Sir Ronald, and the slow progress of soft psychology. Journal of Consulting and Clinical Psychology. 1978;**46**(4):806-834

[47] Wetzels R, Matzke D, Lee MD, Rouder JN, Iverson GJ, Wagenmakers E-J. Statistical evidence in experimental psychology an empirical comparison using 855 t tests. Perspectives on Psychological Science. 2011;**6**(3):291-298

[48] Wagenmakers EJ, Wetzels R, Borsboom D, van der Maas H. Yes, Psychologists Must Change the Way They Analyse Their Data: Clarifications for Bem, Utts, and Johnson (2011). 2011. Available from: http://web.stanford.edu/class/psych201s/psych201s/papers/ClarificationsForBemUttsJohnson.pdf [Accessed: July 26, 2017]

[49] Bem DJ. Feeling the future: Experimental evidence for anomalous retroactive influences on cognition and affect. Journal of Personality and Social Psychology. 2011;**100**(3):407-425

[50] Bem DJ, Utts J, Johnson WO. Must psychologists change the way they analyze their data? Journal of Personality and Social Psychology. 2011;**101**(4):716-719

[51] Bernardo JM. A Bayesian analysis of classical hypothesis testing. Trabajos de estadística y de investigación operativa. 1980;**31**(1):605-647

[52] Lindley DV. The philosophy of statistics. Journal of the Royal Statistical Society: Series D (The Statistician). 2000;**49**:293-337

[53] Jeffreys H. Theory of Probability. Oxford: Clarendon Press; 1961

[54] O'Hagan A, Forster JJ. Kendall's Advanced Theory of Statistics. Vol. 2B. Bayesian Inference. London: Arnold; 2004

[55] Gamerman D, Lopes HF. Markov Chain Monte Carlo: Stochastic Simulation for Bayesian Inference. Boca Raton: CRC Press; 2006

[56] Gilks WR, Richardson S, Spiegelhalter DJ. Introducing Markov Chain Monte Carlo, Markov Chain Monte Carlo in Practice. London: Chapman & Hall; 1996

[57] Lunn D, Spiegelhalter D, Thomas A, Best N. The BUGS project: Evolution, critique and future directions. Statistics in Medicine. 2009;**28**(25):3049-3067

[58] Lunn DJ, Thomas A, Best N, Spiegelhalter D. WinBUGS-a Bayesian modelling framework: Concepts, structure, and extensibility. Statistics and Computing. 2000;**10**(4):325-337

[59] Kruschke JK. Introduction: Credibility, Models, and Parameters, Doing Bayesian Data Analysis: A Tutorial with R, JAGS, and Stan. Boston: Academic Press; 2015. pp. 15-30

[60] Plummer M. JAGS: A program for analysis of Bayesian graphical models using Gibbs sampling. In: Proceedings of the 3rd International Workshop on Distributed Statistical Computing. Vienna: TU Wien; 2003. p. 125

[61] Love J, Selker R, Marsman M, Jamil T, Dropmann D, Verhagen A, Wagenmakers E. JASP (Version 0.7) [Computer Software]. Amsterdam, the Netherlands: JASP Project; 2015

[62] Griffiths TL, Tenenbaum JB, Kemp C. Bayesian inference. In: Holyoak K, Morrison R, editors. The Oxford Handbook of Thinking and Reasoning. New York: Oxford University Press; 2012. pp. 22-35

[63] Samaniego F. A Comparison of the Bayesian and Frequentist Approaches to Estimation. New York: Springer; 2010

[64] Bayes T. An essay toward solving a problem in the doctrine of chances. Philosophical Transactions of the Royal Society of London. 1763;**53**:370-418

[65] Lee MD, Wagenmakers E-J. Bayesian Cognitive Modeling: A Practical Course. Cambridge, New York: Cambridge University Press; 2014

[66] Berger JO, Moreno E, Pericchi LR, Bayarri MJ, Bernardo JM, Cano JA, De la Horra J, Martín J, Ríos-Insúa D, Betrò B. An overview of robust Bayesian analysis. Test. 1994;**3**(1): 5-124

[67] Jackman S. Bayesian Analysis for the Social Sciences. West Sussex: Wiley Chichester; 2009

[68] Myung IJ, Pitt MA. Applying Occam's razor in modeling cognition: A Bayesian approach. Psychonomic Bulletin & Review. 1997;**4**(1):79-95

[69] Bigler ED. Symptom validity testing, effort, and neuropsychological assessment. Journal of the International Neuropsychological Society. 2012;**18**(04):632-640

[70] Lakens D. Equivalence tests: A practical primer for t-tests, correlations, and meta-analyses. Social Psychological and Personality Science. 2017;March 4:1-21

[71] Barbey AK, Sloman SA. Base-rate respect: From ecological rationality to dual processes. Behavioral and Brain Sciences. 2007;**30**(03):241-254

[72] Gigerenzer G, Hoffrage U, Mellers BA, et al. How to improve Bayesian reasoning without instruction: Frequency formats. Psychological Review. 1995;**102**:684-704

[73] Fiedler K, Brinkmann B, Betsch T, Wild B. A sampling approach to biases in conditional probability judgments: Beyond base rate neglect and statistical format. Journal of Experimental Psychology: General. 2000;**129**(3):399-418

[74] Lesage E, Navarrete G, De Neys W. Evolutionary modules and Bayesian facilitation: The role of general cognitive resources. Thinking & Reasoning. 2013;**19**(1):27-53

[75] Ayal S, Beyth-Marom R. The effects of mental steps and compatibility on Bayesian reasoning. Judgment and Decision Making. 2014;**9**(3):226-242

[76] Lindley DV. Bayesian statistics: A review. Society for Industrial and Applied Mathematics; 1972

[77] Gliner JA, Leech NL, Morgan GA. Problems with null hypothesis significance testing (NHST): What do the textbooks say? The Journal of Experimental Education. 2002;**71**(1):83-92

[78] Morey RD, Rouder JN. Bayes Factor: Computation of Bayes Factors for Common Designs. R package version 0.9.12-2. 2015. Available from: https://cran.r-project.org/package=BayesFactor [Accessed: June 21, 2017]

[79] Berry DA. Bayesian clinical trials. Nature Reviews Drug Discovery. 2006;**5**(1):27-36

[80] Briggs AH. A Bayesian approach to stochastic cost-effectiveness analysis. Health Economics. 1999;**8**(3):257-261

[81] Ortega A, Wagenmakers E-J, Lee MD, Markowitsch HJ, Piefke M. A Bayesian latent group analysis for detecting poor effort in the assessment of malingering. Archives of Clinical Neuropsychology. 2012;**27**(4):453-465

[82] Stegmueller D. How many countries for multilevel modeling? A comparison of Frequentist and Bayesian approaches. American Journal of Political Science. 2013;**57**(3):748-761

[83] Morey RD. On verbal categories for the interpretation of Bayes factors. 2015. Available from: http://bayesfactor.blogspot.cl/2015/01/on-verbal-categories-for-interpretation.html [Accessed: June 21, 2017]

Section 3

Applications of Bayesian Inference in Engineering

Chapter 13

Bayesian Inference and Compressed Sensing

Solomon A. Tesfamicael and Faraz Barzideh

Additional information is available at the end of the chapter

http://dx.doi.org/10.5772/intechopen.70308

Abstract

This chapter provides the use of Bayesian inference in compressive sensing (CS), a method in signal processing. Among the recovery methods used in CS literature, the convex relaxation methods are reformulated again using the Bayesian framework and this method is applied in different CS applications such as magnetic resonance imaging (MRI), remote sensing, and wireless communication systems, specifically on multiple-input multiple-output (MIMO) systems. The robustness of Bayesian method in incorporating prior information like sparse and structure among the sparse entries is shown in this chapter.

Keywords: Bayesian inference, compressive sensing, sparse priors, clustered priors, convex relaxation

1. Introduction

In order to estimate parameters in a signal, one can apply wisdoms of the two schools of thoughts in statistics called the classical (also called the frequentist) and the Bayesian. These methods of computing are competitive with each other at times. The definition of probability is where the basic difference arises from. The frequentist define $P(A)$ as a long-run relative frequency with which A occurs in identical repeats of an experiment, whereas Bayesian defines $P(A|B)$ as a real number measure of the probability of a proposition A, given the truth of the information represented by proposition B. So under Bayesian theory, probability is considered as an extension of logic [1, 2]. Probabilities represent the investigator degree of belief—hence it is subjective. But this is not acceptable under classical theory, making it to be not flexible. To add on the differences, under the classical inference, parameters are not random, they are fixed and prior information is absent. But under the Bayesian, parameters are random variables, and prior information is an integral part, and the Bayesian has no excuse for that. Since one is free to invent new estimators or confidence intervals or hypothesis test, adhockery exists and hence frequentist approach lack consistency whereas Bayesian theory is flexible and consistent [1–9]. Therefore, Bayesian inference is our main focus applied to a special paradigm in signal processing in this chapter.

After presenting the theoretical frameworks, Bayesian theory, CS, and convex relaxation methods in Section 2, the use of Bayesian inference in CS problem by considering two priors modeling the sparsity and clusteredness is shown in Section 3. In Section 4, we present three examples of applications that show the connection of the two theories, Bayesian and compressive sensing. In Section 5, the conclusion is given briefly.

2. Theoretical framework

2.1. Bayesian framework

For two random variables A and B, the product rule gives

$$P(A, B) = P(A|B)P(B) = P(B|A)P(A) \tag{1}$$

and the famous Bayes' theorem provides

$$P(B|A) = \frac{P(A|B)P(B)}{P(A)}. \tag{2}$$

Using the same framework, consider model M_j and a vector of parameter $\boldsymbol{\theta}$. We infer what the model's parameter $\boldsymbol{\theta}$ might be, given the data, D, and a prior information I. Using Bayes' theorem, the probability of the parameters θ given model M_j, data D, and information I is given by

$$P(\boldsymbol{\theta}|D, M_j, I) = \frac{P(D|\boldsymbol{\theta}, M_j, I)P(\boldsymbol{\theta}|M_j, I)}{P(D|M_j, I)}, \tag{3}$$

where $P(\boldsymbol{\theta}|D, M_j, I)$, is posterior probability, $P(\boldsymbol{\theta}|M_j, I)$ is the non data information about $\boldsymbol{\theta}$, called prior probability distribution function, while $P(D|\boldsymbol{\theta}, M_j, I)$ is the density of the data conditional on the parameters of the model, called likelihood. $P(D|M_j, I)$ is called the evidence of model M_j, or the normalizing constant, given by:

$$P(D|M_j, I) = \int_{\boldsymbol{\theta}} P(\boldsymbol{\theta}|M_j, I)P(D|\boldsymbol{\theta}, M_j, I)d\boldsymbol{\theta}. \tag{4}$$

$P(\boldsymbol{\theta}|D, M_j, I)$ is the fundamental interest for the first level of inference called model fitting. It is the task of inferring what the model parameters might be given the model and the data. Further, we can do inference on higher level, which is comparing models M_j. In the light of prior information I and data D, a given set of models $\{M_1, M_2, \cdots, M_n\}$ is most likely to be the correct one. Now focusing on the first level of inference, we can ignore the normalizing constant in (3) since it has no relevance at this level of inference about the parameters θ. Hence we get:

$$P(\boldsymbol{\theta}|D, M_j, I) \propto P(D|\boldsymbol{\theta}, M_j, I)P(\theta|M_j, I). \tag{5}$$

The *posterior* probability is proportional to the *Prior* probability times the *Likelihood*. Eq. (5) is called Updating Rule [1, 3], in which the data allow us to update our prior views about $\boldsymbol{\theta}$, and as a result, we get the posterior which combines both the data and non-data information of $\boldsymbol{\theta}$. As an example for a binomial trial, let us have beta distribution as a prior and as a result we get posterior distribution which is beta distribution. **Figure 1** shows that the posterior density is taller and narrower than the prior density. It therefore favors strongly a smaller range of $\boldsymbol{\theta}$ values, reflecting the fact that we now have more information. That is why inference based on the posterior distribution is superior to the one only based on the likelihood.

Now, we first find the maximum of the posterior distribution called maximum a posteriori (MAP). It defines the most probable value for the parameters denoted $\widehat{\boldsymbol{\theta}}_{MP}$. MAP is related to Fisher's methods of maximum likelihood estimation (MLE), $\widehat{\boldsymbol{\theta}}_{ML}$. If f is the sampling distribution of D, then the likelihood function of $D:\boldsymbol{\theta} \mapsto f(D|\boldsymbol{\theta})$ and the maximum likelihood estimation of θ:

$$\boldsymbol{\theta}_{ML}(D) = \arg\max_{\boldsymbol{\theta}} f(D|\boldsymbol{\theta}). \tag{6}$$

But under Bayesian inference, let g be a prior distribution of $\boldsymbol{\theta}$, then the posterior distribution of $\boldsymbol{\theta}$ becomes

$$\boldsymbol{\theta}_{ML} \mapsto \frac{f(D|\boldsymbol{\theta})g(\boldsymbol{\theta})}{f(D)} \tag{7}$$

and the maximum a posteriori estimation of θ:

$$\begin{aligned}\widehat{\boldsymbol{\theta}}_{MP} &= \frac{f(D|\boldsymbol{\theta})g(\boldsymbol{\theta})}{\int_{\boldsymbol{\vartheta}} f(D|\boldsymbol{\vartheta})g(\boldsymbol{\vartheta})d\boldsymbol{\vartheta}} \\ &= \arg\max_{\boldsymbol{\theta}} f(D|\boldsymbol{\theta})g(\boldsymbol{\theta})\end{aligned} \tag{8}$$

Inference based on the posterior is not an easy task since it involves multiple integral, which are cumbersome to solve at times. However, it can be computed in several ways: Numerical optimization (like Conjugate gradient method, Newton method,...), modification of an expectation-maximization algorithm and others. As we can see it from (22) and (8), the difference between MLE and MAP is the prior distribution. The latter can be considered as a

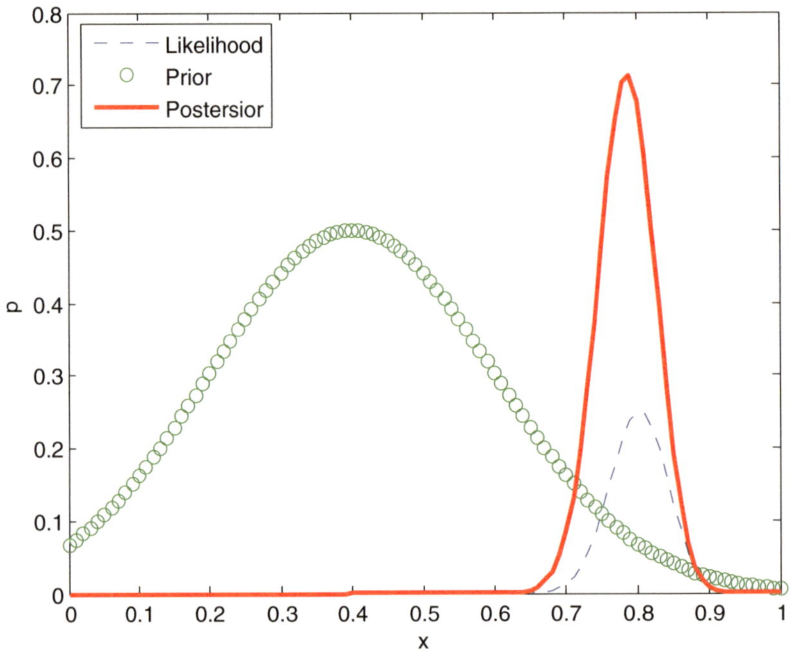

Figure 1. Figure showing the updating rule: the posterior synthesizes and compromises by favoring values between the maximum of the prior density and likelihood. The prior we had is challenged to shift by the arrival of little amount of data.

regularization of the former. Here we can summarize the posterior distribution by the value of the best fit parameters θ_{MP} and error bars (confidence intervals) on the best fit parameters. Error bars can be found from the curvatures of the posterior. To proceed further, we replace the random variable D and θ by vectors $\mathbf{y}$ and $\mathbf{x}$ and we assume prior distributions on $\mathbf{x}$ in the next section.

2.2. Compressive sensing

Compressive sensing (CS) is a paradigm to capture information at lower rate than the Nyquist-Shannon sampling rate when signals are sparse in some domain [10–13]. CS has recently gained a lot of attention due to its exploitation of signal sparsity. Sparsity, an inherent characteristic of many natural signals, enables the signal to be stored in a few samples and subsequently be recovered accurately.

As a signal processing scheme, CS follows a similar fashion: encoding, transmission/storing, and decoding. Focusing on the encoding and decoding of such a system with noisy measurement, the block diagram is given in **Figure 2**. At the encoding side, CS combines the sampling and compression stages of a traditional signal processing into one step by measuring few samples that contain maximum information about the signal. This measurement/sampling is done by linear projections using random sensing transformations as shown in the landmark papers by the authors mentioned above. Having said this, let us define the CS problem formally as follows:

http://dx.doi.org/10.5772/intechopen.70308

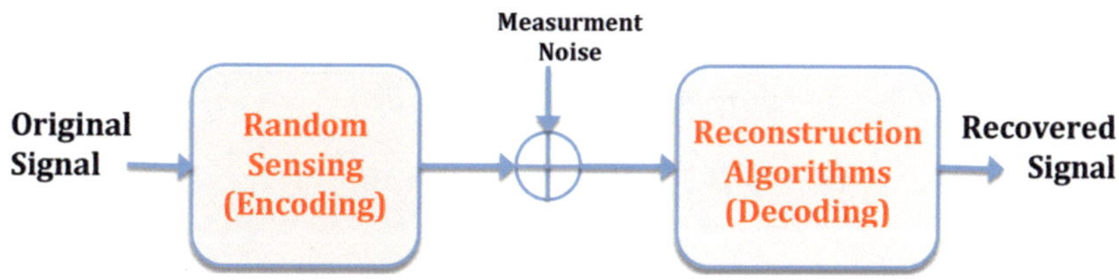

Figure 2. Blockdiagram for CS-based reconstruction.

Definition 1. (*The standard CS problem*)

Find the k-sparse signal vector $\mathbf{x} \in \mathbb{R}^N$ *provided the measurement vector* $\mathbf{y} \in \mathbb{R}^M$*, the measurement matrix* $\mathbf{A} \in \mathbb{R}^{M \times N}$ *and the under-determined set of linear equations as*

$$\mathbf{y} = \mathbf{A}\mathbf{x}, \tag{9}$$

where $k \ll M \ll N$.

One can ask again two of the questions here, in relation to the standard CS problem. First, how should we design the matrix **A** to ensure that it preserves the information in the signal **x**? Second, how can we recover the original signal **x** from measurements **y** [14]? To address the first question, the solution for the CS problem presented here is dependent on the design of **A**. This matrix can be considered as a transformation of the signal from the signal space to the measurement space, **Figure 3** [15]. There have been different criteria that matrix **A** should satisfy to have meaningful reconstruction. One of the main criteria is given in [11]. The authors defined the sufficient condition that matrix **A** should satisfy for the reconstruction of the signal **x**. It is called the Restricted Isometric Property (RIP) and it is defined below.

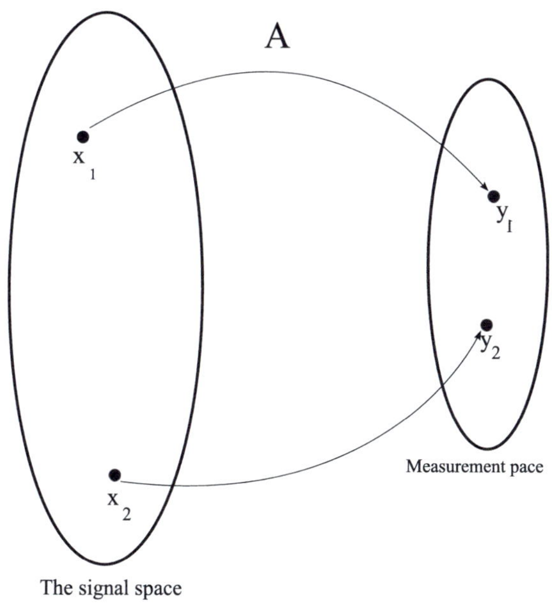

Figure 3. Transformation from the signal-space to the measurement-space.

Definition 2. (*Restricted Isometry Property*)

For all $\mathbf{x} \in \mathbb{R}^N$ *so that* $\| \mathbf{x} \|_0 \leq k$, *if there exists* $0 \leq \delta_k < 1$ *such that*

$$(1-\delta_k)\|\mathbf{x}\|_2^2 \leq \|\mathbf{A}\mathbf{x}\|_2^2 \leq (1+\delta_k)\|\mathbf{x}\|_2^2 \tag{10}$$

is satisfied, then $\mathbf{A}$ *fulfills RIP of order k with radius* δ_k.

An equivalent description of the RIP is to say that all subsets of k columns taken from $\mathbf{A}$ are nearly orthogonal (the columns of $\mathbf{A}$ cannot be exactly orthogonal since we have more columns than rows) [16]. For example, if a matrix $\mathbf{A}$ satisfies the RIP of order $2k$, then we can interpret (10) saying that $\mathbf{A}$ approximately preserves the distance between any pair of k-sparse vectors. For random matrix $\mathbf{A}$, the following theorem is one of the results in relation to RIP for the noiseless CS problem, provided that the entries of the random matrix $\mathbf{A}$ are drawn from some distributions which are given later.

Theorem 1. (*Perfect Recovery Condition, Candes and Tao* [13])

If $\mathbf{A}$ *satisfies the RIP of order 2k with radius* δ_{2k}, *then for any k-sparse signal* $\mathbf{x}$ *sensed by* $\mathbf{y}=\mathbf{A}\mathbf{x}$, $\mathbf{x}$ *is with high probability perfectly recovered by the ideal program*

$$\begin{aligned} \widehat{\mathbf{x}} &= arg \min_{\mathbf{x}} \quad \|\mathbf{x}\|_0 \\ &subject\ to \quad \mathbf{y} = \mathbf{A}\mathbf{x} \end{aligned} \tag{11}$$

and it is unique, where $\| \mathbf{x} \|_0 = k \equiv \# \{i \in \{1, 2, \cdots, N\} | x_i \neq 0\}$.

This means, if $\mathbf{A}$ satisfies the RIP of order k with radius δ_k, then for any $k' < k$, $\mathbf{A}$ satisfies the RIP of order k' with constant $\delta_{k'} < \delta_k$ [?]. Note that, this theorem is stated for the noiseless CS problem and it is possible to extend it for the noisy CS system. The proof of these theorems is deferred to the literature mentioned, [13], in for the sake of space.

Under conventional sensing paradigm, the dimension of the original signal and the measurement should be at least equal. But in CS, the measurement vector can be far less than the original. While at the decoding side, reconstruction is done using nonlinear schemes. Eventually, the reconstruction is more cumbersome than the encoding which was only projections from a large space to a smaller space. On the other hand, finding a unique solution that satisfies the constraint that the signal itself is sparse or sparse in some domain is complex in nature. Fortunately, there are many algorithms to solve the CS problem, such as iterative methods such as *greedy iterative algorithms* [17] and *iterative thresholding algorithms* [18]. This chapter focuses merely on the *convex relaxations* methods [12, 13]. The regularizing terms in these methods can be reinterpreted as prior information under Bayesian inference. We consider a noisy measurement and apply convex relaxation algorithms for robust reconstruction.

2.3. Convex relaxation methods for CS

Various methods for estimating $\mathbf{x}$ may be used. We have the least square (LS) estimator in which no prior information is applied:

http://dx.doi.org/10.5772/intechopen.70308

$$\widehat{\mathbf{x}} = \left(\mathbf{A}^T\mathbf{A}\right)^{-1}\mathbf{A}^T\mathbf{y}, \tag{12}$$

which performs very badly for the CS estimation problem we are considering. In order to incorporate the methods called convex relaxation, let us define an important concept first.

Definition 3. (*Unit Ball*)

A unit ball in l_p-space of dimension N can be defined as

$$\mathcal{B}_p \equiv \left\{\mathbf{x} \in \mathbb{R}^N : \quad \|\mathbf{x}\|_p \le 1\right\}. \tag{13}$$

Unit balls corresponding to $p = 0$, $p = 1/2$, $p = 1$, $p = 2$, $p = \infty$, and $N = 2$, the balls are shown in **Figure 4**.

The exact solution for the noiseless CS problem is given by

$$\min_{\mathbf{x}} \|\mathbf{x}\|_0, \quad \text{such that } \mathbf{y} = \mathbf{A}\mathbf{x}. \tag{14}$$

However, minimizing l_0-norm is a non-convex optimization problem which is NP-hard [19]. By relaxing the objective function to convexity, it is possible to get good approximation. That is, replacing the l_0-norm by the l_1-norm, one can find a problem which is tractable. Note that it is also possible to use other l_p-norms to relax the condition given by l_0. However, keeping our focus on l_1-norm, consider the minimization problem instead of (14).

$$\min_{\mathbf{x}} \quad \|\mathbf{x}\|_1, \text{ such that } \mathbf{y} = \mathbf{A}\mathbf{x} \tag{15}$$

The solution of the relaxed problem (15) gives the same as that of (14) and this equivalence was provided by Donoho and Huo in [20].

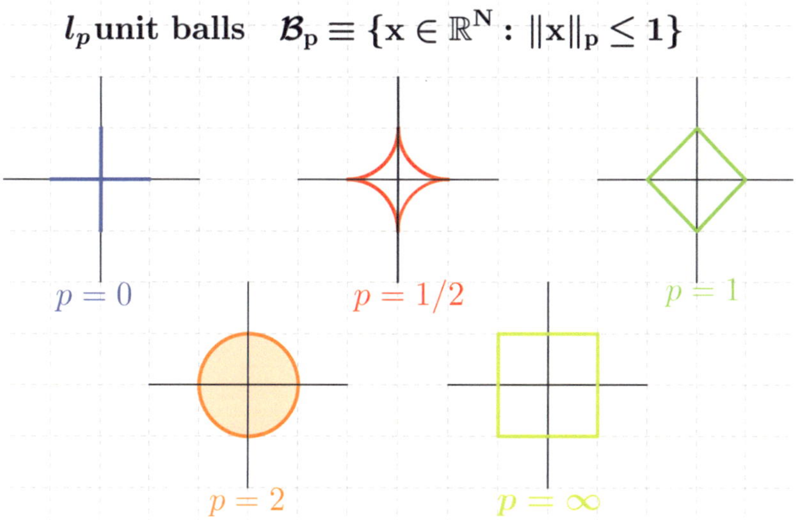

Figure 4. Different l_p-balls in different l_p-spaces for N=2, only balls with p≥1 are convex.

Theorem 2. (l_0–l_1 *Equivalence* [13])

If **A** *satisfies the RIP of order 2k with radius* $\delta_{2k} < \sqrt{2} - 1$, *then*

$$\begin{aligned} \widehat{\mathbf{x}} = arg \ \underset{x}{min} \ \|\mathbf{x}\|_1 \\ subject \ \ to \ \ \mathbf{y} = \mathbf{A}\mathbf{x} \end{aligned} \tag{16}$$

is equivalent to (11) and will find the same unique $\widehat{\mathbf{x}}$.

Justified by this theorem, (15) is an optimization problem which can be solved in polynomial time and the fact that it gives the exact solution for the problem (14) under some circumstance has been one of the main reasons for the recent developments in CS. There is a simple geometric intuition on why such an approach gives good approximations. Among the l_p-norms that can be used in the construction of CS related optimization problems, only those which are convex give rise to a convex optimization problem which is more feasible than the non-convex counter parts, which means l_p-norms with only $p \geq 1$ satisfy such a condition. On the other hand, l_p-norms with $p > 1$ do not favor sparsity, for example, l_2-norm minimization tends to spread reconstruction across all coordinates even if the true solution is sparse. But l_1-norm is able to enforce sparsity. The intuition is that l_1-minimization solution is most likely to occur at corners or edges, not faces [21, 45]. That is why l_1-norm became famous for CS. Further, in CS literature, convex relaxation is presented as either l_2-penalized l_1-minimization called Basis Pursuit Denoising (BPDN) [22] or l_1-penalized l_2-minimization called least absolute shrinkage and selection operator (LASSO) [45], which are equivalent and effective in estimating a high-dimensional data.

Usually real world systems are contaminated with noise, **w**, and in this chapter, the focus is on such problems. The noisy recovery problem becomes a simple extension of (15),

$$\min_{\mathbf{x}} \|\mathbf{x}\|_1, \text{ such that } \|\mathbf{y} - \mathbf{A}\mathbf{x}\|_2 \leq \epsilon \tag{17}$$

where ϵ is a bound on $\|\mathbf{w}\|_2$. The real problem for (17) is stability. Introducing small changes in the observations should result in small changes in the recovery. We can visualize this using the balls shown in **Figure 5**.

Both the l_0 and l_1-norms give exact solutions for the noise-free CS problem while giving a close solution for the noisy problem. However, the l_2-norm gives worst approximation in both cases compared to the other l_p-norms with $p < 2$ (see **Figure 5**). Moreover, (17) is equivalent to an unconstrained quadratic programming problem as

$$\min_{x} \frac{1}{2}\|\mathbf{y} - \mathbf{A}\mathbf{x}\|_2^2 + \gamma\|\mathbf{x}\|_1, \tag{18}$$

as it will be shown later as LASSO, where γ is a tuning parameter. The equivalency of (17) and (18) is shown in [23, 24]. In this chapter, the generalized form of the minimization problem in (18) with different l_p-norm regularization is considered, that is,

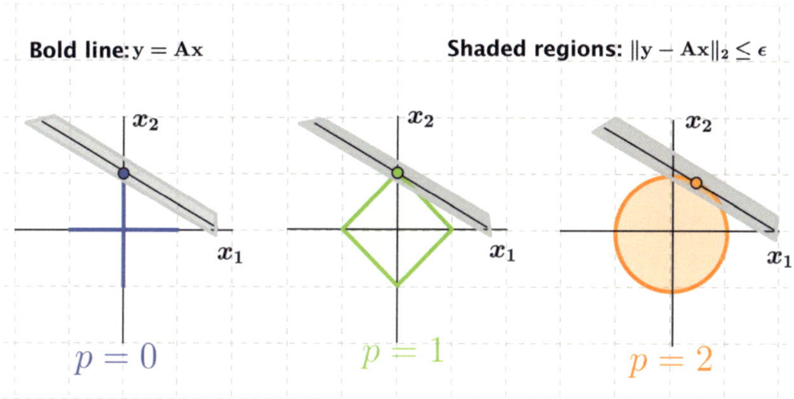

Figure 5. l_p-norm approximations: the constraints for the noise-free CS problem is given by the bold line while the shaded region is for the noisy one.

$$\min_{x} \frac{1}{2}\|\mathbf{y} - \mathbf{A}\mathbf{x}\|_2^2 + \gamma\|\mathbf{x}\|_p. \tag{19}$$

Further, this chapter provides the use of Bayesian framework in compressive sensing by incorporating two different priors modeling the sparsity and the possible structure among the sparse entries in a signal. Basically, it is the summary of the recent works [2, 25–27].

3. Bayesian inference used in CS problem

Under Bayesian inference, consider two random variables $\mathbf{x}$ and $\mathbf{y}$ with probability density function (pdf) $p(\mathbf{x})$ and $p(\mathbf{y})$, respectively. The product rule gives us $p(\mathbf{x},\mathbf{y}) = p(\mathbf{x}|\mathbf{y})p(\mathbf{y}) = p(\mathbf{y}|\mathbf{x})p(\mathbf{x})$ and Bayes' theorem provides

$$p(\mathbf{x}|\mathbf{y}) = \frac{p(\mathbf{y}|\mathbf{x})p(\mathbf{x})}{p(\mathbf{y})}. \tag{20}$$

Further, the maximum a posteriori (MAP), $\widehat{\mathbf{x}}_{MP}$, is defined as

$$\begin{aligned}\widehat{\mathbf{x}}_{MP} &= \arg\max_{\mathbf{x}} \frac{p(\mathbf{y}|\mathbf{x})p(\mathbf{x})}{\int_{\tilde{\mathbf{x}}} p(\mathbf{y}|\tilde{\mathbf{x}})p(\tilde{\mathbf{x}})d\tilde{\mathbf{x}}} \\ &= \arg\max_{\mathbf{x}} p(\mathbf{y}|\mathbf{x})p(\mathbf{x})\end{aligned} \tag{21}$$

MAP is related to Fisher's methods of maximum likelihood estimation (MLE), $\widehat{\mathbf{x}}_{ML}$:

$$\widehat{\mathbf{x}}_{ML} = \arg\max_{\mathbf{x}} p(\mathbf{y}|\mathbf{x}). \tag{22}$$

As we can see it from (21) and (22), the difference between MAP and MLE is the prior distribution. The former can be considered as a regularized form of the latter. Since we apply Bayesian inference we assume further different prior distributions on **x**.

3.1. Sparse prior

The estimators of **x** resulting from (19) for the sparse problem we consider in this chapter, can be presented as a maximum a posteriori (MAP) estimator under the Bayesian framework as in [28]. We show this by defining a prior probability distribution for **x** on the form

$$p(\mathbf{x}) = \frac{e^{-uf(\mathbf{x})}}{\int_{\mathbf{x}\in\mathbb{R}^N} e^{-uf(\mathbf{x})}d\mathbf{x}} \tag{23}$$

where the regularizing function f: $\chi \rightarrow R$ is some scalar-valued, non negative function with $\chi \subseteq \mathbb{R}$ which can be expanded to a vector argument by

$$f(\mathbf{x}) = \sum_{i=1}^{N} f(x_i), \tag{24}$$

such that for sufficiently large u,

$$\int_{\mathbf{x}\in\mathbb{R}^N} \exp(-uf(\mathbf{x}))d\mathbf{x}$$

is finite. Furthermore, let the assumed variance of the noise be given by $\sigma^2 = \frac{\lambda}{u}$, where λ is the system parameter which can be taken as $\lambda = \sigma^2 u$.

Since the pdf of the noise **w** is gaussian, the likelihood function of **y** given **x** is given by

$$p_{\mathbf{y}|\mathbf{x}}(\mathbf{y}|\mathbf{x}) = \frac{1}{(2\pi\sigma)^{N/2}} e^{-\frac{1}{2\sigma^2}\|\mathbf{y}-A\mathbf{x}\|_2^2}. \tag{25}$$

Together with (20) and (23), this now gives

$$p_{\mathbf{x}|\mathbf{y}}(\mathbf{x}|\mathbf{y};\mathbf{A}) = \frac{e^{-u\left(\frac{1}{2}\|\mathbf{y}-\mathbf{A}\mathbf{x}\|_2^2+\lambda f(\mathbf{x})\right)}}{(2\pi\sigma)^{N/2}\int_{\mathbf{x}\in\mathbb{R}^N} e^{-u\left(\frac{1}{2\lambda}\|\mathbf{y}-\mathbf{A}\mathbf{x}\|_2^2+\lambda f(\mathbf{x})\right)}d\mathbf{x}}.$$

The MAP estimator, (21), becomes

$$\widehat{\mathbf{x}}_{MP} = \arg\min_{\mathbf{x}\in\mathbb{R}^N} \frac{1}{2}\|\mathbf{y}-\mathbf{A}\mathbf{x}\|_2^2 + \lambda f(\mathbf{x}). \tag{26}$$

Now, as we choose different regularizing function, we get different estimators as listed below [28]:

http://dx.doi.org/10.5772/intechopen.70308

1. Linear estimators: when $f(\mathbf{x}) = \|\mathbf{x}\|_2^2$ (26) reduces to

$$\widehat{\mathbf{x}}_{\text{Linear}} = \mathbf{A}^T\left(\mathbf{A}\mathbf{A}^T + \lambda\mathbf{I}\right)^{-1}\mathbf{y}, \tag{27}$$

   which is the LMMSE estimator. But we ignore this estimator in our analysis since the results are not sparse. However, the following two estimators are more interesting for CS problems since they enforce sparsity into the vector **x**.

2. LASSO estimator: when $f(\mathbf{x}) = \|\mathbf{x}\|_1$ we get the LASSO estimator and (26) becomes,

$$\hat{x}\text{LASSO} = \arg\min_{x\in\mathbb{R}^N} \frac{1}{2}\|\mathbf{y} - \mathbf{A}\mathbf{x}\|_2^2 + \lambda\|\mathbf{x}\|_1. \tag{28}$$

3. Zero-norm regularization estimator: when $f\mathbf{x} = \|\mathbf{x}\|_0$, we get the zero-norm regularization estimator and (26) becomes

$$\hat{x}\text{Zero-Norm} = \arg\min_{x\in\mathbb{R}^N} \frac{1}{2}\|\mathbf{y} - \mathbf{A}\mathbf{x}\|_2^2 + \lambda\|\mathbf{x}\|_0. \tag{29}$$

As mentioned earlier, (29) is the best solution for estimation of the sparse vector **x**, but is NP-complete. The worst approximation for the sparse problem considered is the L2-regularization solution given by (27). However, the best approximation is given by Eq. (28) and its equivalent forms. We have used some of the algorithms in literature in our simulation which are considered as equivalent to this approximations such as Bayesian compressive sensing (BCS) [29] and L1-norm regularized least-squares (L1-LS) [11–13].

3.2. Clustering prior

The entries of the sparse vector **x** may have some special structure (clusteredness) among themselves. This can be modeled by modifying the previous prior distribution.[1] We use another penalizing parameter γ to represent clusteredness in the data. For that we define the clustering using the distance between the entries of the sparse vector **x** by

$$D \equiv \sum_{i=1}^{N} |x_i - x_{i-1}|,$$

and we use a regularizing parameter γ. Hence, we define the clustering prior to be

$$q(\mathbf{x}) = \frac{e^{-\gamma D(\mathbf{x})}}{\int_{\mathbf{x}\in\mathbb{R}^N} e^{-\gamma D(\mathbf{x})} d\mathbf{x}}. \tag{30}$$

The new posterior involving this prior under the Bayesian framework is proportional to the product of the three pdfs:

---

[1]In [30] a hierarchical Bayesian generative model for sparse signals is found in which they have applied full Bayesian analysis by assuming prior distributions to each parameter appearing in the analysis. We follow a different approach.

$$p(\mathbf{x}|\mathbf{y}) \propto p(\mathbf{y}|\mathbf{x})p(\mathbf{x})q(\mathbf{x}). \tag{31}$$

By similar argument s as used in 3.1, we arrive at the clustered LASSO estimator

$$\hat{x}\text{Clu-Lasso} = \arg\min_{x \in \mathbb{R}^N} \frac{1}{2}\|\mathbf{y} - \mathbf{A}\mathbf{x}\|_2^2 + \lambda\|\mathbf{x}\|_1 + \gamma\sum_{i=1}^{N}|x_i - x_{i-1}|. \tag{32}$$

Here λ, γ are our tuning parameters for the sparsity in $\mathbf{x}$ and the way the entries are clustered, respectively.

4. Bayesian inference in CS applications

Compressed sensing paradigm has been applied to many signal processing areas [31–41]. However, at this time, building the hardware that can translate the CS theory into practical use is very limited. Nonetheless, the demand for cheaper, faster, and efficient devices will motivate the use of CS paradigm in real-time systems in the near future.

So far, in image processing, one can mention the single-pixel imaging via compressive sampling [31], in magnetic resonance imaging (MRI) for reducing scan time and improved image quality [32], in seismic images [33], and in radar systems for simplifying hardware design and to obtain high resolution [34, 35]. In communication and networks, CS theory has been studied for sparse channel estimation [36], for under water acoustic channels which are inherently sparse [37], spectrum sensing in cognitive radio networks [38], for large wireless sensor networks (WSNs) [39], as a channel coding scheme [40], localization [41] and so on. A good CS application literature review is provided in [21], which basically is the summary of the bulk of literatures given at http://dsp.rice.edu/cs.

In this chapter, there are examples of CS theory applications using Bayesian inference in imaging, like magnetic resonance imaging (MRI) and, in communication, i.e., multiple-input multiple-output (MIMO) systems, and in remote sensing. First, let us see the impact of the estimators derived above, LASSO and clustered LASSO, in MRI.

4.1. Magnetic resonance imaging (MRI)

MRI images are usually very weak due to the presence of noise and due to the weak nature of the signal itself. Compressed sensing (CS) paradigm can be applied in order to boost such signal recoveries. We applied CS paradigm via Bayesian framework, that is, incorporating the different prior information such as sparsity and the special structure that can be found in such sparse signal recovery method is applied on different MRI images.

4.1.1. Angiogram image

Angiogram images are already sparse in the pixel representation. An angiogram image taken from the University Hospital Rechts der Isar, Munich, Germany [42] is used for our analysis.

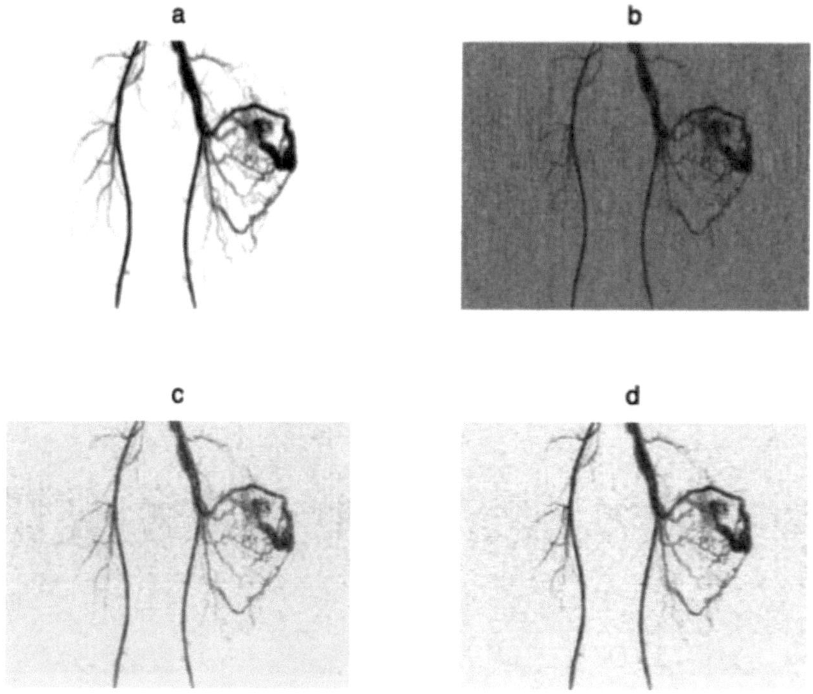

Figure 6. Comparison of reconstruction schemes together with performance comparison using mean square error (MSE) in dB: (a) original image **x**; (b) LMMSE (−35.1988 dB); (c) LASSO (−53.6195 dB); and (d) clustered Lasso (−63.6889 dB).

The image we took is sparse and clustered even in the pixel domain. The original signal after vectorization is $\mathbf{x}$ of length $N = 960$. By taking 746 measurements, and maximum number of non-zero elements $k = 373$, we applied different reconstruction schemes and the results are shown in **Figure 6**.

4.1.2. *Phantom image*

Another MRI image considered is the Shepp-Logan phantom which is not sparse in spatial domain. However, we sparsified it in K-space by zeroing out small coefficients. We then measured the sparsified image and added noise. The original signal after vectorization is $\mathbf{x}$ of length $N = 200$. By taking 94 measurements, that is, $\mathbf{y}$ is of length $M = 94$, and maximum number of non-zero elements $k = 47$, we applied different reconstruction algorithms used above. The result shows that clustered LASSO does well compared to the others as can be seen in **Figure 7**.

4.1.3. *fMRI image*

Another example to apply the clustered LASSO based image reconstruction using Bayesian framework to medical images is a functional magnetic resonance imaging (fMRI), a non-invasive technique of brain mapping, which is crucial in the study of brain activity. Taking many slices in fMRI data, we saw how these data sets are sparse in the Fourier domain. This is

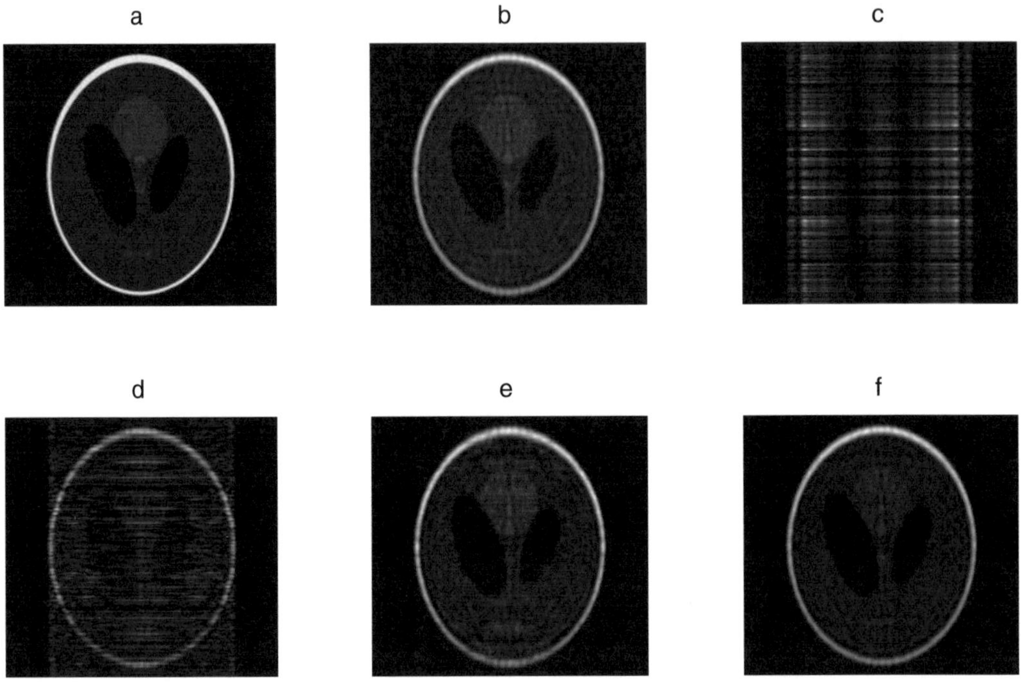

Figure 7. Comparison of reconstruction schemes together with performance comparison using mean square error (MSE) in dB: (a) original image **x**; (b) sparcified image; (c) least square (LS) (−21.3304 dB); (d) LMMSE (−27.387 dB); (e) LASSO (−37.9978 dB); and (f) clustered LASSO (−40.0068 dB).

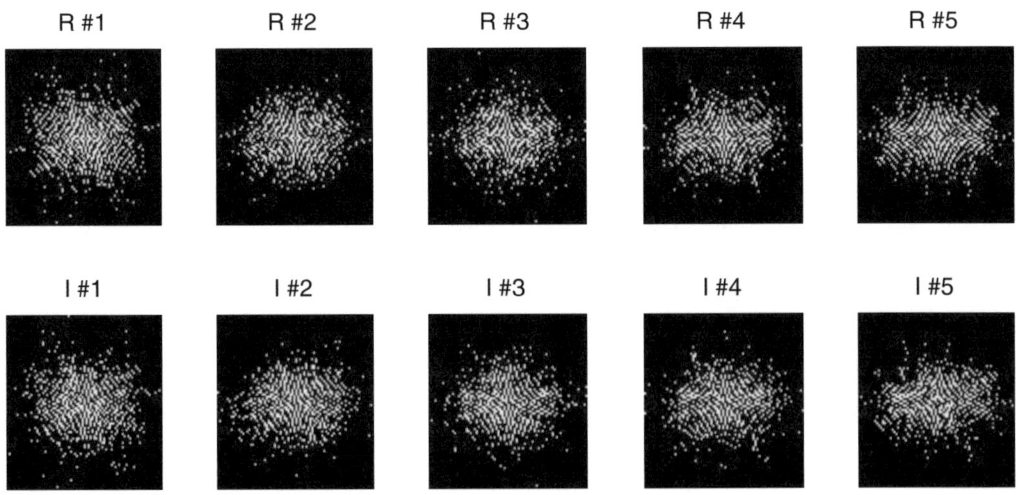

Figure 8. The five column images represent the real and imaginary parts of the Fourier transform representation of the data set we have chosen to present further, which in general shows that the fMRI image have sparse and clustered representation.

shown in **Figure 8**. We observed the whole data in this domain for the whole brain image. They all share the characteristics we have based our analysis, i.e., sparsity and clusteredness. Then we took some slices which are consecutive in the slice order and took different N, k, and M=2k, on these slices. We can see the numbers at the top of **Figure 9**, in which the two numbers represent k and N, respectively.

In fMRI, results are compared using image intensity which gives a good ground for a health practitioner to observe and decide in accordance to the available information. The more one have prior knowledge on how the brain regions work in human beings or pets the better priors that one incorporate to analyze the data. So this is an interesting tool for researchers in the future.

4.2. MIMO systems

Multiple-input multiple-output (MIMO) systems are integrated in modern wireless communications due to their advantage in improving performance with respect to many performance metrics. One such advantage is the ability to transmit multiple streams using spatial multiplexing, but channel state information (CSI) at the transmitter is needed to get optimal system performance.

Consider a frequency division duplex (FDD) MIMO system consisting of N_t transmit and N_r receive antennas. Assume that the channel is a flat-fading, temporally correlated channel denoted by a matrix $\mathbf{H}[n]\in\mathbb{C}^{N_r\times N_t}$ where n indicates a channel feedback time index with block fading assumed during the feedback interval. The singular value decomposition (SVD) of $\mathbf{H}[n]$ gives

$$\mathbf{H}[n] = \mathbf{U}[n]\Sigma[n]\mathbf{V}^H[n],$$

where $\mathbf{U}\in\mathbb{C}^{N_r\times r}$ and $\mathbf{V}\in\mathbb{C}^{N_t\times r}$ are unitary matrices and $\mathbf{\Sigma}\in\mathbb{C}^{r\times r}$ is a diagonal matrix consisting of $r = \min(N_t, N_r)$ singular values. In the presence of perfect channel state information (CSI), a MIMO system model can be given by the equation

$$\tilde{\mathbf{y}} = \mathbf{U}^H[n]\mathbf{H}[n]\mathbf{V}[n]\tilde{\mathbf{x}} + \mathbf{U}^H[n]\mathbf{n} \tag{33}$$

where $\tilde{\mathbf{x}}\in\mathbb{C}^{r\times 1}$ is transmitted vector, $\mathbf{V}[n]$ is used as precoder at the transmitter, $\mathbf{U}^H[n]$ is used as decoder at the receiver, $\mathbf{n}\in\mathbb{C}^{N_r\times 1}$ denotes a noise vector whose entries are i.i.d. and distributed according to $\mathcal{CN}(0,1)$ and $\tilde{\mathbf{y}}\in\mathbb{C}^{N_r\times 1}$ is the received vector.

Channel adaptive transmission requires knowledge of channel state information at the transmitter. In temporally correlated MIMO channels, the correlation can be utilized to reduce feedback overhead and improve performance. CS methods and rotative quantization are used to compress and feedback the CSI for MIMO systems [43]. This was done as an extension work of [44]. It is shown that the CS-based method reduces feedback overhead while delivering the same performance as the direct quantization scheme, using simulation.

Three methods are compared in the simulations, perfect CSI, without CS and with CS using matched filter (MF) and minimum mean square error estimator (MMSE) receivers for different total feedback bits $B = 10$ and $B = 5$. In **Figure 10**, sum rates are compared against signal-to-noise-ratio (SNR). Using CS, half of the number of bits can be saved. In **Figure 11**, where the bit-error-rate is plotted against SNR, the CS method has a better bit error rate performance using same number of bits for the CS and without CS cases. These two figures demonstrate the clear advantage of using CS in feedback of singular vectors in rotative based method. The detail is deferred to [43].

4.3. Remote sensing

Remote sensing satellites provide a repetitive and consistent view of the Earth and they offer a wide range of spatial, spectral, radiometric, and temporal resolutions. Image fusion is applied to extract all the important features from various input images. These images are integrated to form a fused image which is more informative and suitable for human visual perception or computer processing. Sparse representation has been applied to fuse image to improve the quality of fused image [45].

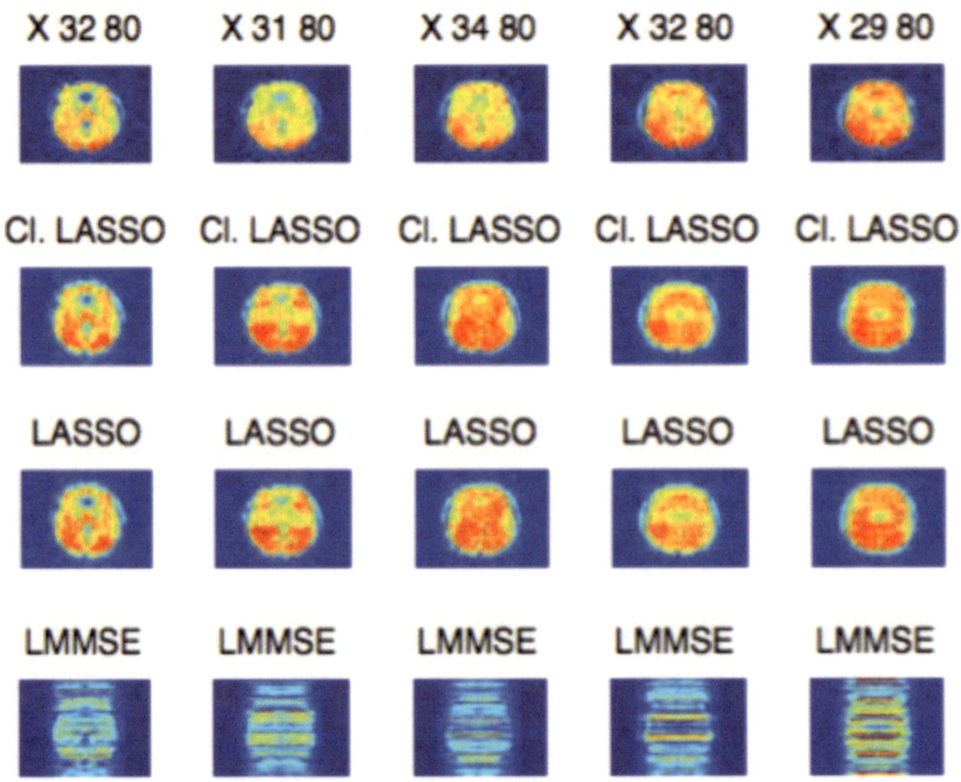

Figure 9. Application of sparse and cluster prior, LASSO and clustered LASSO (CL. LASSO), on a fMRI data analysis for $N = 80$, $k > 50$ for $\sigma^2 = 0.1$ and $\lambda = 0.1$, where LMMSE is with L2-regularised one.

http://dx.doi.org/10.5772/intechopen.70308

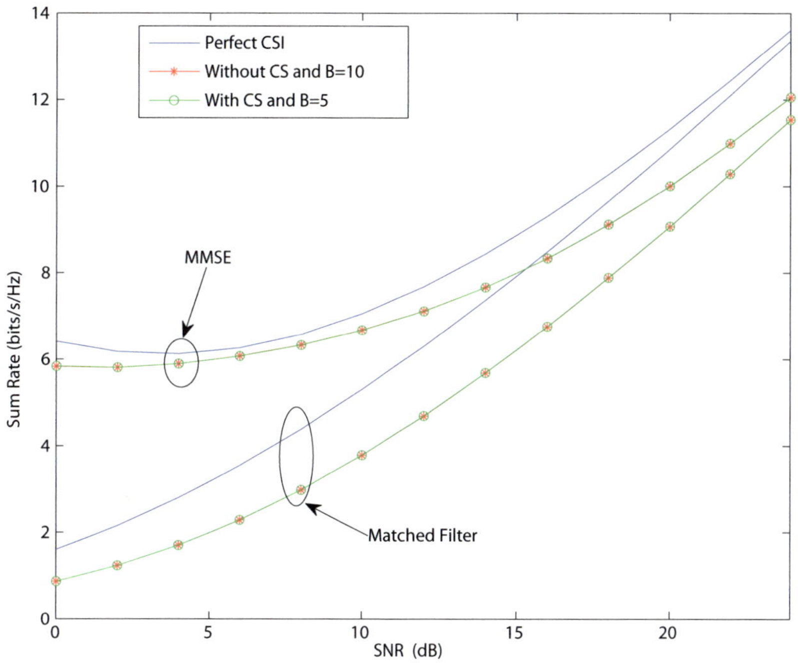

Figure 10. Sum rate vs. SNR for a 2×2 MIMO system with and without CS with two streams. We can observe that the performance of the CS method is almost equal to that of the method without using CS while saving half the number of bits.

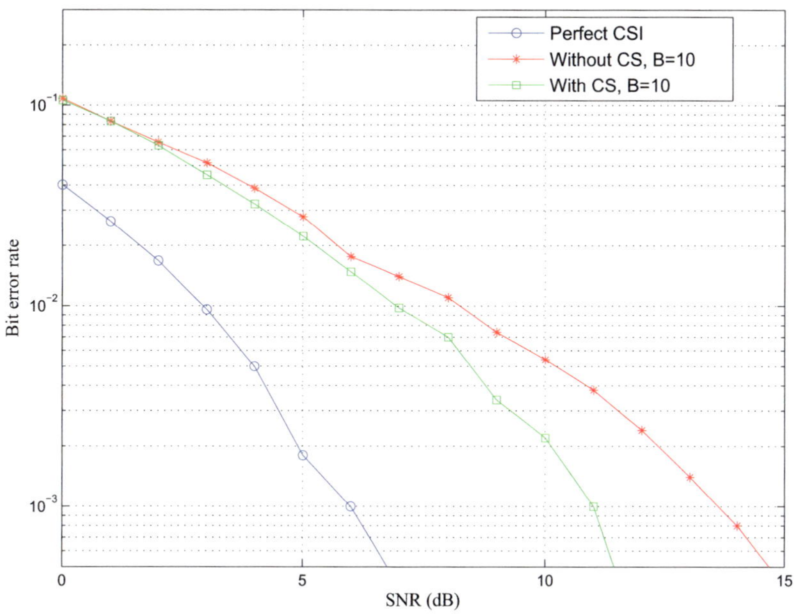

Figure 11. Bit error rate vs. SNR using matched filter receiver for a 2×2 MIMO system with one stream.

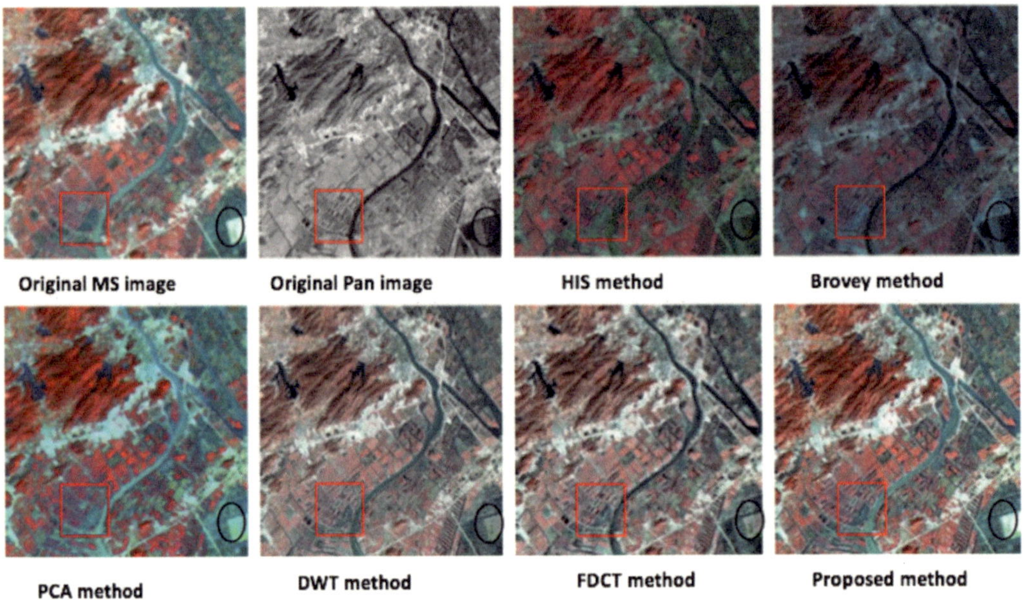

Figure 12. Comparison of image fusion methods for remote sensing applications using Brovey, DWT, PCA, FDCT, and the sparse representation methods [46].

To improve the quality of the fused image, a remote sensing image fusion method based on sparse representation is proposed in [46]. In these methods, the source images were represented with sparse coefficients first. Then, the larger values of sparse coefficients of panchromatic (Pan) image are set to 0. Thereafter, the coefficients of panchromatic (Pan) and multispectral (MS) image are combined with the linear weighted averaging fusion rule. Finally, the fused image is reconstructed from the combined sparse coefficients and the dictionary. The proposed method is compared with intensity-hue-saturation (IHS), Brovey transform (Brovey), discrete wavelet transform (DWT), principal component analysis (PCA) and fast discrete curvelet transform (FDCT) methods on several pairs of multifocus images. The proposed method using sparse representation outperforms, see **Figure 12**, better than the usual methods listed here. We believe that our method of clustered compressed sensing can also further improve this result.

5. Conclusions

In this chapter, a Bayesian way of analyzing data on CS paradigm is presented. The method assumes prior information like the sparsity and clusteredness of signals in the analysis of the data. Among the different reconstruction methods, the convex relaxation methods are redefined using Bayesian inference. Further, three CS applications are presented: MRI imaging, MIMO systems, and remote sensing. For MRI imaging, the two different priors are incorporated, while for MIMO systems and remote sensing, only the sparse prior is applied in

http://dx.doi.org/10.5772/intechopen.70308

the analysis. We suggest that including the special structure among the sparse elements of the data can be included in the analysis to further improve the results.

Author details

Solomon A. Tesfamicael[1]* and Faraz Barzideh[2]

*Address all correspondence to: solomon.a.tesfamicael@ntnu.no

1 Department of Education, Norwegian University of Science and Technology (NTNU), Trondheim, Norway

2 Department of Electrical Engineering and Computer Science, University of Stavanger (UiS), Stavanger, Norway

References

[1] Jaynes ET. Probability Theory: The Logic of Science. Cambridge University Press; 2003

[2] Tesfamicael SA, Barzideh F. Clustered compressed sensing in fMRI data analysis using a Bayesian framework. International Journal of Information and Electronics Engineering. 2014;**4**(2):74-80

[3] Mackay DJC. Information Theory, Inference, and Learning Algorithms. Cambridge University Press; 2003. ISBN: 978-0-521-64298-9

[4] O'Hagen A, Forster J. Kendall's Advanced Theory of statistics, volume 2B. Bayesian Inference. Arnold, a member of the Hodder Headline Group; 2004. ISBN: 0 340 807520

[5] Berger JO. Bayesian and Conditional Frequentist Hypothesis Testing and Model Selection. VIII C:L:A:P:E:M; La Havana, Cuba; November 2001

[6] Efron B. Modern Science and the Bayesian-Frequentist Controversy; 2005-19B/233. January 2005

[7] Botje MRA. Fisher on Bayes and Bayes' Theorem, Bayesian Analysis; 2008

[8] Moreno E, Javier Giron F. On the Frequentist and Bayesian approaches to hypothesis testing. January-June 2006;3-28

[9] Friston KJ, Penny W, Phillips C, Kiebel S, Hinton G, Ashburner J, Classical and Bayesian inference in neuroimaging: Theory. NeuroImage. June 2002;**16**(2):465-483

[10] Donoho D. Compressed sensing. IEEE Transactions on Information Theory. 2006;**52**(4): 12891306

[11] Candes EJ, Tao T. Decoding by linear programming. IEEE Transactions on Information Theory. December 2005;**51**(12)

[12] Cand'es E, Romberg J, Tao T. Robust uncertainty principles: Exact signal reconstruction from highly incomplete frequency information. IEEE Transactions on Information Theory. February 2006;**52**(2):489509

[13] Cande's EJ, Tao T. Near-optimal signal recovery from random projections: Universal encoding strategies? IEEE Transactions on Information Theory. December 2006;**52**:54065425

[14] Eldar YC, Kutyniok G. Compressed Sensing: Theory and Applications. Cambridge University Press; 2012

[15] Eldar YC, Kutyniok G. Compressed Sensing: Algorithms and Applications. KTHKTH, Communication Theory. ACCESS Linnaeus Centre; 2012

[16] Candes EJ. The restricted isometry property and its implications for compressed sensing. Comptes Rendus Mathematique. 2008

[17] Guan X, Gao Y, Chang J, Zhang Z. Advances in Theory of Compressive Sensing and Applications in Communication. 2011 First International Conference on Instrumentation, Measurement, Computer, Communication and Control; 2011

[18] Blumensath T, Davies ME. Iterative hard thresholding for compressed sensing. Applied and Computational Harmonic Analysis. 2009

[19] Natarajan BK. Sparse approximate solutions to linear systems. SIAM Journal of Computing. 1995

[20] Natarajan BK. Uncertainty principles and ideal atomic decomposition. IEEE Transactions on Information Theory. January 2001;**47**:2845-2862

[21] Qaisar S, Bilal RM, Iqbal W, Naureen M, Lee S. Compressive sensing: From theory to applications, a survey. Journal of Communications and Networks. October 2013;**15**(5):443-456

[22] Figueiredo MAT, Nowak RD, Wright SJ. Gradient projection for sparse reconstruction: Application to compressed sensing and other inverse problems. Journal of Selected Topics in Signal Processing. 2007;**1**(4):586-597

[23] Schniter P, Potter LC, Ziniel J. Subspace pursuit for compressive sensing signal reconstruction. Information Theory and Applications Workshop. February 2008. pp. 326-333

[24] Teixeira FCA, Bergen SWA, Antoniou A. Robust signal recovery approach for compressive sensing using unconstrained optimization. Proceedings of 2010 IEEE International Symposium on Circuits and Systems (ISCAS). May 2010. pp. 3521-3524

[25] Tesfamicael SA, Barzideh F. Clustered compressed sensing via Bayesian framework. IEEE UKSim-AMSS 17th International Conference on Computer Modelling and Simulation. UKSim2015-19.S.Image, Speech and Signal Processing; Cambridge, United Kingdom. 2015. pp. 25-27

[26] Tesfamicael SA, Barzideh F. Clustered compressive sensing: Application on medical imaging. International Journal of Information and Electronics Engineering. 2015;**5**(1):48-50

[27] Tesfamicael SA. Compressive sensing in signal processing: Performance analysis and applications [doctoral thesis]. NTNU; 2016. p. 182

[28] Rangan S, Fletcher AK, Goyal VK. Asymptotic Analysis of MAP Estimation via the Replica Method and Applications to Compressed Sensing. 2009. arXiv:0906.3234v1

[29] Ji S, Xue Y, Carin L. Bayesian compressive sensing. IEEE Transactions on Signal Processing. June 2008;**56**(6):2346-2356

[30] Yu L, Sun H, Pierre Barbot J, Zheng G. Bayesian compressive sensing for clustered sparse signals. IEEE International Conference on Acoustics, Speech and Signal Processing (ICASSP). 2011

[31] Duarte MF, Davenport MA, Takhar D, Laska JN, Sun T, Kelly KF, Baraniuk RG. Single-pixel imaging via compressive sampling. IEEE Signal Processing Magazine. March 2008;**25**(2):83-91

[32] Lustig M, Donoho D, Pauly JM. Sparse MRI: The application of compressed sensing for rapid MR imaging. Magnetic Resonance in Medicine. 2007;**58**(6):1182-1195

[33] Lustig M, Donoho D, Pauly JM. Simply denoise: Wavefield reconstruction via jittered undersampling. Geophysics. 2007;128-133

[34] Baraniuk R, Steeghs P. 2007 IEEE Compressive Radar Imaging Radar Conference. 19-28 April, 2007

[35] Herman MA, Strohmer T. High-resolution radar via compressed sensing. IEEE Transactions on Signal Processing. June 2009;**57**(6):2275-2284

[36] Bajwa WU, Haupt J, Sayeed AM, Nowak R. Compressed channel sensing: A new approach to estimating sparse multipath channel. Proceedings of the IEEE. June 2010;**98**(6): 1058-1107

[37] Berger CR, Zhou S, Preisig JC, Willett P. Sparse channel estimation for multicarrier underwater acoustic communication: From subspace methods to compressed sensing. IEEE Transactions on Signal Processing. March 2010;**58**(3):1708-1721

[38] Zeng F, Zhi T, Chen L. Distributed compressive wideband spectrum sensing in cooperative multi-hop cognitive networks. IEEE International Conference on Communications (ICC). 1-5 May, 2010

[39] Qing L, Zhi T. Decentralized sparse signal recovery for compressive sleeping wireless sensor networks. IEEE Transactions on Signal Processing. July 2010;**58**(7):3816-3827

[40] Charbiwala Z, Chakraborty S, Zahedi S, Kim Y, Srivastava MB, He T, Bisdikian C. Compressive oversampling for robust data transmission in sensor networks. IEEE INFOCOM Proceedings. 2010. pp. 1-9

[41] Chen F, Au WSA, Valaee S, Zhenhui T. Compressive sensing based positioning using RSS of WLAN access points. IEEE INFOCOM Proceedings. 1-9 March 2010

[42] Image. Depiction of Vessel Diseases with a Wide Range of Contrast and Non-Contrast Enhanced Techniques. Munich, Germany: University Hospital Rechts der Isar; 2014

[43] Tesfamicael SA, Lundheim L. Compressed sensing based rotative quantization in temporally correlated MIMO channels. Recent Developments in Signal Processing; 2013

[44] Godana SBE, Ekman T. Rotative quantization using adaptive range for temporally correlated MIMO channels. 2013 IEEE 24th International Symposium on Personal Indoor and Mobile Radio Communications (PIMRC). 2013. pp. 1233-1238

[45] Tibshirani R. Compressive sensing: From theory to applications, a survey. Journal of the Royal Statistical Society. Series B (Methodological). 1996;**58**(1):267-288

[46] Yu X, Gao G, Xu J, Wang G. Remote sensing image fusion based on sparse representation. 2014 IEEE Geoscience and Remote Sensing Symposium

Chapter 14

Sparsity in Bayesian Signal Estimation

Ishan Wickramasingha, Michael Sobhy and
Sherif S. Sherif

Additional information is available at the end of the chapter

http://dx.doi.org/10.5772/intechopen.70529

Abstract

In this chapter, we describe different methods to estimate an unknown signal from its linear measurements. We focus on the underdetermined case where the number of measurements is less than the dimension of the unknown signal. We introduce the concept of signal sparsity and describe how it could be used as prior information for either regularized least squares or Bayesian signal estimation. We discuss compressed sensing and sparse signal representation as examples where these sparse signal estimation methods could be applied.

Keywords: inverse problems, signal estimation, regularization, Bayesian methods, signal sparsity

1. Introduction

In engineering and science, a system typically refers to a physical process whose outputs are generated due to some inputs [1, 2]. Examples of systems include measuring instruments, imaging devices, mechanical and biomedical devices, chemical reactors and others. A system could be abstracted as a block diagram,

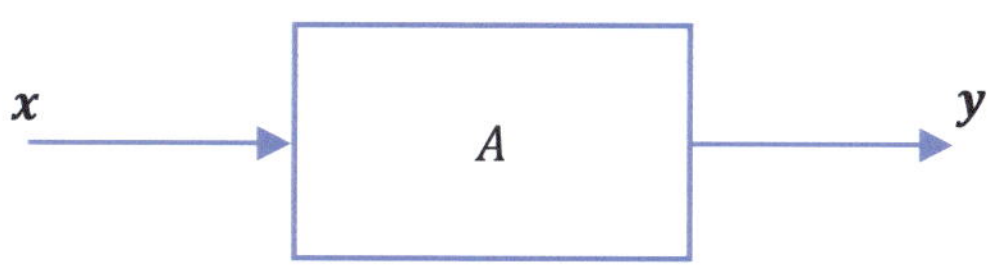

where x and y represent the inputs and outputs of the system, respectively. The block, A, formalizes the relation between these inputs and the outputs using mathematical equations [2, 3]. Depending on the nature of the system, the relation between its inputs and outputs could be either linear or nonlinear. For a linear relation, the system is called a *linear system* and it would be represented by a set of linear equations [3, 4]

$$y = Ax. \tag{1}$$

In this chapter, we will restrict our attention to linear systems, as they could adequately represent many actual systems in a mathematically tractable way.

When dealing with systems, two typical types of problems arise, forward and inverse problems.

1.1. Forward problems

In a *forward problem*, one would be interested in obtaining the output of a system due to a particular input [5, 6]. For linear systems, this output is the result of a simple matrix-vector product, Ax. Forward problems usually become more difficult as the number of equations increases or as uncertainties about the inputs, or the behavior of the system, are present [6].

1.2. Inverse problems

In an *inverse problem*, one would be interested in inferring the inputs to a system x that resulted in observed outputs, i.e., measured y [5, 6]. Another formulation of an inverse problem is to identify the behavior of the system, i.e., construct A, from knowledge of different input and output values. This problem formulation is known as *system identification* [1, 7, 8]. In this chapter, we will only consider the input inference problem. The nature of the input x to be inferred further leads to two broad categories of this problem: *estimation*, and *classification*. In input estimation, the input could assume an infinite number of possible values [4, 9], while in input classification the input could assume only a finite number (usually small) of possible values [4, 9]. Accordingly, in input classification, one would like to only assign an input to a predetermined signal class. In this chapter, we will only focus on estimation problems, particularly on restoring an input signal x from noisy data y that is obtained using a linear measuring system represented by a matrix A.

2. Signal restoration as example of an inverse problem

To solve the above signal restoration problem, we need to estimate input signal x through the inversion of matrix A. This could be a hard problem because in many cases the inverse of A might not exist, or the measurement data, y, might be corrupted by noise. The existence of the inverse of A depends on the number of acquired independent measurements relative to the dimension of the unknown signal [5, 10]. The conditions for the existence of a stable solution of any inverse problem, i.e., for an inverse problem to be well-posed, have been addressed by Hadamard as:

- *Existence*: for measured output y there exists at least one corresponding input x.
- *Uniqueness*: for measured output y there exists only one corresponding input x.
- *Continuity*: as the input x changes slightly, the output y changes slightly, i.e., the relation between x and y is continuous.

These conditions could be applied to linear systems as conditions on the matrix $\boldsymbol{A}$. Let the matrix $\boldsymbol{A} \in \mathbb{R}^{n \times m}$, such that $\mathbb{R}^{n \times m}$ denotes the set of matrices of dimension $n \times m$ with its elements being real values. The matrix equation, $\boldsymbol{y}_{n \times 1} = \boldsymbol{A}_{n \times m} \boldsymbol{x}_{m \times 1}$, is equivalent to n linear equations with m unknowns. The matrix $\boldsymbol{A}$ is a linear transformation that maps input signals from its domain $\mathfrak{D}(\boldsymbol{A}) = \mathbb{R}^m$ to its range $\mathcal{R}(\boldsymbol{A}) = \mathbb{R}^n$ [4, 5, 10]. For any measured output signal $\boldsymbol{y} \in \mathbb{R}^n$, we could identify three cases based on the values of n and m.

2.1. Underdetermined linear systems

In this case, $n < m$, i.e., the number of equations is less than the number of unknowns,

$$\boldsymbol{A} = \begin{bmatrix} a_{11} & a_{12} & \cdots & a_{1m} \\ \vdots & \vdots & \ddots & \vdots \\ a_{n1} & a_{n2} & \cdots & a_{nm} \end{bmatrix}. \tag{2}$$

If these equations are consistent, *Hadamard's Existence* condition will be satisfied. However, Hadamard's *Uniqueness* condition is not satisfied because the *Null Space*($\boldsymbol{A}$) $\neq \{\boldsymbol{0}\}$, i.e., there exist $\boldsymbol{z} \neq \boldsymbol{0} \in$ *Null Space*($\boldsymbol{A}$) such that,

$$\boldsymbol{A}(\boldsymbol{x} + \boldsymbol{z}) = \boldsymbol{y}. \tag{3}$$

This linear system is called *under-determined* because its equations, i.e., system constraints, are not enough to uniquely determine $\boldsymbol{x}$ [4, 5]. Thus, the inverse of $\boldsymbol{A}$ does not exist.

2.2. Overdetermined linear systems

In this case, $m > n$, the number of equations is more than the number of unknowns,

$$\boldsymbol{A} = \begin{bmatrix} a_{11} & \cdots & a_{1m} \\ a_{21} & \cdots & a_{2m} \\ \vdots & & \vdots \\ & \ddots & \\ \vdots & & \vdots \\ a_{n1} & \cdots & a_{nm} \end{bmatrix}. \tag{4}$$

If these equations are consistent, Hadamard's *Existence* condition will not be satisfied. However, *Hadamard*'s *Uniqueness* condition will be satisfied, if $\boldsymbol{A}$ has full rank. In this case, *Null Space* ($\boldsymbol{A}$) = $\{\boldsymbol{0}\}$, i.e.,

$$\boldsymbol{A}(\boldsymbol{x} + \boldsymbol{0}) = \boldsymbol{A}\boldsymbol{x} = \boldsymbol{y}. \tag{5}$$

This linear system is called *over-determined*, because its equations, i.e., system constraints, are too many for $\boldsymbol{x}$ to exist [4, 5]. Also, the inverse of $\boldsymbol{A}$ does not exist.

2.3. Square linear systems

The case where $m = n$, the number of equations is equal to the number of unknowns,

$$A = \begin{bmatrix} a_{11} & \cdots & a_{1n} \\ \vdots & \ddots & \vdots \\ a_{n1} & \cdots & a_{nn} \end{bmatrix}. \quad (6)$$

If A has full rank, its *Null Space*(A) = {**0**} and both Hadamard's *Existence* and *Uniqueness* conditions will be satisfied. In addition, if A has a small condition number, the relation between x, y will be continuous, and Hadamard's *Continuity* condition will be satisfied [4, 5, 10]. In this case, the inverse problem formulated by this system of linear equations is well-posed.

3. Methods for signal estimation

In this section, we will focus on the estimation of an input signal x from a noisy measurement y of the output of a linear system A.

The linear system shown in **Figure 1**, could be modeled as,

$$y = Ax + v. \quad (7)$$

where v is additive Gaussian noise. As a consequence of the *Central Limit Theorem*, this assumption of Gaussian distributed noise is valid for many output measurement setups.

Statistical Estimation Theory allows one to obtain an estimate $\hat{x}$ of a signal x that is input to a known system A from measurement y (see **Figure 2**) [11, 12]. However, this estimate $\hat{x}$ is not unique, as it depends on the choice of the used estimator from the different ones available. In addition to measurement y, if other information about the input signal is available, it could be

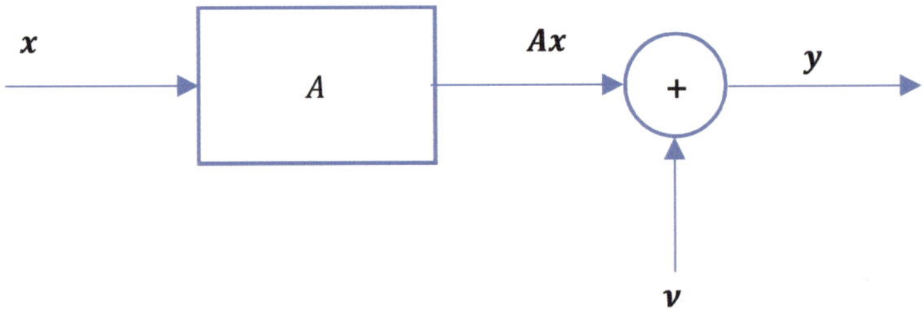

Figure 1. Linear system with noisy output measurement.

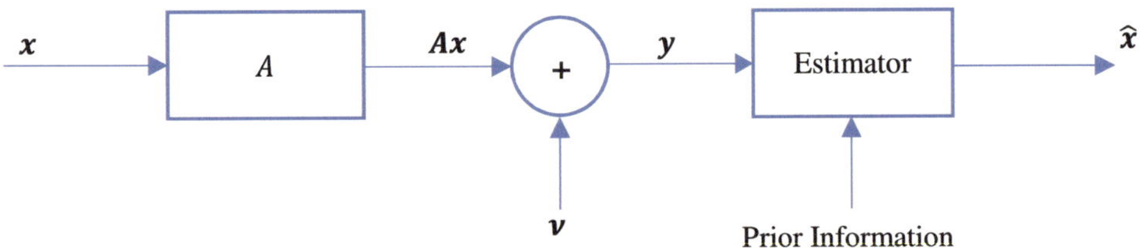

Figure 2. Signal estimation using prior information.

http://dx.doi.org/10.5772/intechopen.70529

used as prior information to constrain the estimator to produce a better estimate of $\boldsymbol{x}$. Signal estimation for overdetermined systems could be achieved without any prior information about the input signal. However, for underdetermined systems, prior information is necessary to ensure a unique estimate.

3.1. Least squares estimation

If there is no information available about the statistics of the measured data,

$$\boldsymbol{y} = \boldsymbol{A}\boldsymbol{x} + \boldsymbol{v}, \tag{8}$$

least squares estimation could be used. The least squares estimate is obtained by minimizing the square of the L_2 norm of the error between the measurement and the linear model, $\boldsymbol{v} = \boldsymbol{y} - \boldsymbol{A}\boldsymbol{x}$. It is given by

$$\widehat{\boldsymbol{x}} = \arg\min_{x} \|\boldsymbol{y} - \boldsymbol{A}\boldsymbol{x}\|_2^2. \tag{9}$$

The L_2 norm is a special case of the p-norm of a vector, where $p = 2$, that is defined as $\|\boldsymbol{x}\|_p = \left(\sum_{i=1}^{m} |x_i|^p\right)^{\frac{1}{p}}$. In Eq. (9), the unknown $\boldsymbol{x}$ is considered deterministic, so its statistics are not required. The noise $\boldsymbol{v}$ in this formulation is implicitly assumed to be white noise with variance σ^2 [13, 14]. Least squares estimation is typically used to estimate input signals $\boldsymbol{x}$ in overdetermined problems. Since $\widehat{\boldsymbol{x}}$ is unique in this case, no prior information, additional constraints, for $\boldsymbol{x}$ is necessary.

3.2. Weighted least squares estimation

If the noise $\boldsymbol{v}$ in Eq. (8) is not necessarily white and its second order statistics, i.e., mean and covariance matrix, are known, then weighted least squares estimation could be used to further improve the least squares estimate. In this estimation method, measurement errors are not weighted equally, but a weighting matrix $\boldsymbol{C}$ explicitly specifies such the weights. The weighted least squares estimate is given by

$$\widehat{\boldsymbol{x}} = \arg\min_{x} \|\boldsymbol{C}^{-1/2}(\boldsymbol{y} - \boldsymbol{A}\boldsymbol{x})\|_2^2. \tag{10}$$

We note that the least squares problem, Eq. (9), is a special case of the weighted least squares problem, Eq. (10), when $\boldsymbol{C} = \sigma^2\boldsymbol{I}$.

3.3. Regularized least squares estimation

In underdetermined problems, the introduction of additional constraints on $\boldsymbol{x}$, also known as *regularization*, could ensure the uniqueness of the obtained solution. Standard least squares estimation could be extended, through regularization, to solve underdetermined estimation problems. The regularized least squares estimate is given by

$$\arg\min_{x} \|y - Ax\|_2^2 + \lambda \|Lx\|_2, \tag{11}$$

where L is a matrix specifying the additional constraints and λ is a *regularization parameter* whose value determines the relative weights of the two terms in the objective function. If the combined matrix $\begin{bmatrix} A \\ L \end{bmatrix}$ has full rank, the regularized least squares estimate $\widehat{x}$ is unique [4]. In this optimization problem, the unknown x is once again considered deterministic, so its statistics are not required. It is worthwhile noting that while *regularization* is necessary to solve underdetermined inverse problems, it could also be used to improve numerical properties, e.g., condition number, of either linear overdetermined or linear square inverse problems.

3.4. Maximum likelihood estimation

If the probability distribution function (pdf) of the measurement y, parameterized by an unknown deterministic input signal x, is available, then the *maximum likelihood* estimate of x is given by,

$$\widehat{x} = \arg\max_{x} f(y|x). \tag{12}$$

This maximum likelihood estimate $\widehat{x}$ is obtained by assuming that measurement y is the most likely measurement to occur given the input signal x. This corresponds to choosing the value of x for which the probability of the observed measurement y is maximized. In maximum likelihood estimation, the negative log of the likelihood function, $f(y|x)$, is typically used to transform Eq. (12) into a simpler minimization problem. When, $f(y|x)$ is a Gaussian distribution, $N(Ax, C)$, minimizing the negative log of the likelihood function is equivalent to solving the weighted least squares estimation problem.

3.5. Bayesian estimation

If the conditional pdf of the measurement y, given an unknown random input signal x, is known, in addition to the marginal pdf of x, representing prior information about x, is given, then a Bayesian estimation method would be possible. The first step to obtain one of the many possible Bayesian estimates of x is to use Bayes rule to obtain the *a posteriori* pdf,

$$f(x|y) = \frac{f(y|x)f(x)}{\int f(y|x)f(x)}. \tag{13}$$

Once this *a posteriori* pdf is known, different Bayesian estimates $\widehat{x}$ could be obtained. For example, the *minimum mean square error* estimate is given by,

$$\widehat{x}_{MMSE} = E_x[f(x|y)] = E_x\left[\frac{f(y|x)f(x)}{\int f(y|x)f(x)}\right], \tag{14}$$

while the *maximum a priori* (MAP) estimate is given by,

$$\widehat{x}_{MAP} = \arg\max_{x} f(x|y) = \arg\max_{x} f(y|x)f(x). \tag{15}$$

We note that the maximum likelihood estimate, Eq. (12), is a special case of the MAP estimate, when $f(x)$ is a uniform pdf over the entire domain of x. The use of prior information is essential to solve underdetermined inverse problems, but it also improves the numerical properties, e.g., condition number, of either linear overdetermined or linear square inverse problems.

3.5.1. Bayesian least squares estimation

In least squares estimation, the vector x is assumed to be an unknown deterministic variable. However, in Bayesian least squares estimation, it is considered a vector of scalar random variables that satisfies statistical properties given by an *a priori* probability distribution function [5]. In addition, in least squares estimation, the L_2 norm of the measurement error is minimized, while in Bayesian least squares estimation, it is the estimation error, $e = \widehat{x} - x$, not measurement error, that is used [5]. Since x is assumed to be a random vector, the estimation error e will also be a random vector. Therefore, the Bayesian least squares estimate could be obtained by minimizing the condtional mean of the square of the estimation error, given measurement, y,

$$\widehat{x} = \arg\min_{x} E\left[(\widehat{x} - x)^T(\widehat{x} - x)|y\right]. \tag{16}$$

When x has a Gaussian distribution and A represents a linear system, then measurement y will also have a Gaussian distribution. In this case, the Bayesian least squares estimate given by Eq. (16) could be reinterpreted as a regularized least squares estimate given by,

$$\widehat{x} = \arg\min_{x} \|y - Ax\|_2^2 + \|\mu - x\|, \tag{17}$$

where μ is the mean of the *a priori* distribution of x [5]. Therefore, a least squares Bayesian estimate is analogous to a regularized least squares estimate, where *a priori* information about x is expressed as additional constraints on x in the form of a *regularization* term.

3.5.2. Advantages of Bayesian estimation over other estimation methods

Bayesian estimation techniques could be used, given that a reliable *a priori* distribution is known, to obtain an accurate estimate of a signal x, even if the number available measurements is smaller than the dimension of the signal to estimated. In this underdetermined case, Bayesian estimation could accurately estimate a signal while un-regularized least squares estimation or maximum likelihood estimation could not. The use of prior information in Bayesian estimation could also improves the numerical properties, e.g., condition number, of either linear overdetermined or linear square inverse problems. This could be understood by keeping in mind the mathematical equivalence between obtaining one scalar measurement related to x, and specifying one constraint that x has to satisfy. Therefore, as the number of available measurements significantly increases, both Bayesian and maximum likelihood estimates would converge to the same estimate.

Bayesian estimation also could be easily adapted to estimate dynamic signals that change over time. This is achieved by sequentially using past estimates of a signal, e.g., x_{t-1}, as prior information to estimate its current value x_t. More generally, Bayesian estimation could be easily adapted for *data fusion*, i.e., combination of multiple partial measurements to estimate a complete signal in remote sensing, stereo vision and tomographic imaging, e.g., Positron emission tomography (PET), Magnetic resonance imaging (MRI), computed tomography (CT) and optical coherence tomography (OCT). Bayesian methods could also easily fuse all available prior information to provide an estimate based on measurements, in addition to all known information about a signal.

Bayesian estimation techniques could be extended in straight forward ways to estimate output signals of nonlinear systems or signals that have complicated probability distributions. In these cases, numerical Bayesian estimates are typically obtained using Monte Carlo methods.

3.5.3. Sparsity as prior information for underdetermined Bayesian signal estimation

Sparse signal representation means the representation of a signal in a domain where most of its coefficients are zero. Depending on the nature of the signal, one could find an appropriate domain where it would be sparse. This notion could be useful in signal estimation because assuming that the unknown signal x is sparse could be used as prior information to obtain an accurate estimate of it, even if only a small number of measurements are available. The rest of this chapter will focus on using signal sparsity as prior information for underdetermined Bayesian signal estimation.

4. Sparse signal representation

As shown in **Figure 3**, a sinusoid is a dense signal in the time domain. However, it could be represented by a single value, i.e., it has a sparse representation, in the frequency domain.

We note that any signal could have a sparse representation in a suitable domain [15]. A sparse signal representation means a representation of the signal in a domain where most of its coefficients are zero. Sparse signal representations have many advantages including:

1. A sparse signal representation requires less memory for its storage. Therefore, it is a fundamental concept for signal compression.

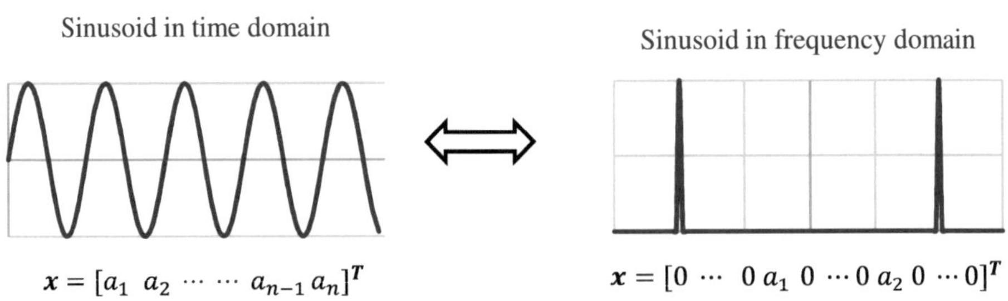

Figure 3. A sinusoid in time and frequency domains.

2. A sparse signal representation could lead to simpler signal processing algorithms. For example, signal denoising could be achieved by simple thresholding operations in a domain where the signal is known to be sparse.

3. Sparse signal representations have fewer coefficients than dense signal representations. Therefore, the computational cost for sparse representations would be lower than for dense representations.

4.1. Signal representation using a dictionary

A *dictionary* $\mathcal{D}$ is a collection of vectors $\{\phi_n\}_{n\epsilon\Gamma}$, indexed by a parameter $n \epsilon \Gamma$ equal to the dimension of a signal *f*, where we could represent *f* as a linear combination [16],

$$f = \sum_{n\epsilon\Gamma} c_n \phi_n. \tag{18}$$

If the vectors $\{\phi_n\}_{n\epsilon\Gamma}$ are linearly independent, then such dictionary is called a *basis*. Representing a signal as a linear combination of sinusoids, i.e., using a *Fourier* dictionary, is very common. *Wavelet* dictionaries and *Chirplet* dictionaries are also common dictionaries for signal representation. Dictionaries could be combined together to obtain a larger dictionary, where $n \epsilon \Gamma$ is larger than the dimension the signal *f*, that is called an *overcomplete* dictionary or a frame.

4.1.1. Signal representation using a basis

A set of vectors form a basis for $\mathbb{R}^n$ if they span $\mathbb{R}^n$ and are linearly independent. A basis in a vector space *V* is a set *X* of linearly independent vectors such that every vector in *V* is a linear combination of elements in *X*. A vector space *V* is finite dimensional if it has a finite number of basis vectors [17].

Depending on the properties of $\{\phi_n\}_{n\epsilon\Gamma}$, bases could be classified into different types, e.g., orthogonal basis, orthonormal basis, biorthogonal basis, global basis and local basis. For an orthogonal basis, its basis vectors in the vector space *V* are mutually orthogonal,

$$\langle \phi_m, \phi_n \rangle = 0 \text{ for } m \neq n. \tag{19}$$

For an orthonormal basis, its basis vectors in the vector space *V* are mutually orthogonal and have unit length,

$$\langle \phi_m, \phi_n \rangle = \delta(m - n), \tag{20}$$

where $\delta(m - n)$ is the Kronecker delta function. For a biorthogonal basis, its basis vectors are not orthogonal to each other, but they are orthogonal to vectors in another basis, $\left\{\widetilde{\phi}_n\right\}_{n\epsilon\Gamma}$, such that

$$\left\langle \phi_m, \widetilde{\phi}_n \right\rangle = \delta(m - n). \tag{21}$$

In addition, depending on the domain (support) on which these basis vectors are defined, we could also classify a basis as either global or local. Sinusoidal basis vectors used for the discrete Fourier transform are defined on the entire domain (support) of f, so they are considered a *global* basis. Many wavelet basis vectors used for the discrete wavelet transform are defined on only part of the domain (support) of f, so they are considered a *local* basis.

4.1.2. Signal representation using a frame

A frame is a set of vectors $\{\phi_n\}_{n\epsilon\Gamma}$ that spans $\mathbb{R}^n$ and could be used to represent a signal f from the inner products $\{\langle f, \phi_n\rangle\}_{n\epsilon\Gamma}$. A frame allows the representation of a signal as a set of frame coefficients, and its reconstruction from these coefficients in a numerically stable way

$$f = \sum_{n\epsilon\Gamma} \langle f, \phi_n\rangle \phi_n. \tag{22}$$

Frame theory analyzes the completeness, stability, and redundancy of linear discrete signal representations [18]. A frame is not necessarily a basis, but it shares many properties with bases. The most important distinction between a frame and a basis is that the vectors that comprise a basis are linearly independent, while those comprising frame could be linearly dependent. Frames are also called *overcomplete* dictionaries. The redundancy in the representation of a signal using frames could be used to obtain sparse signal representations.

4.2. Sparse signal representation as a regularized least squares estimation problem

If designed to concentrate the energy of a signal in a small number of dimensions, an orthogonal basis would be the minimum-size dictionary that could yield a sparse representation of this signal [15]. However, finding an orthogonal basis that yields a highly sparse representation for a given signal is usually difficult or impractical. To allow more flexibility, the orthogonality constraint is usually dropped, and *overcomplete* dictionaries (frames) are usually used. This idea is well explained in the following quote by Stephane Mallat:

> *"In natural languages, a richer dictionary helps to build shorter and more precise sentences. Similarly, dictionaries of vectors that are larger than bases are needed to build sparse representations of complex signals. Sparse representations in redundant dictionaries can improve pattern recognition, compression, and noise reduction but also the resolution of new inverse problems. This includes super resolution, source separation, and compressed sensing" [15].*

Thus representing a signal using a particular *overcomplete* dictionary has the following goals [16]

- Sparsity—this representation should be more sparse than other representations.
- Super resolution—the resolution of the signal when represented using this dictionary should be higher than when represented in any other dictionary.
- Speed—this representation should be computed in $O(n)$ or $O(n\ log(n))$ time.

A simple way to obtain an *overcomplete* dictionary $\boldsymbol{A}$ is to use a union of basis $\boldsymbol{A}_i$ that would result in the following representation of a signal $\boldsymbol{y}$,

$$(y) = \underbrace{([A_1][A_2][A_3][A_4][A_5])}_{A}(x) \Rightarrow y = Ax, \tag{23}$$

where A is a $n \times m$ matrix representing the dictionary and x are the coefficients representing y in the domain defined by A. Since A represents an *overcomplete* dictionary, the number of its rows will be less than the number of its columns. Eq. (23) is a formulation of the signal representation problem as an underdetermined inverse problem.

To obtain a sparse solution for Eq. (23) one needs to find an $m \times 1$ coefficient vector $\hat{x}$, such that,

$$\hat{x} = \arg\min_{x} \|y - Ax\|_2^2 + \lambda\|x\|_0, \tag{24}$$

where $\|x\|_0$ is the cardinality of vector x, i.e., its number of nonzero elements, and $\lambda > 0$ is a regularization parameter that quantifies the tradeoff between the signal representation error, $\|y - Ax\|_2^2$, and its sparsity level, $\|x\|_0$ [19]. The cardinality of vector x is sometimes referred to as the L_0 norm of x, even though $\|x\|_0$ is actually a *pseudo norm* that does not satisfy the requirements of a norm in $\mathbb{R}^m$. This sparse signal representation problem, Eq. (24), has a form similar to the regularized least squares estimation problem, Eq. (11), that would be *underdetermined* in the case of an *overcomplete* dictionary. Because of the correspondence between regularized least squares estimation and Bayesian estimation, the problem of finding a sparse representation of a signal could be formulated as a Bayesian estimation problem.

5. Compressed sensing

Compressed sensing involves the estimation of a signal using a number of measurements that are significantly less than its dimension [20]. By assuming that the unknown signal is sparse in the domain where the measurements were acquired, one could use this sparsity constraint as prior information to obtain an accurate estimate of the signal from relatively few measurements.

Compressed sensing is closely related to *signal compression* that is routinely used for efficient storage or transmission of signals. Compressed sensing was inspired by this question: instead of the typical signal acquisition followed by signal compression, is there a way to acquire (sense) the compressed signal in the first place? If possible, it would significantly reduce the number of measurements and the computation cost [20]. In addition, this possibility would allow acquisition of signals that require extremely high, hence impractical, sampling rates [21]. As an affirmative answer to this question, compressed sensing was developed to combine signal compression with signal acquisition [20]. This is achieved by designing the measurement setup to acquire signals in the domain where the unknown signal is assumed to be sparse.

In compressed sensing, we consider the estimation of an input signal $x \in \mathbb{R}^n$ from m linear measurements, where $m \ll n$. As discussed above, this problem could be written as an underdetermined linear system,

$$y = Ax, \tag{25}$$

where $y \in \mathbb{R}^m$ and $A \in \mathbb{R}^{m \times n}$ represent the measurements and measurement (sensing) matrix, respectively.

Assuming that the unknown signal x is s-sparse, i.e., $x \in \sum_s$ has only s nonzero elements, in the domain specified by the measurement (sensing) matrix A, and assuming that A satisfies the restricted isometry property (RIP) of order 2s, i.e., there exists a constant $\delta_{2s} \in (0, 1)$ such that,

$$(1 - \delta_{2s})||z||_2^2 \leq ||Az||_2^2 \leq (1 + \delta_{2s})||z||_2^2, \tag{26}$$

for all $z \in \sum_{2s}$, then x could be reconstructed from $m \geq s$ measurements by different optimization algorithms [20]. When the measurements y are noiseless, x could be exactly estimated from,

$$\min_x ||x||_0 \text{ subject to } Ax = y. \tag{27}$$

However, when the measurements y are contaminated by noise,x could be obtained as the regularized least squares estimate,

$$\hat{x} = \arg\min_x ||Ax - y||_2^2 + \lambda ||x||_0. \tag{28}$$

This minimization problem could also be mathematically reformulated and solved as a Bayesian estimation problem.

6. Obtaining sparse solutions for signal representation and signal estimation problems

From Sections 4 and 5 we note that the problem of obtaining a sparse signal representation, Eq. (24) and the problem of sparse signal estimation in compressed sensing, Eq. (28), both have the same mathematical form [11, 22],

$$\hat{x} = \arg\min_x \|y - Ax\|_2^2 + \lambda\|x\|_0. \tag{29}$$

In this section, we describe different approaches to solving this minimization problem. From Eq. (29), we note that the first term of its RHS, $\|y - Ax\|_2^2$, represents either signal reconstruction error (sparse signal representation problem) or measurement fitting error (sparse signal estimation in compressed sensing problem), while the second term of its RHS,$\|x\|_0$, represents the cardinality (number of nonzero coefficients) of the unknown signal. The regularization parameter λ specifies the tradeoff between these two terms in the objective function. The selection of an appropriate value of λ to balance the reconstruction, or fitting error, and signal sparsity is very important. Regularization theory and Bayesian approaches could provide ways to determine optimal values of λ [23–26].

Convex optimization problems is a class of optimization problems that are significantly easier to solve compared to nonconvex problems [34]. Another advantage of convex optimization problems is that any local solution, e.g., a local minimum, is guaranteed to be a global solution. We note that obtaining an exact solution for the minimization problem in Eq. (29) is difficult because it is nonconvex. Therefore, one could either seek an approximate solution to this nonconvex problem or approximate this problem by a convex optimization whose exact solution could be obtained easily.

Considering the general regularized least squares estimation problem,

$$\widehat{x} = \arg\min_{x} \|y - Ax\|_2^2 + \lambda \|x\|_p, \tag{30}$$

we note that it is a nonconvex optimization problem for $0 \le p < 1$ and a convex optimization problem for $p \ge 1$. One alternative to approximate Eq. (29) by a convex optimization problem, one could relax the strict condition of minimizing the cardinality of the signal, $\|x\|_0$, by replacing by it by the sparsity-promoting condition of minimizing the L_1 norm of the signal, $\|x\|_1$. Another alternative to approximate Eq. (29) by another nonconvex optimization problem that is easier to solve than the original problem using a Bayesian formulation, is to replace $\|x\|_0$ by $\|x\|_p$, $0 < p < 1$. The minimization of Eq. (30) using $\|x\|_p$, $0 < p < 1$ would result in a higher degree of signal sparsity compared to when $\|x\|_1$ is used. This could be understood visually by examining **Figure 4**, that shows the shapes of two-dimensional unit balls using (pseudo)norms with different values of p.

We explain further details in the following subsections.

6.1. Obtaining a sparse signal solution using L_0 minimization

The sparsest solution of the regularized least squares estimation problem, Eq. (29) would be obtained when $p = 0$ in $\|x\|_p$. As shown in **Figure 5**, the solution of the regularized least squares problem, $\widehat{x}$, is given by the intersection of the circles, possibly ellipses, representing the

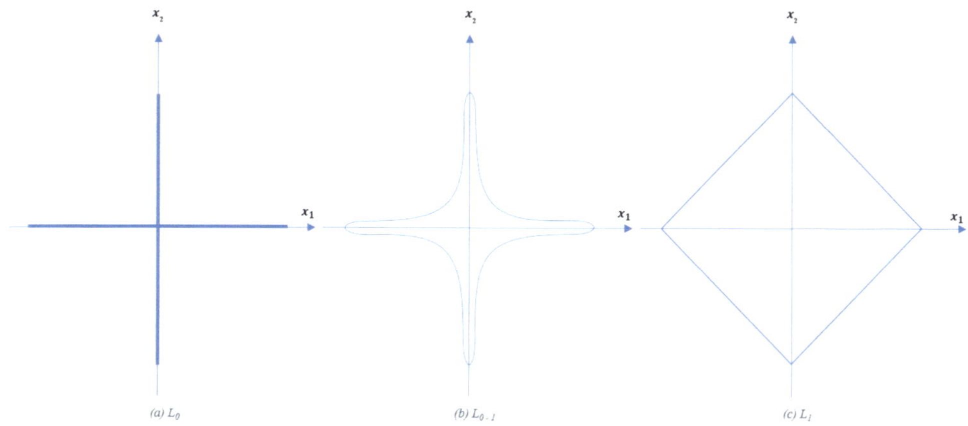

Figure 4. Two-dimensional unit ball using different (pseudo)norms. (a) L_0, (b) L_{0-1}, and (c) L_1.

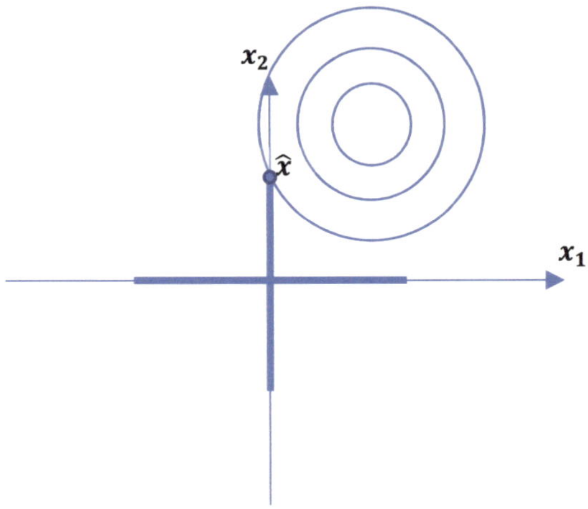

Figure 5. Regularized least squares using L_0.

solution of the unconstrained least squares estimation problem and the unit ball using L_0 representing the constraint of minimizing L_0. In this case of minimizing L_0, the unconstrained least squares solution will always intersect the unit ball at an axis, this yielding the most possible sparse solution. However, as mentioned earlier, this L_0 minimization problem is difficult to solve because it is nonconvex. Approximate solutions for this problem could be obtained using greedy optimization algorithms, e.g., Matching Pursuits [27] and Least Angle Regression (LARS) [28].

6.2. Obtaining a sparse signal solution using L_1 minimization

On relaxing the nonconvex regularized least squares using L_0 minimization problem, by setting $p = 1$, we obtain the convex L_1 minimization problem. As shown in **Figure 4(c)**, the unit ball using the L_1 norm covers a larger area than the unit ball using the L_0 pseudo norm, shown in **Figure 4(a)**. Therefore, as shown in **Figure 6**, the solution for the regularized least squares problem using the L_1 minimization would be sparse, but it should not be expected to be as sparse as the L_0 minimization problem.

This L_1 minimization problem could be solved easily using various algorithms, e.g., Basis Pursuits [16], Method of frames (MOF) [29], Lasso [30, 31], and Best Basis Selection [32, 33]. A Bayesian formulation of this L_1 minimization problem is also possible by assuming that the *a priori* probability distribution of x is Laplacian, $x \sim e^{-|x|}$.

6.3. Obtaining a sparse signal solution using L_{0-1} minimization

As discussed above, solving the regularized least squares problem with L_0 minimization should yield the sparsest signal solution. However, only approximate solutions are available for this difficult nonconvex problem. Alternatively, solving the regularized least squares problem with L_1 minimization should yield an exact sparse solution that would be less sparse than in the L_0 case, but it is considerably easier to obtain.

http://dx.doi.org/10.5772/intechopen.70529

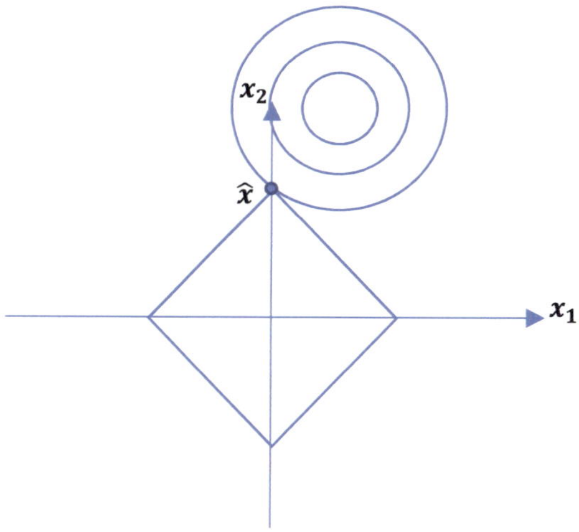

Figure 6. Regularized least squares using L_1.

The regularized least squares problem could also be formulated as an L_{0-1} minimization problem. As $\|x\|_p$, $0 < p < 1$,that we abbreviate as L_{0-1}, is not an actual norm, this optimization problem would be nonconvex [34]. The advantage of using L_{0-1} minimization is that, as shown in **Figure 4(b)**, compared to unit ball using the L_1 norm, the unit ball using the L_{0-1} pseudo norm has a narrower area that is concentrated around the axes. Therefore, as shown in **Figure 7**, the L_{0-1} minimization problem should yield a sparser solution compared to the L_1 minimization problem.

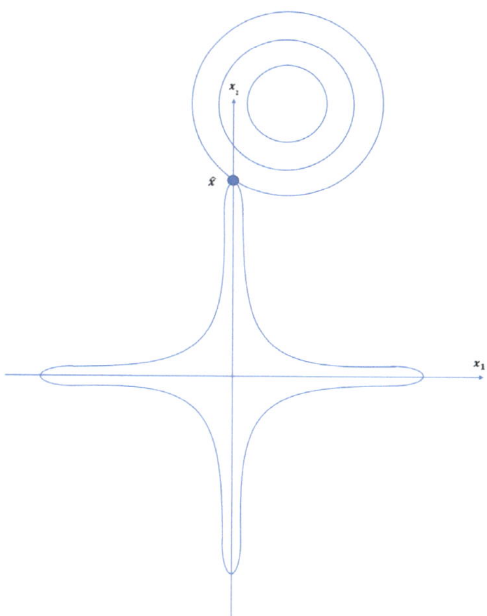

Figure 7. Regularized least squares using L_{0-1}.

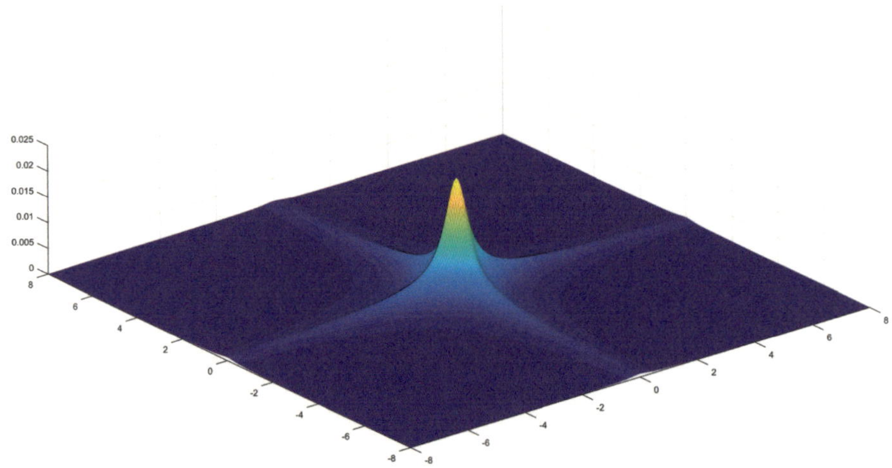

Figure 8. Product of two *student-t* probability distributions.

Another advantage of using L_{0-1} minimization is that this nonconvex optimization problem could be easily formulated as a Bayesian estimation problem that could be solved using Markov Chain Monte Carlo (MCMC) methods. As shown in **Figure 8**, the product of *student-t* probability distributions has a shape similar to the unit ball using the L_{0-1} pseudo norm, so *student-t* distributions could be used as *a priori* distributions to approximate the L_{0-1} pseudo norm.

6.4. Bayesian method to obtain a sparse signal solution using L_{0-1} minimization

As mentioned in Section 3.5, the first step to obtaining one of the many possible Bayesian estimates of x is to use Bayes rule to obtain the *a posteriori* pdf,

$$f(\boldsymbol{x}|\boldsymbol{y}) = \frac{f(\boldsymbol{y}|\boldsymbol{x})f(\boldsymbol{x})}{\int f(\boldsymbol{y}|\boldsymbol{x})f(\boldsymbol{x})}. \tag{31}$$

Using this *a posteriori* distribution, one could obtain a sparse signal solution using L_{0-1} minimization, as the *maximum a posteriori* (MAP) estimate given by Eq. (15). Compared to other Bayesian estimates, the MAP estimate could be easier to obtain because the calculation of the normalizing constant, $\int f(\boldsymbol{y}\,|\,\boldsymbol{x})f(\boldsymbol{x})$,would not be needed. The maximization of the product of conditional probability distribution of $\boldsymbol{y}$ given $\boldsymbol{x}$ and the *a priori* distribution of $\boldsymbol{x}$ is equivalent to the minimizing of the sum of their negative logarithms,

$$\widehat{\boldsymbol{x}}_{MAP} = \arg\min_{\boldsymbol{x}} \left[-\log p(\boldsymbol{y}|\boldsymbol{x}) - \log p(\boldsymbol{x})\right]. \tag{32}$$

In the case of white Gaussian measurement noise, $p(\boldsymbol{y}\,|\,\boldsymbol{x}) \sim N_{\boldsymbol{x}}(\boldsymbol{A}\boldsymbol{x}, \sigma^2\boldsymbol{I})$ where $-\log p(\boldsymbol{y}|\boldsymbol{x}) \propto \|\boldsymbol{y} - \boldsymbol{A}\boldsymbol{x}\|_2^2$, which the first term of the RHS of Eq. (30). As discussed in the previous section, the *a priori* probability $p(\boldsymbol{x})$ corresponding to L_{0-1} minimization could be represented as a product of univariate *student-t* probability distribution functions [14],

$$p(\boldsymbol{x}) = \prod_{i=1}^{M} stud_{\boldsymbol{x}_i}[0, 1, \vartheta] = \prod_{i=1}^{M} \frac{\Gamma\left(\frac{\vartheta+1}{2}\right)}{\sqrt{\vartheta\pi}\Gamma\left(\frac{\vartheta}{2}\right)} \left(1 + \frac{\boldsymbol{x}_i^2}{\vartheta}\right)^{-\frac{(\vartheta+1)}{2}}, \tag{33}$$

where Γ is the Gamma function, and ϑ is the number of degrees of freedom of the *student-t* distribution. Since this *a priori* distribution function is not an exponential function, we would use Eq. (15) instead of Eq. (32) to obtain the MAP estimate.

Because the prior is not a Gaussian distribution, there is no simple closed form expression for the posterior, $p(\boldsymbol{x}|\boldsymbol{y})$ with a *student-t a priori* probability distribution. However, we could express each *student-t* distribution as an infinite weighted sum of Gaussian distributions, where the hidden variables h_i determine their variances [14].

$$p(\boldsymbol{x}) = \prod_{i=1}^{M} \int N_{\boldsymbol{x}_i}(0, 1/h_i)\, Gam_{h_i}[\vartheta/2, \vartheta/2] dh_i = \int N_{\boldsymbol{x}}\left(\boldsymbol{0}, \boldsymbol{H}^{-1}\right) \prod_{i=1}^{M} Gam_{h_i}[\vartheta/2, \vartheta/2] d\boldsymbol{H}, \tag{34}$$

where the matrix $\boldsymbol{H}$ contains the hidden variables $\{h_i\}_{i=1}^{M}$ on its diagonal and has zeros elsewhere, and $Gam_{h_i}[\vartheta/2, \vartheta/2]$ is the gamma probability distribution function with parameters ($\vartheta/2$, $\vartheta/2$). Using this approximation, the *a posteriori* pdf could be written as

$$\begin{aligned} p(\boldsymbol{x}|\boldsymbol{y}) \propto p(\boldsymbol{y}|\boldsymbol{x})p(\boldsymbol{x}) &= N_{\boldsymbol{x}}\left(\boldsymbol{A}\boldsymbol{x}, \sigma^2\boldsymbol{I}\right) \int N_{\boldsymbol{x}}\left(\boldsymbol{0}, \boldsymbol{H}^{-1}\right) \prod_{i=1}^{M} Gam_{h_i}\left[\frac{\vartheta}{2}, \frac{\vartheta}{2}\right] d\boldsymbol{H} \\ &= \int N_{\boldsymbol{x}}\left(\boldsymbol{A}\boldsymbol{x}, \sigma^2\boldsymbol{I}\right) N_{\boldsymbol{x}}\left(\boldsymbol{0}, \boldsymbol{H}^{-1}\right) \prod_{i=1}^{M} Gam_{h_i}\left[\frac{\vartheta}{2}, \frac{\vartheta}{2}\right] d\boldsymbol{H}. \end{aligned} \tag{35}$$

The product of two Gaussian distributions is also a Gaussian distribution [35],

$$N_{\boldsymbol{x}}\left(\boldsymbol{\mu}_1, \boldsymbol{\Sigma}_1\right) N_{\boldsymbol{x}}\left(\boldsymbol{\mu}_2, \boldsymbol{\Sigma}_2\right) = k.N_{\boldsymbol{x}}(\boldsymbol{\mu}, \boldsymbol{\Sigma}), \tag{36}$$

where the mean and covariance ($\boldsymbol{\mu}$, $\boldsymbol{\Sigma}$) of the new Gaussian distribution in Eq. (36) is given by,

$$\boldsymbol{\mu} = \left(\boldsymbol{\Sigma}_1^{-1} + \boldsymbol{\Sigma}_2^{-1}\right)^{-1}\left(\boldsymbol{\Sigma}_1^{-1}\boldsymbol{\mu}_1 + \boldsymbol{\Sigma}_2^{-1}\boldsymbol{\mu}_2\right) \text{and}\, \boldsymbol{\Sigma} = \left(\boldsymbol{\Sigma}_1^{-1} + \boldsymbol{\Sigma}_2^{-1}\right)^{-1}, \tag{37}$$

and k is a constant. Therefore, we could simplify the product of two the Gaussian distributions given in Eq. (35) as,

$$N_{\boldsymbol{x}}\left(\boldsymbol{A}\boldsymbol{x}, \sigma^2\boldsymbol{I}\right).N_{\boldsymbol{x}}\left(\boldsymbol{0}, \boldsymbol{H}^{-1}\right) = k.N_{\boldsymbol{x}}\left(\left(\sigma^{-2}I + \boldsymbol{H}\right)^{-1}\left(\sigma^{-2}\boldsymbol{A}\boldsymbol{x}\right), \left(\sigma^{-2}I + \boldsymbol{H}\right)^{-1}\right). \tag{38}$$

From Eqs. (35) and (38) we could write $p(\boldsymbol{x}|\boldsymbol{y})$ as,

$$p(\boldsymbol{x}|\boldsymbol{y}) = k\int N_{\boldsymbol{x}}\left(\left(\sigma^{-2}I + \boldsymbol{H}\right)^{-1}\left(\sigma^{-2}\boldsymbol{A}\boldsymbol{x}\right), \left(\sigma^{-2}I + \boldsymbol{H}\right)^{-1}\right) \prod_{i=1}^{M} Gam_{h_i}\left[\frac{\vartheta}{2}, \frac{\vartheta}{2}\right] d\boldsymbol{H}. \tag{39}$$

We still could not compute the integral in Eq. (39) in closed form. However, we could maximize the RHS of Eq. (39) over the hidden variables $\boldsymbol{H}$ to obtain an approximation for the *a posteriori* probability distribution function

$$p(\mathbf{x}|\mathbf{y}) \approx \arg\max_{\mathbf{H}} \left[N_x \left(\left(\sigma^{-2}I + \mathbf{H}\right)^{-1} \left(\sigma^{-2}\mathbf{A}\mathbf{x}\right), \left(\sigma^{-2}I + \mathbf{H}\right)^{-1} \right) \prod_{i=1}^{M} Gam_{h_i}\left[\frac{\vartheta}{2}, \frac{\vartheta}{2}\right] \right]. \quad (40)$$

Eq. (40) would be a good approximation of $p(\mathbf{x}|\mathbf{y})$, if the actual distribution over the hidden variables is concentrated tightly around its mode [14]. When h_i has a large value, its corresponding ith component of the *a priori* probability distribution function $p(\mathbf{x})$ would have a small variance, $\frac{1}{h_i}$, so that this ith component of $p(\mathbf{x})$ could be set to zero. Therefore, this ith dimension of the prior $p(\mathbf{x})$ would not contribute to the solution of Eq. (30), thus increasing its sparsity.

Since both Gaussian and gamma pdfs in Eq. (40) are members of the exponential family of probability distributions, we could obtain $\widehat{\mathbf{x}}_{MAP}$ by maximizing the sum of their logarithms. Section 3.5 in [11] and Section 8.6 in [14] describe an iterative optimization method to obtain $\widehat{\mathbf{x}}_{MAP}$ from the approximate *a posteriori* probability distribution function given by Eq. (40).

7. Conclusion

In this chapter, we described different methods to estimate an unknown signal from its linear measurements. We focused on the underdetermined case where the number of measurements is less than the dimension of the unknown signal. We introduced the concept of signal sparsity and described how it could be used as prior information for either regularized least squares or Bayesian signal estimation. We discussed compressed sensing and sparse signal representation as examples where these sparse signal estimation methods could be applied.

Author details

Ishan Wickramasingha[1], Michael Sobhy[2] and Sherif S. Sherif[1]*

*Address all correspondence to: sherif.sherif@umanitoba.ca

1 Department of Electrical and Computer Engineering, University of Manitoba, Winnipeg, Canada

2 Biomedical Engineering Graduate Program, University of Manitoba, Winnipeg, Canada

References

[1] Keesman KJ. System Identification: An Introduction. London: Springer Science & Business Media; 2011

[2] Von Bertalanffy L. General system theory. New York. 1968;**41973**(1968):40

[3] Chen C-T. Linear System Theory and Design. New York, NY: Oxford University Press, Inc.; 1999

[4] Moon TK, Stirling WC. Mathematical Methods and Algorithms for Signal Processing. Upper Saddle River, NJ: Prentice Hall; 2000

[5] Fieguth P. Statistical Image Processing and Multidimensional Modeling. New York, NY: Springer Science+Business Media, LLC; 2011

[6] Tarantola A. Inverse problem theory and methods for model parameter estimation. Philadelphia, PA: Society for Industrial and Applied Mathematics; 2005. p. 1-37

[7] Ljung L. Perspectives on system identification. Annual Reviews in Control. 2010 Apr 30;**34**(1):1-2

[8] Wellstead PE. Non-parametric methods of system identification. Automatica. 1981 Jan 1;**17**(1):55-69

[9] Shanmugan KS, Breipohl AM. Random Signals: Detection, Estimation, and Data Analysis. New York, NY: Wiley; 1997

[10] Saad Y. Iterative Methods for Sparse Linear Systems. Philadelphia, PA: Society for Industrial and Applied Mathematics; 2003

[11] Bishop CM. Pattern Recognition and Machine Learning. New York, NY: Springer; 2006

[12] Mendel JM. Lessons in Estimation Theory for Signal Processing, Communications, and Control. Englewood Cliffs, N.J.: Prentice-Hall; 1995

[13] Sorenson HW. Least-squares estimation: From Gauss to Kalman. IEEE Spectrum. 1970 Jul;**7**(7):63-68

[14] Prince SJ. Computer Vision: Models, Learning, and Inference. Cambridge: Cambridge University Press; 2012

[15] Mallat S. A Wavelet Tour of Signal Processing: The Sparse Way. Amsterdam: Academic Press; 2009

[16] Chen SS, Donoho DL, Saunders MA. Atomic decomposition by basis pursuit. SIAM Review. 2001;**43**(1):129-159

[17] Paul R. Halmos. Finite-Dimensional Vector Spaces. Mineola, UNITED STATES: Dover Publications; 2017

[18] Mallat S. A Wavelet Tour of Signal Processing. San Diego: Academic Press; 1999

[19] Shannon CE. A mathematical theory of communication. ACM SIGMOBILE Mobile Computing and Communications Review. 2001;**5**(1):3-55

[20] Eldar YC, Kutyniok G, editors. Compressed Sensing: Theory and Applications. Cambridge: Cambridge University Press; 2012

[21] Asif MS. Dynamic compressive sensing: Sparse recovery algorithms for streaming signals and video [Doctoral dissertation]. Georgia Institute of Technology

[22] Huang K, Aviyente S. Sparse representation for signal classification. In: NIPS. Vol. 19; 2006. pp. 609-616

[23] Poggio T, Torre V, Koch C. Computational vision and regularization theory. Nature. 1985 Sep 26;**317**(6035):314-319

[24] Tikhonov AN, Arsenin VI. Solutions of Ill-posed Problems. Washington, DC: Winston; 1977 Jan

[25] Wahba G, Wendelberger J. Some new mathematical methods for variational objective analysis using splines and cross validation. Monthly Weather Review. 1980 Aug;**108**(8): 1122-1143

[26] Lin Y, Lee DD. Bayesian L1-Norm Sparse Learning. In: 2006 IEEE International Conference on Acoustics Speech and Signal Processing Proceedings. Toulouse, France: vol. 5; 2006. p. V–V.

[27] Mallat SG, Zhang Z. Matching pursuits with time-frequency dictionaries. IEEE Transactions on Signal Processing. 1993 Dec;**41**(12):3397-3415

[28] Efron B, Hastie T, Johnstone I, Tibshirani R. Least angle regression. The Annals of Statistics. 2004 Apr;**32**(2):407-499

[29] Daubechies I. Time-frequency localization operators: A geometric phase space approach. IEEE Transactions on Information Theory. 1988 Jul;**34**(4):605-612

[30] Tibshirani R. Regression shrinkage and selection via the lasso. Journal of the Royal Statistical Society. Series B (Methodological). 1996 Jan 1;**58**(1):267–288

[31] Yuan M, Lin Y. Model selection and estimation in regression with grouped variables. Journal of the Royal Statistical Society: Series B (Statistical Methodology). 2006 Feb 1;**68**(1): 49-67

[32] Coifman RR, Wickerhauser MV. Entropy-based algorithms for best basis selection. IEEE Transactions on Information Theory. 1992 Mar;**38**(2):713-718

[33] Rao BD, Kreutz-Delgado K. An affine scaling methodology for best basis selection. IEEE Transactions on Signal Processing. 1999 Jan;**47**(1):187-200

[34] Boyd S, Vandenberghe L. Convex Optimization. New York: Cambridge University Press; 2004.

[35] Bromiley P. Products and convolutions of Gaussian probability density functions. Tina-Vision Memo. 2003;**3**(4):1–13.

Chapter 15

Dynamic Bayesian Network for Time-Dependent Classification Problems in Robotics

Cristiano Premebida, Francisco A. A. Souza and
Diego R. Faria

Additional information is available at the end of the chapter

http://dx.doi.org/10.5772/intechopen.70059

Abstract

This chapter discusses the use of dynamic Bayesian networks (DBNs) for time-dependent classification problems in mobile robotics, where Bayesian inference is used to infer the class, or category of interest, given the observed data and prior knowledge. Formulating the DBN as a time-dependent classification problem, and by making some assumptions, a general expression for a DBN is given in terms of classifier priors and likelihoods through the time steps. Since multi-class problems are addressed, and because of the number of time slices in the model, additive smoothing is used to prevent the values of priors from being close to zero. To demonstrate the effectiveness of DBN in time-dependent classification problems, some experimental results are reported regarding semantic place recognition and daily-activity classification.

Keywords: dynamic Bayesian network, Bayesian inference, probabilistic classification, mobile robotics, social robotics

1. Introduction

Bayesian inference finds applications in many areas of engineering, and mobile robotics is not an exception. When time is a variable to be considered, the dynamic Bayesian network (DBN) [1–5] is a powerful approach to be considered. Due to its graphical representation and modelling versatility, DBN facilitates the problem-solving process in probabilistic time-dependent applications. Therefore, DBNs provide an effective way to model time-based (dynamic) probabilistic problems and also enable a very suitable and intuitive representation by means of a graph-based tree.

Depending on the structure of the DBN, the joint probabilistic distribution that governs a given system can be decomposed by a tractable product of probabilities, where the conditional terms only depend on their directly linked nodes. This chapter concentrates on inference problems using DBN where the variable to be inferred from a feature vector (data) represents a set of semantic classes $\mathcal{C} = \{c_1, c_2, \ldots, c_{nc}\}$ or categories, in the context of intelligent perception systems for mobile robotics applications. Namely, we will address problems where $\mathcal{C}$ denotes semantic places in a given indoor environment [6, 7] e.g. $\mathcal{C} = \{$ *'corridor', 'office',…,'kitchen'*$\}$ and also when the classes of interest are daily-live activities $\mathscr{C} = \{$*'drinking','talking',…,'walking'*$\}$ [8, 9].

The principle of Bayesian inference basically depends on two elements: the prior and the likelihood; in practical problems, the evidence probability acts 'only' as a normalization to guarantees that the posterior sums to one. In this chapter, we will deal with the problems of the classical Bayesian form $posterior \propto likelihood \cdot prior$, but the incorporation of (past) time will be explicitly modelled in a discrete-time basis, and the *past* information is assumed to be contained in the prior probabilities. Inference will be considered beyond the first-order Markov assumption, which means that a DBN with a finite number of time slices (T) will be addressed. Current time step t and previous/past time steps will be considered in the formulation of the DBN; thus, the time interval is $\{t, t-1, \ldots, t-T\}$.

The observed data enters the DBN in the form of a vector of features X calculated from sensory data; examples of sensors are laser scanners (or 2D Lidar) and RGB-D camera, as shown in **Figure 1**. Later, in the formulation of the DBN, we will consider that the feature vector at a given time step (X^t) is conditionally independent of previous time steps; therefore, $P(X^t \mid X^{t-1}) = P(X^t)$.

The use of Bayesian inference in mobile robotics for purpose of localization, simultaneous localization and mapping (SLAM), object detection, path planning and navigation, has been addressed in many scientific works; see Ref. [10] for a review. The majority of those applications involve stochastic filtering, such as Kalman filter (KF), particle filter (PF), Monte Carlo techniques and hidden Markov model (HMM) [11, 12]. However, when the parameter of interest has to be inferred from multidimensional feature vectors (e.g. feature vectors with hundreds of elements) and also when the distribution that the observed data were drawn is not known (in unseen/knew or testing scenarios) then, a DBN can be used to handle such complex problems. In robotics, semantic place classification [6, 7] and activity recognition [8, 9] are examples of such problems and belong to the research area of pattern recognition. For

Figure 1. Sensors commonly used in mobile robotics for perception systems.

these application cases, the class-conditional probabilities (or likelihoods) can be modelled using machine learning techniques, for example, naive Bayes classifier (NBC), support vector machines (SVMs) and artificial neural networks (ANNs) [13, 14].

The remainder of this chapter is organized as follows: a brief review of the DBN is given in Section 2. Section 3 addresses inference in DBN, formulated for purposes of pattern recognition in robotics, followed by the use of additive smoothing on the prior distributions. In Section 4, experimental results on semantic place classifications and activity recognition are presented. Finally, Section 5 presents our conclusions.

2. Preliminaries on DBN

Basically, a DBN is used to express the joint probability of events that characterizes a time-based (dynamic) system, where the relationships between events are expressed by conditional probabilities. Given evidence (observations) about events of the DBN, and prior probabilities, statistical inference is accomplished using the Bayes theorem. Inference in pattern recognition applications is the process of estimating the probability of the classes/categories given the observations, the class-conditional probabilities, and the priors [15, 16]. When time is involved, usually the system is assumed to evolve according to the first-order Markov assumption and, as consequence, a single time slice is considered.

In this chapter, we address DBN structures with more than one time slice. Moreover, the conditional probabilities of the DBN will be modelled by supervised machine learning techniques (also known as classifier or classification method). Two case studies will be particularly discussed: activity recognition for human-robot interaction and semantic place classification for mobile robotics navigation.

The observed data variable, denoted by $X = \{X_1, \ldots, X_{nx}\}$, enters into the DBN in the form of conditional probabilities $P(X|C)$, where the values of X are feature vectors. To give an idea of the dimensionality of X, in semantic place classification [6], the number of features can be $nx = 50$, while in activity recognition we have 51 features [8]. Given such dimensionalities, which can be even higher, it becomes infeasible to estimate the probability distribution that characterizes $P(X|C)$ without the use of advanced algorithms. Although a simple Naïve Bayes classifier can be incorporated in a DBN to model $P(X|C)$, more powerful solutions, such as the ensemble of classifiers in the DBMM approach introduced in Ref. [8], tend to achieve higher classification performance.

In summary, DBN is a direct acyclic graph (DAG) that consists of a finite set of events (the nodes or vertices) connected through edges (or arcs) that model the dependencies among the events and also the time variable. Here, the nodes are given by the variables $\{X, C\}$, and the dynamic (time-based) behaviour of the BDN is considered to be governed by the current time t and by a finite set of previous time slices $\{t-1, t-2, \ldots, t-T\}$. So, future time slices will be not considered. **Figure 2** shows the structure of the DBN, with $T + 1$ time slices, that will be considered in the problem formulation presented in the sequel.

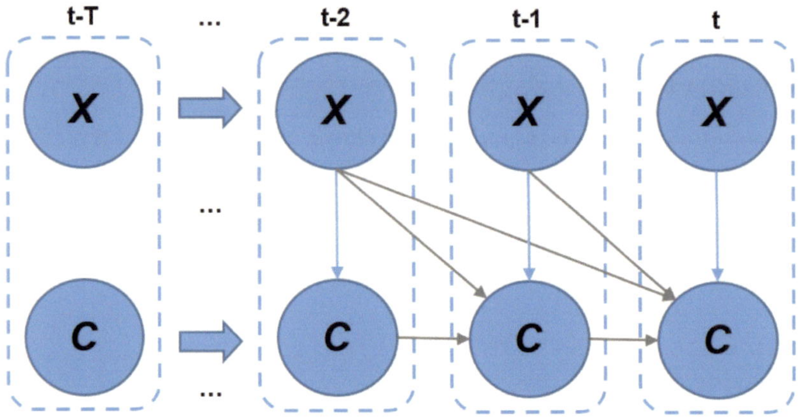

Figure 2. An example of a DBN with $T+1$ time slices and two nodes $\{C, X\}$.

3. Inference with DBN

The problem is formulated by considering $P(X^t, X^{t-1}, \ldots, X^{t-T}, C^t, C^{t-1}, \ldots, C^{t-T})$ i.e., the joint distribution of the nodes over the time up to T. The goal is to infer the current-time value of the class C^t given the data $X^{t:t-T} = \{X^t, X^{t-1}, \ldots, X^{t-T}\}$ and the prior knowledge of the class, which is attained by the a-posteriori probability $P(C^t \mid C^{t-1:t-T}, X^{t:t-T})$. The superscript notation denotes the set of values over a time interval: $\{t:t-T\} = \{t, t-1, t-2, \ldots, t-T\}$.

The simplest case is for a single time slice where the posterior reduces to $P(C^t \mid X^t) \propto P(X^t \mid C^t)P(C^t)$. For two time slices, we have

$$P(C^t \mid C^{t-1}, X^{t:t-1}) \propto P(X^t \mid X^{t-1}, C^{t:t-1})P(X^{t-1} \mid C^{t:t-1})P(C^t \mid C^{t-1})P(C^{t-1}). \quad (1)$$

As the number of time slices increases, the problem of inferring the class becomes more complex; therefore, some assumptions can be made in order to find a tractable solution. As a first assumption, let the nodes be independent of later (subsequent in time) nodes. As a consequence, and taking as the example for $T = 1$, the probability $P(X^{t-1} \mid C^{t:t-1}) = P(X^{t-1} \mid C^{t-1})$ that is, the node X^{t-1} does not depend on the node C^t which is after a time-slice. The second assumption, more strong, is that the feature-vector node X is independent for all time slices hence, and following the previous example, $P(X^t \mid X^{t-1}, C^{t:t-1})$ becomes $P(X^t \mid C^{t:t-1})$. Given these two assumptions, we can state the general problem of calculating the posterior probability of a DBN with $T+1$ time slices by the expression

$$P(C^t \mid C^{t-1:t-T}, X^{t:t-T}) = \frac{1}{\beta} \prod_{k=t}^{t-T} \left\{ P(X^k \mid C^k)P(C^k) \right\}. \quad (2)$$

where β is the scale (normalization) factor to guarantee that the values of the a-posteriori sum to one. The class-conditional probabilities $P(X^k \mid C^k)$ come from a supervised classifier or from an ensemble of classifiers as in Ref. [8], while $P(C^k)$ assumes the value of the previous posterior probability; thus, $P(C^t) \leftarrow posterior^{t-1}$.

This strategy for 'updating' the values of the prior by taking the values of previous posteriors is a very common and effective technique used in Bayesian sequential systems. The steps involved in the calculation of the posterior probability, as expressed in Eq. (2), are illustrated in **Figure 3**.

Selection of the class-conditional model to express $P(X \mid C)$ is an important part of the approach and can be achieved by well-known probabilistic machine learning methods. Although generative methods (e.g. Naïve Bayes, GMM and HMM) provide direct probabilistic interpretation and, therefore, constitute appropriate choices, discriminative methods (e.g. SVM, random forest and ANN) tend to have better classification performance. However, to be a suitable model, a given discriminative method has to be of a probabilistic form; this implies, at least, that the outcomes from the classifier sum to one. A more advanced method can be used to model $P(X \mid C)$ in a DBN, as the dynamic Bayesian mixture model (DBMM) [8], where a mixture of n classifiers is used to model the conditional probability which assumes the form $P(X \mid C) = \sum_{j=1}^{n} \omega_j P(X \mid C)_j$, $j = 1, \ldots, n$; where ω_j are the weighting parameters and $P(X \mid C)_j$ are the probabilities from the classifiers. Further details are provided in Ref. [6].

The product of likelihoods and priors, in the expression of the a-posteriori Eq. (2), has the consequence of penalizing the classes that are less likely to occur. In other words, the classes with low probability, i.e. close to zero, will have an even more low values of posterior; this effect is intensified as the number of time slices increases. Because the priors are recursively assigned by assuming the values of the previous posteriors, we suggest to use additive smoothing to avoid values of priors to be very close to zero.

Additive smoothing, also called Lidstone smoothing, adds a term (α) to the prior distribution and can be expressed as

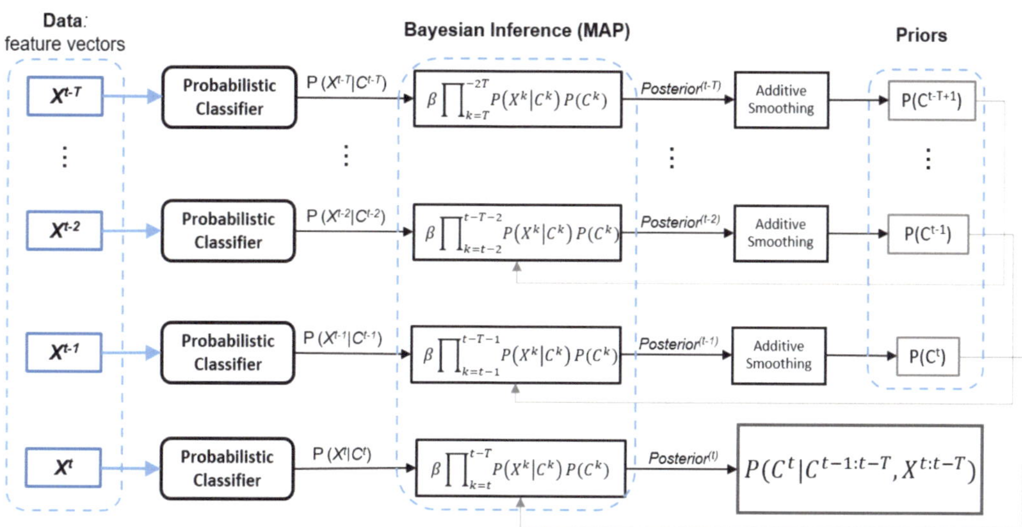

Figure 3. This figure illustrates the DBN, with $T + 1$ time slices, as formulated according to the assumptions presented in Section 3. The product of likelihoods and priors, over the time interval $[t - T, t]$, becomes the posterior probability as expressed in Eq. (2).

$$\hat{P}(C_i) = \frac{P(C_i) + \alpha}{1 + \alpha \cdot (nc)}, \ i = 1, \ldots, nc \quad (3)$$

where α is the additive smoothing factor and nc is the number of classes. The influence of α on the smoothed prior $\hat{P}(C_i)$ has to be such that the values of $\hat{P}(C_i)$ are greater than zero ($\hat{P}(C_i) > 0, \forall i$) and, moreover, the prior distribution $\hat{P}(C_i)$ should be consistent (the values of $\hat{P}(C_i)$ must of course sum to one). A practical range is $0 < \alpha < 0.1$.

Figure 4 provides an example of the impact of α on a given prior, with values of α equal to {0, 0.01, 0.05 and 0.1}. As the value of α increases, the prior distribution tends to lose its initial definiteness due to the uniform 'bias' introduced by α. In the example shown in **Figure 4**, we have considered a five-class case ($nc = 5$).

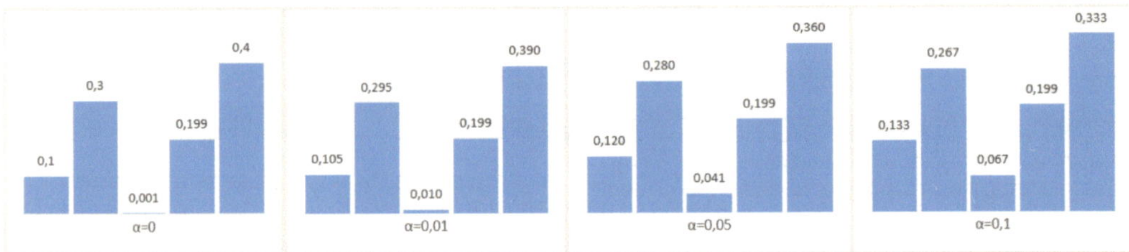

Figure 4. An example of the influence of the additive factor (α) in a given $P(C_i)$, $i = 1, \ldots, 5$.

4. Experiments on classification: mobile robotics case studies

In order to demonstrate the use of the DBN as formulated above, we will consider two classification problems that find applications in mobile robotics: semantic place recognition [6] and activity classification [8].

4.1. Semantic place recognition

Figure 5 illustrates a probabilistic system for semantic place recognition where data comes from a laser scanner sensor. In a practical application, the sensor is mounted on-board a mobile robot [6, 7]. Based on **Figure 5**, we can make a direct correspondence with the DBN discussed above by verifying that the feature vector is X, the probabilistic classifier outputs the class-conditional probability $P(X|C)$ and the priors transmit the time-based information through the network.

As an example of the DBN application in semantic place classification, let us report some results from Ref. [6], where a DBN was applied on the image database for robot localization (IDOL) dataset: available at http://www.cas.kth.se/IDOL/. In this context, the problem of semantic place classification can be stated as follows: 'given a set of features, calculated on

http://dx.doi.org/10.5772/intechopen.70059

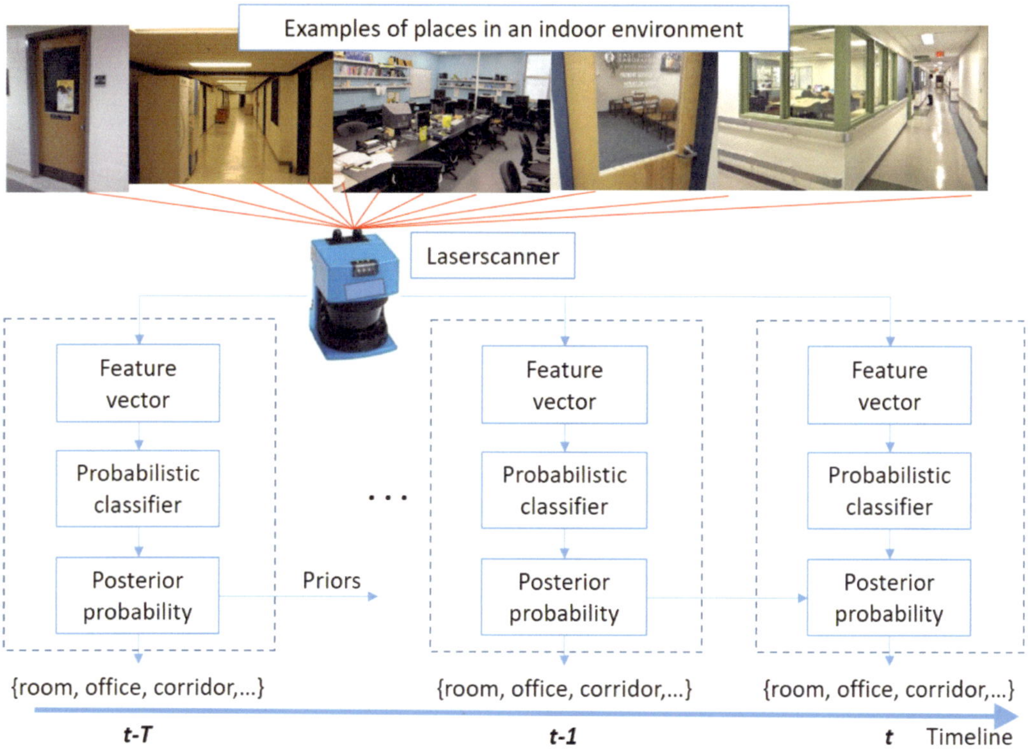

Figure 5. Illustration for a time-dependent probabilistic system applied in semantic place recognition. In this system, data obtained from a laser scanner.

data from laser scanner sensors (installed on-board a mobile robot), determine the semantic robot location ('corridor', 'room', 'office', etc) by using a classification method'. The experiments in Ref. [6] use a mixture of classifiers to model the class-conditional probability in the DBN; such approach is called DBMM [8].

Figure 6 shows recognition results in a sequence of nine frames from the IDOL dataset, where the first row depicts images of indoor places as captured by a camera mounted on-board a mobile robot. The second row provides classification results without time slices (i.e. time-base prior probabilities are not incorporated into the DBN), and the subsequent rows show classification probabilities for a DBN with time-slices up to three. In the figure, the vertical line (in red) indicates the transition between classes: from the class 'kitchen' (KT) to the class 'corridor' (CR).

4.2. Activity classification

In the case of the activity classification problem described here, the objective is to classify the human's daily activity based on spatiotemporal skeleton-based features. In such a case, mobile robots mounted with appropriated cameras can make use of such classification models to

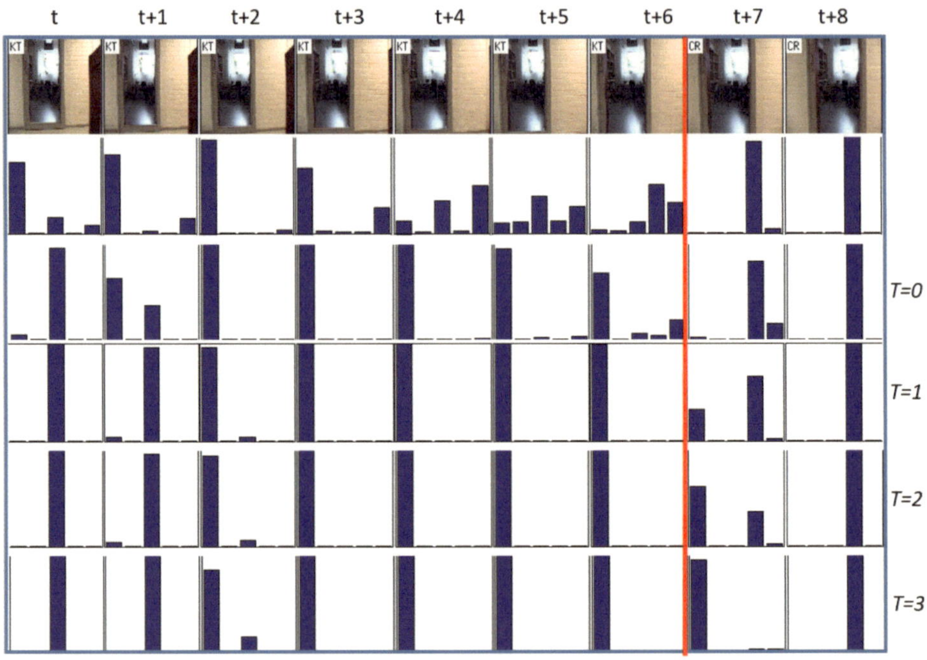

Figure 6. Classification results on a five-class semantic place recognition problem, extracted from reference [6], using a DBN with mixture models of three classifiers (DBMM [6, 8]).

improve the quality of life of, for example, old-age people, by assisting them in their daily life or detecting anomalous situations. Similar to semantic place recognition problem, the activity classification problem can also be seen as a time-dependent probabilistic system, where the feature vector X is the skeleton-based features. From Ref. [8], we report some results on the activity classification.

Figure 7 exhibits an activity classification framework, based on Ref. [8], which uses a DBN with mixture models (the DBMM approach as previously described in the semantic place classification problem), where the data is acquired by using an RGB-D sensor, followed by the skeleton detection step and the feature extraction process, where the latter is based on geometrical features. From the training stage, global weights are computed using an uncertainty measure (e.g. entropy) as a confidence level for each base classifier based on their performance on the training set. During the test, given the input data (i.e. skeleton features for the current activity), base classifiers are used and merged as mixture models with time slices (using previous time instant classification) to reinforce the current classification.

The well-known dataset for activity recognition Cornell Activity Dataset (CAD60) [9, 17] was used to evaluate the proposed framework in Refs. [8, 18]. The CAD-60 dataset comprises video sequences and skeleton data of human daily activities acquired from a RGB-D sensor. There are 12 human' daily activities performed by four different subjects (two male and two female, one of them being left-handed) grouped in five different environments: office, kitchen,

http://dx.doi.org/10.5772/intechopen.70059

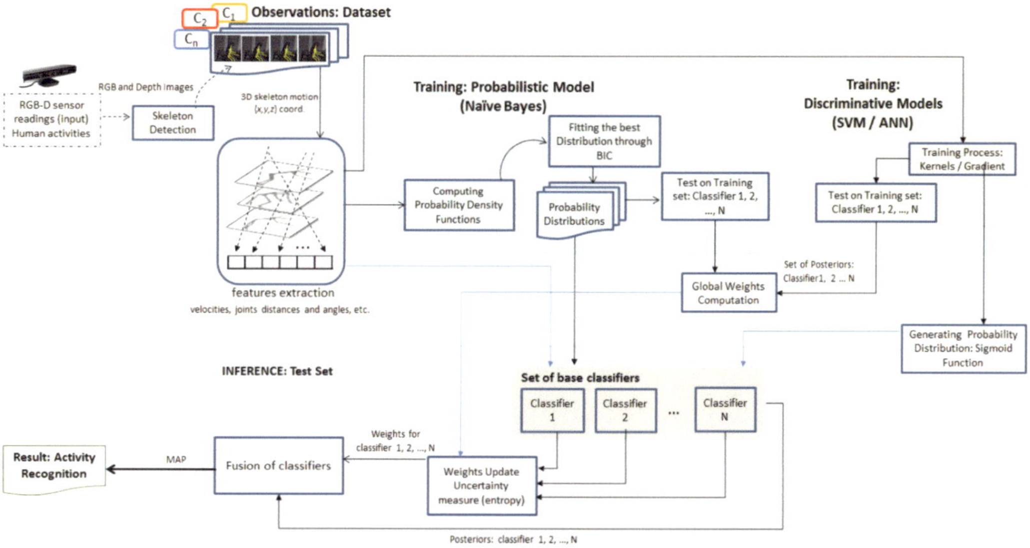

Figure 7. Illustration for a time-dependent probabilistic system applied to activity classification. In this system, data obtained from a RGB-D camera, which provides the spatiotemporal skeleton-based features.

bedroom, bathroom and living room. Additionally, the CAD-60 dataset has two more activities (random movements and still), which are used for classification assessment on test sets, in order to evaluate precision and generalization capacity of the approaches since these activities encompass similar movements to some other activities. We have adopted the same strategy described in Ref. [17], so that we present the classification results in terms of precision (Prec) and recall (Rec) for each scenario. The evaluation criterion was carried out using leave-one-out cross-validation. The idea is to verify the generalization capacity of the classifier by using the strategy of 'new person', i.e. learning from different persons and testing with an unseen person. The classification is made frame-by-frame to account for the accuracy of the frames correctly classified.

Results show the DBMM approach obtained better classification performance compared to other state-of-the-art methods presented in the ranked table in Ref. [17]. The overall results were precision: 94.83%; recall: 94.74% and accuracy: 94.74%. **Figure 8** presents the classification performance (i.e. precision and recall) for the 'new person' tested in each scenario. For comparison purposes, **Table 1** summarizes the results in terms of accuracy of state-of-the-art single classifiers and a simple averaged ensemble compared with the proposed DBMM for the bedroom (scenario with more misclassification), showing that our approach outperforms other classifiers. The classification performance in terms of overall accuracy, precision and recall has shown that our proposed framework outperforms state-of-the-art methods that use the same datasets [17].

In this section, we have shown the DBMM [8, 18] performance using an offline dataset. Additionally, further tests using a mobile platform with an RGB-D sensor on-board running on-the-fly in an assisted living context was also successfully validated with accuracy above

Figure 8. Performance on the CAD-60 ('new person'). Results are reported in terms of precision (Prec) and recall (Rec) and an average (AV) per scenario. Overall AV: precision 94.83%; recall: 94.74%. Activities in (a): Act1—rinsing water; Act2—brushing teeth; Act3—wearing lens; Act4—random + still; activities in (b): Act1—talking on phone; Act2—drinking water; Act3—opening container; Act4—random + still; activities in (c): Act1—talking on phone; Act2—drinking water; Act3—talking on coach; Act4—relaxing on coach; Act5—random + still; activities in (d) Act1—drinking water; Act2—cooking chopping; Act3—cooking stirring; Act4—opening container; Act5—random + still; activities in (e): Act1—talking on phone; Act2—writing on whiteboard; Act3—drinking water; Act4—working on computer; Act5—random + still.

Location	Activity	Bayes	ANN	SVM	AV	DBMM
Bedroom	1	79.90%	74.70%	74.90%	76.50%	84.10%
	2	72.70%	76.60%	81.40%	76.90%	86.40%
	3	79.60%	91.10%	93.10%	87.90%	98.30%
	4	65.70%	93.50%	92.60%	83.90%	97.40%
	Average	**74.48%**	**83.98%**	**85.50%**	**81.30%**	**91.55%**

Activity: 1—talk.on phone, 2—drink.water, 3—open.container, 4—random + still.

Table 1. Results in terms of accuracy on the bedroom scenario of the CAD-60 dataset ('new person') using single classifiers, a simple averaged ensemble (AV) and the DBMM.

90%, as reported in Ref. [18]. More details about the DBMM using a mobile robot for activity recognition and a video showing the classification performance can be found in Ref. [18].

5. Conclusion

In this chapter, the authors have presented a DBN formulation for classification of time-dependent problems together with experimental results on applications of two mobile robots. The first one regarding the semantic place classification and the second one based on activity classification. In both formulations, the DBN was used as basis to compose the DBMM [6, 8, 18], a more complex structure used to handle more complex scenarios. In both applications, the DBMM has shown to be a powerful choice in modelling of time-dependent scenarios.

When it comes to semantic place classification, the model could detect classes' transitions during the robot navigation, thanks to the different time slices (i.e. higher than 2) and the additive smoothing used in the model. In the case of activity recognition, since the activities in the dataset do not have classes' transitions, i.e. only one activity is performed during a task, in this case, a simple version of the DBMM using only one time slice is enough to correct classify all activities. For real-time applications using a mobile robot and in accordance with experimental results reported in Ref. [6], it is suggested to use more than two time slices in the mode.

Author details

Cristiano Premebida[1]*, Francisco A. A. Souza[1] and Diego R. Faria[2]

*Address all correspondence to: cpremebida@isr.uc.pt

1 Institute of Systems and Robotics, University of Coimbra, Coimbra, Portugal

2 School of Engineering & Applied Science, Aston University, Birmingham, UK

References

[1] Friedman N, Murphy K, Russell S. Learning the structure of dynamic probabilistic networks. In: Proceeding of the Fourteenth Conference on Uncertainty in Artificial Intelligence (UAI'98); 24-26 July 1998; Madison, Wisconsin. San Francisco, CA, USA: Morgan Kaufmann Publishers Inc.; 1998

[2] Korb KB, Nicholson AE. Bayesian Artificial Intelligence. 2nd ed. Boca Raton, FL: CRC Press, Inc. 2010

[3] Koller D, Friedman N. Probability graphical models: Principles and techniques. In: Adaptive Computation and Machine Learning. The MIT Press; Cambridge, MA, USA; 2009

[4] Murphy KP. Dynamic Bayesian networks: Representation, inference and learning. Ph.D. Dissertation. University of California, Berkeley; 2002

[5] Mihajlovic V, Petkovic M. Dynamic Bayesian Networks: A State of the Art. Technical Report, Computer Science Department, University of Twente, Netherlands; 2001

[6] Premebida C, Faria D, Nunes U. Dynamic Bayesian network for semantic place classification in mobile robotics. Autonomous Robots (AURO), Springer; 2016

[7] Rottmann A, Mozos OM, Stachniss C, Burgard W. Semantic place classification of indoor environments with mobile robots using boosting. In: Proceeding of the 20th National Conference on Artificial Intelligence (AAAI'05); 9-13 July 2005; Pittsburgh, Pennsylvania: AAAI Press; 2005

[8] Faria DR, Premebida C, Nunes C. A probabilistic approach for human everyday activities recognition using body motion from RGB-D images. In: Proceedings of the IEEE RO-MAN'14: International Symposium on Robot and Human Interactive Communication; 25-29 August. 2014; Edinburgh, UK. IEEE; Cambridge, MA, USA; 2014

[9] Sung J, Ponce C, Selman B, Saxena A. Unstructured human activity detection from RGBD images. In: Proceedings of the IEEE International Conference on Robotics and Automation (ICRA), Saint Paul, MN, New York, NY, USA, May 2012; pp. 842-849

[10] Thrun S, Burgard W, Fox D. Probabilistic Robotics. MIT Press; New Jersey, NJ, USA; 2005

[11] Li T, Prieto J, Corchado JM, Bajo J. On the use and misuse of Bayesian filters. In: Proceeding of the IEEE 18th Int. Conference on Information Fusion (Fusion); 6-9 July 2015; Washington, DC, USA. IEEE; 2015

[12] Chen Z. Bayesian filtering: From Kalman filters to particle filters and beyond. Statistics. 2003;**182**(1):1-69

[13] Bishop CM. Pattern recognition. Machine Learning. 2006;**128**:1-58

[14] Duda RO, Hart PE, Stork DG. Pattern Classification. John Wiley & Sons; New Jersey, NJ, USA

[15] Neapolitan RE. Learning Bayesian Networks. Upper Saddle River, NJ, USA: Prentice-Hall, Inc.; 2003

[16] Russell S, Norvig P. Artificial Intelligence: A Modern Approach. 3rd ed. Prentice Hall; New Jersey, NJ, USA; 2010

[17] Cornell Activity Datasets CAD-60 [Internet]. Available from: http://pr.cs.cornell.edu/humanactivities/data.php [Accessed: January 2017

[18] Faria DR, Vieira M, Premebida C, Nunes U. Probabilistic human daily activity recognition towards robot-assisted living. In: Proceeding of the IEEE RO-MAN'15: IEEE International Symposium on Robot and Human Interactive Communication; Kobe, Japan; New York, NY, USA; September 2015

Section 4

Applications of Bayesian Inference in Economics

Chapter 16

A Bayesian Model for Investment Decisions in Early Ventures

Anamaria Berea and Daniel Maxwell

Additional information is available at the end of the chapter

http://dx.doi.org/10.5772/intechopen.70051

Abstract

In this research, we present a Bayesian model to aid the investment decision in early stage start-ups and ventures. This model addresses both the venture and the angel investing markets. The model is informed both by previous academic literature on entrepreneurship and by venture capital investment practices. The model is validated through an anonymized experiment where reviewers with previous experience in entrepreneurship or investment or both scored a list of 20 anonymous real companies for which we knew the outcome a priori. The experiment revealed that the model and online scoring platform that we built provide an accuracy of 83% in identifying companies that would later on fail and where the investments would be lost. The model also performs fairly well in identifying companies where the investors would not lose their money but they would either have to wait for a very long time on their returns or they would not receive large return on investment (ROI), and we also show that the model performs modestly in identifying "big exit" companies or companies where the investors would receive high ROI and in a fairly short amount of time.

Keywords: Bayesian networks, investment, start-up, entrepreneurship, decision models

1. Introduction

One of the biggest challenges facing early stage investors is a lack of actionable data and effective analytics. Most investment decisions are made based on the instinct (heuristics) of the investor who may or may not have experience in the sector and decisions are often inherently biased. In investment environment is increasingly complex, and investors cannot process all of the factors that are critical to the success of a potential investment and make a well-informed decision. Research suggests that well-built analytic models make better decisions than human experts across virtually every field [1].

Some of the newest data on the returns on angel investment show that these are about 2.5 times the value of the initial investment and the average period of recovery of investment is 3.6 years [2].

In general, there is little literature with respect to automated techniques or models of investment decision. A very recently published paper shows an interesting risk analysis model that would reduce the risk of investing in early entrepreneurs [3]. This research takes a similar approach—reduce the "bad" investment decisions—but it uses a different model, based on a Bayesian model, which performs well in identifying the future failures of new ventures.

While there is understandably little academic literature on forecasting future star-up success and its relationship to investment decision-making, due to the confidentiality of the data in this business, the decision-making practice in the venture capital and angel investment industries rely heavily on the experience of the investors and on the "collective" thinking of the investors that gather together to rate or assess the pitches or business proposals for various funding rounds of investment.

Therefore, this chapter presents a model for investment decision-making that is informed mainly by the practitioners and is intended to be applied in to investment practice. Its aim is to be a tool that helps the process of rating seed and start-up ventures become more informative and transparent both for investors and for entrepreneurs.

The model built for this research is mainly informed by the interviews and discussions conducted with investors during the summer of 2014. The nodes of the model and the dependencies between the nodes have been created based on these interviews, while the distributions of the prior probabilities have been informed by the academic literature where such information could be found, otherwise they are normal.

This research describes the model in general terms, how it has been implemented in practice and the results of two experiments that have been run to provide validity of its forecasting accuracy. The construction, implementation, and validation of the model, as well as a discussion of findings are presented in the following sections below.

The rest of this chapter is structured as follows: Section 2 describes the model and the rationale behind building it; Section 3 describes the experiments that were conducted using this model, mainly with the purpose of validating its accuracy; Section 4 presents the results from the experiments and an analysis of the accuracy of the model; and Section 5 summarizes succinctly the conclusions of this research.

2. The Bayesian investment decision model

We used Bayesian networks modeling to build a probabilistic assessment model of early stage companies or ventures. We based our selection of nodes/factors on a series of interviews and working closely with practitioners in venture capital funding. We afterwards implemented this model on an online platform, available at www.exogenius.net (see **Figure 1**).

http://dx.doi.org/10.5772/intechopen.70051

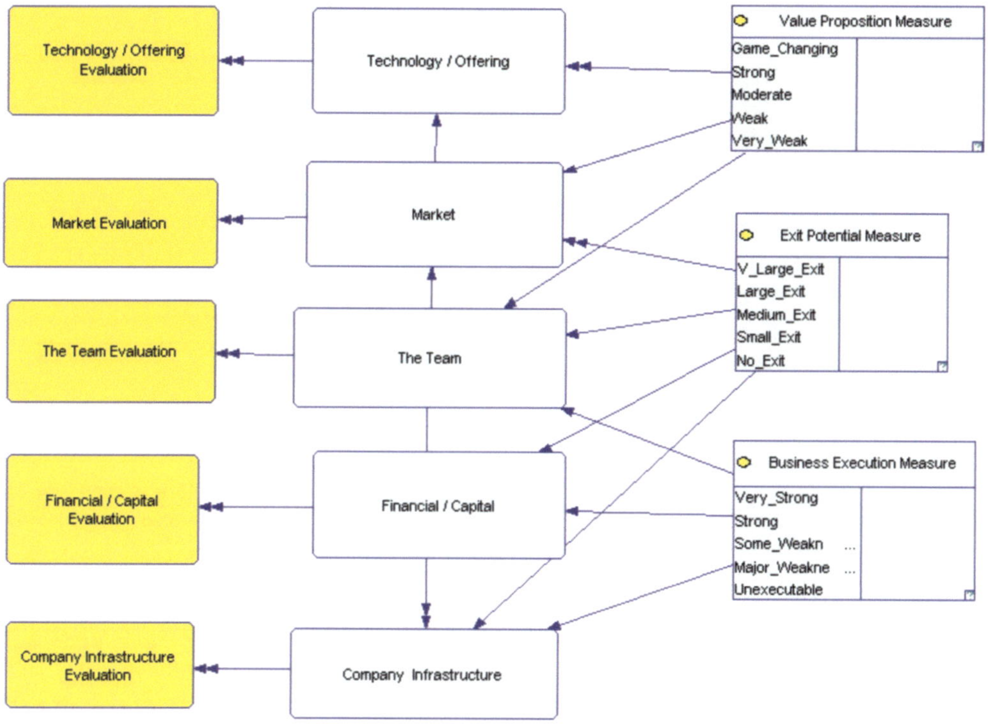

Figure 1. The Bayesian model for investment decision.

The Bayesian model scores on a scale of [0, 100] the potential performance of a company/start-up by identifying three key measures: *business execution, value proposition,* and *exit potential* (see **Figure 2**). These measures are aggregated (nonlinearly) into an overall score of performance. Each of these three important measures scores the future potential of a project or start-up in regard to their proposition (which may be a technological innovation, a social value, or any business value that the entrepreneur presents as the core proposition), their ability to sustain, carry out, and fulfill their proposition (business execution) and the potential of this new venture to exit (either through IPO, buy-out, or in any manner that would be satisfactory for the investor).

Each of these three measures is a child of five subnetworks in the model, which are represented by more granular parent-children nodes each. These five subnetworks are business/entrepreneurship factors or indicators that are measuring the new venture on the following aspects of the business proposal: technical difficulty, uniqueness of innovation, readiness for market, customer engagement, team performance, entrepreneurial and managerial experience, founders and incorporation of the company, and many more. Each of the granular nodes in the model is represented by three to five states and they are informed either by the evidence from published literature (as described below) or otherwise by a uniform distribution priors [4].

The conditional tables of each node have been readjusted after sensitivity analysis was performed, based on data and facts previously published in the entrepreneurship and high-growth companies literature [5–7].

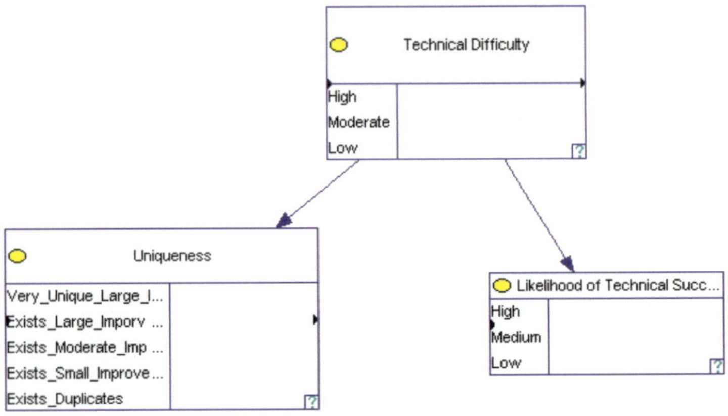

Figure 2. Example of one of five subnetworks of the model—the technology offering is represented by three granular nodes.

For example, the states of the technology (marginal versus breakthrough) node are defined according to the literature on entrepreneurship [7–9]; the number of founders is also determined based on these prior findings, i.e., the state of 2–4 founders has the highest positive impact on the final score, while the other states have low impact or negative impact (more than 5 founders lower the chances of success significantly) [5, 6].

The nodes representing the team complementarity, coordination, and learning are based on the findings of the Startup Genome Project, which was run at Berkley and Stanford Universities [5, 10, 11]. In other words, since the findings show that team complementarity and learning are critically important for the success of the early ventures, the team node in the model reflects these findings through the distribution of prior in its states.

Similarly, the nodes that are assessing the infrastructure of the start-up (broadly construed as not only physical requirements to develop the proposed technology, but also legislative, financial, or logistic infrastructure), are informed by the currently published probabilistic values in previous studies on organizational emergence [12].

The placement of the new venture in the current market is also assessed, and this is done based on the assessment of the projected growth of the company relative to the projected growth of the market or of the industry [7, 12].

For the development of the model, we used both UnBBayes [13] and GeNIe/SMILE [14] open-source softwares dedicated to Bayesian modeling. After the model was built, tested, and developed, it was migrated on the online platform, with easy to use user interface, and where we ran our experiments.

The implementation of the model on an online platform facilitated experimentation for forecasting accuracy. The nodes of the model that provide new evidence, specific to each venture, are represented as a series of 23 questions in a user-friendly interface. For example, the evidence node in the model that represents the uniqueness of the offering became the question "How unique is the proposed offering (idea/innovation/technology/product/service)?" in the online platform. The nodes that were not evidence in the model have obviously not been represented as questions in the online implementation. The reviewers/users have the possibility to see the progression of the three key scores (value proposition, business execution, and exit potential) as well as the final score as they go through answering the individual assessment questions.

3. The experimental design for model validation

In order to validate the accuracy of the model scores, an anonymized experiment was designed, where 20 case studies of companies were recreated from real, historical companies. These case studies included the state of funding and potential of various companies while they were start-ups, before their first or second seed funding and the aim of the experiment was to show whether the exit or the overall scores of the model align statistically with what happened in real life.

In the experiment, there were randomly picked 20 historical cases for which we know the ground truths about their financial history (how they started, how much was their initial funding, and how much was their exit), by using publicly available information from Crunch-Base website, Wikipedia and various failed start-ups, and postmortems case studies. The companies in the sample for the experiment had either high exits (were bought for more than $500 million), medium exits (were bought for 100–1000K or they took a very long time to exit, i.e., 20 years), or no exits (they shut down or went bankrupt soon after their launch).

Each of these 20 case studies in the sample were recreated as anonymous business proposals, given the information at the time when they were seeking initial funding (i.e., 2010). Therefore, each of these anonymized case studies included the following information: the year when the reviewer had to "travel back in time" (i.e., 2010), with a hyperlink toward published most important business and technological events of that year (i.e., the economist), the company location, the number of founders, the type of incorporation, anonymized information about the founders experience, information about the market and industry at that time, information about the customers, the team, the infrastructure, about the financial past of the company if it existed and, most importantly, information about the product or technology without disclosing its brand name. The reviewers were also free to look for additional information on the web regarding the state of technology and business at that particular time in the past. The oldest case study was placed in 1999 and the newest one in 2014.

In other words, all the possible information about a company that could be included prior to the time of their initial funding request was we included, as long as it could be anonymized.

We conducted two experiments: one with experts in business or investing and other with MBA students at the University of Maryland.

The first experiment was carried by 24 volunteer reviewers, who reviewed five of these anonymous case studies each, by answering the questions from online platform at the forefront of our model for each of their assigned five case studies. The reviewers in the experiment are experienced as either entrepreneurs or investors; therefore, they are a panel of experts that completed the experiment.

The second experiment was carried by MBA students at the University of Maryland, in a 1 h long session. The students were also randomly assigned five case studies each and answered the same questions from the online platform as the experts did.

4. Results and accuracy analysis

The first experiment started on March 22, 2016 and by April 13, 2016, 54% of reviewers completed their reviews. We collected 68 (reviews) X-4 (scores) data points. The second experiment was carried out during 1 day in October 2016.

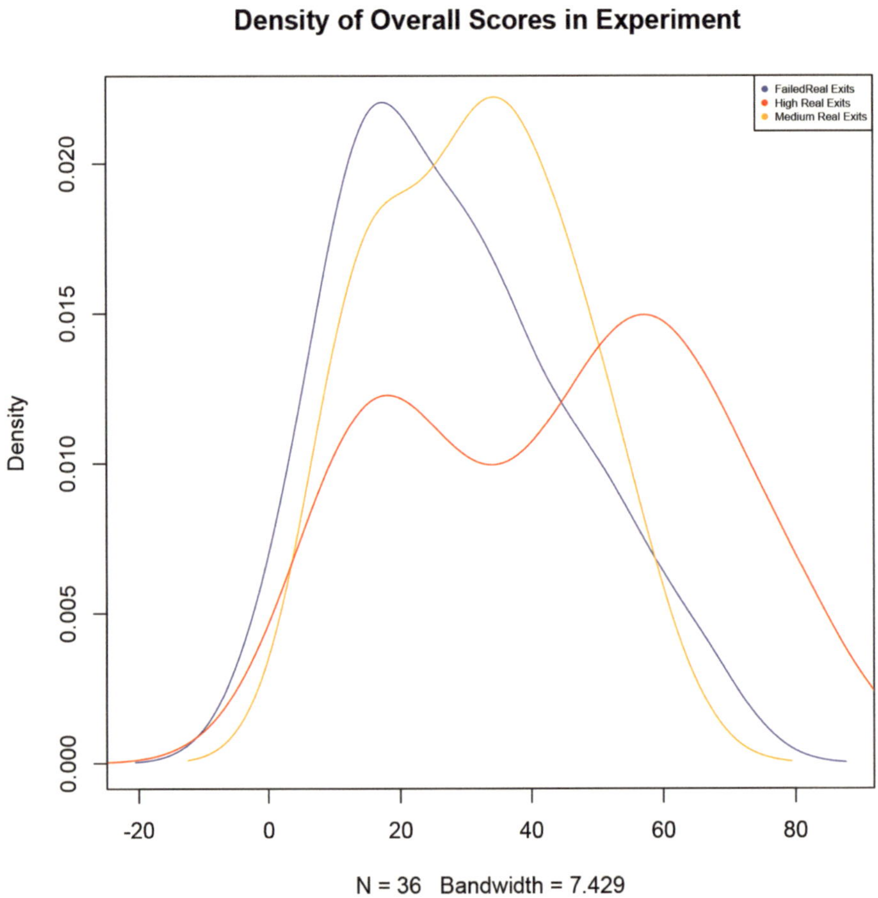

Figure 3. The distribution of overall reviewing scores in the expert experiment. This figure shows the scores on a scale of 0–100 that were given by the professional reviewers (investors and entrepreneurs) in the overall rating for the companies in each of the three groups—high-exits; medium exits; and no exits. The distributions of the reviewers scores show that low exits were scored between 0 and 40 with most scores around a value of 20; medium exits were scored between 0 and 80 with most scores around 40; and the high exits were scored either with scores around 20 or scores around 60.

http://dx.doi.org/10.5772/intechopen.70051

A reviewer provides the observations for the evidence nodes/questions in the model. The model then provides a distribution on all scores as output, conditional on these observations. Thus, the Bayesian model here is a three-layer model where the metrics are at the top level in the network and the observations (market evaluation, team evaluation, etc.) are at the bottom layer of granular nodes.

Both the measures in the model and the observations are discrete.

The data from the anonymized experiments were rematched with the ground truth data from the real case studies and compared the experiments with the evidence on three groups of companies (high exits, medium exits, no exits). The distributions of the exit scores and the overall scores from the experiment for each of these groups are plotted on the following figures (see **Figures 3–7**).

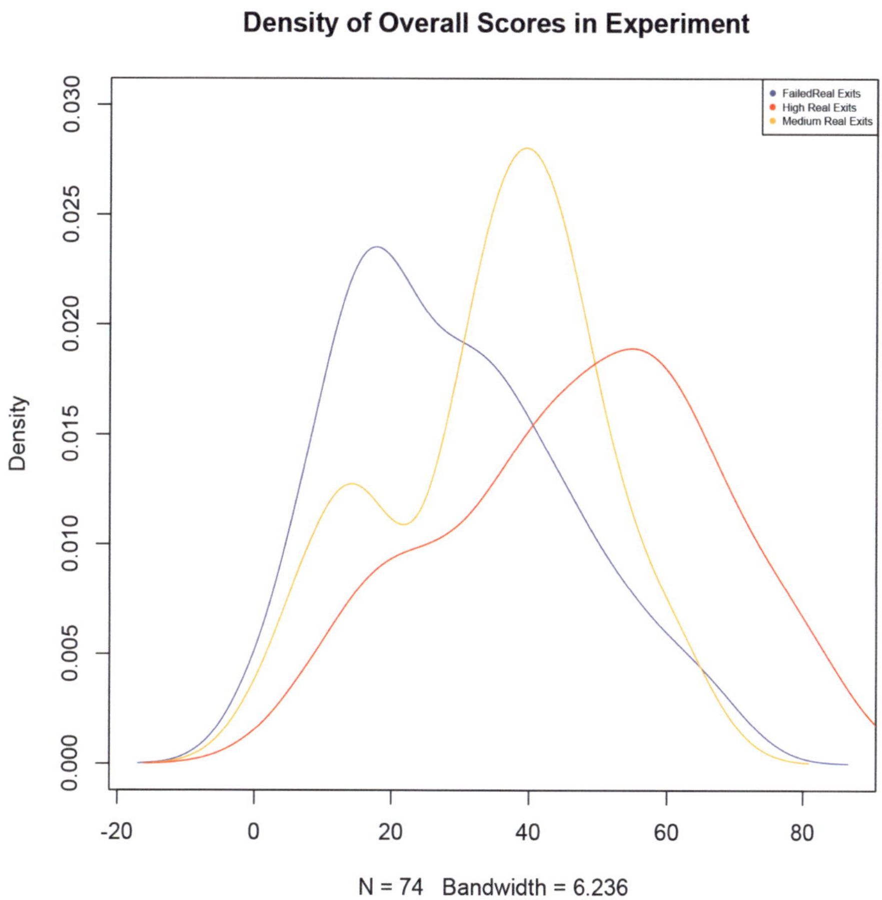

Figure 4. The distribution of overall reviewing scores in the MBA students experiment. Similarly to the plot above, this figure shows the scores on a scale of 0–100 that were given by the University of Maryland students in the overall rating for the companies in each of the three groups—high exits; medium exits; and no exits. The distributions of the reviewers scores show that low exits were scored between 0 and 40 with most scores around a value below 20; medium exits were scored between 0 and 80 with most scores around either 20 or 40; and the high exits were scored either with scores between 20 and 60.

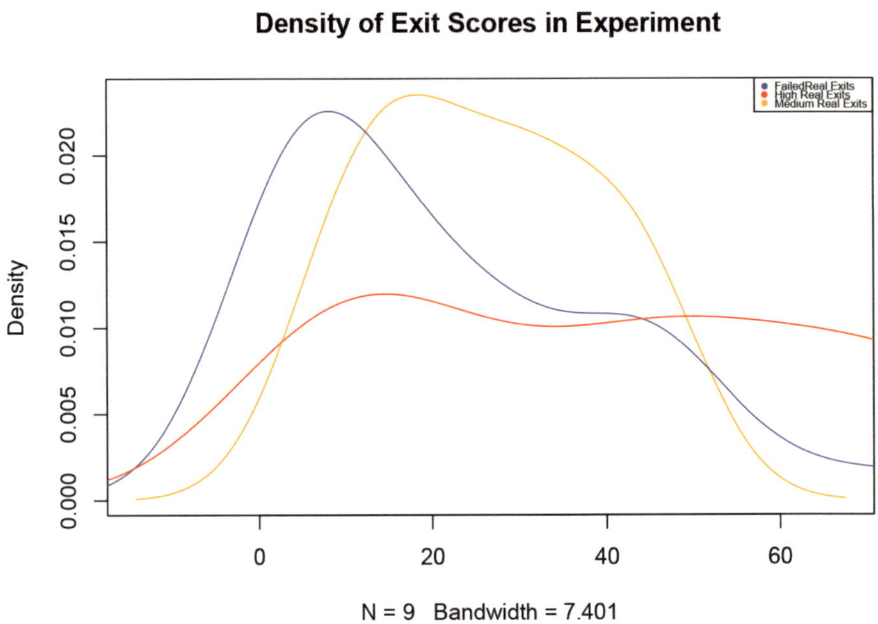

Figure 5. The distribution of exit reviewing scores in the expert experiment. This figure is similar to **Figure 3**, except that these are the scores of the professional reviewers for the exit node and not the overall score. The low exits were scores mainly with values close to 0, medium exits with scores between 10 and 60, and high exits scores were very close to a uniform distribution.

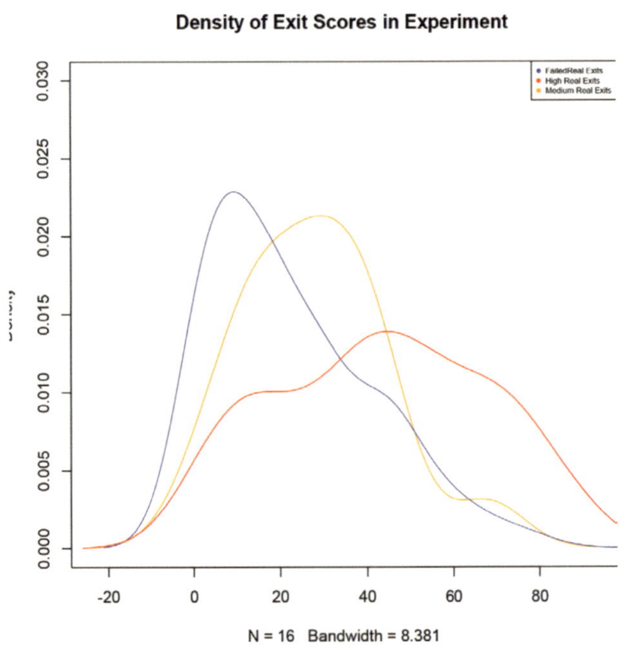

Figure 6. The distribution of exit reviewing scores in the MBA students experiment. Similarly as above, this figure shows the distribution of the exit scores for the student reviewers. The scores of the low exit companies were close to zero, the ones of the medium exits around 30 and the ones of the high exits exhibit a much larger range of scores, from 0 to 100.

http://dx.doi.org/10.5772/intechopen.70051

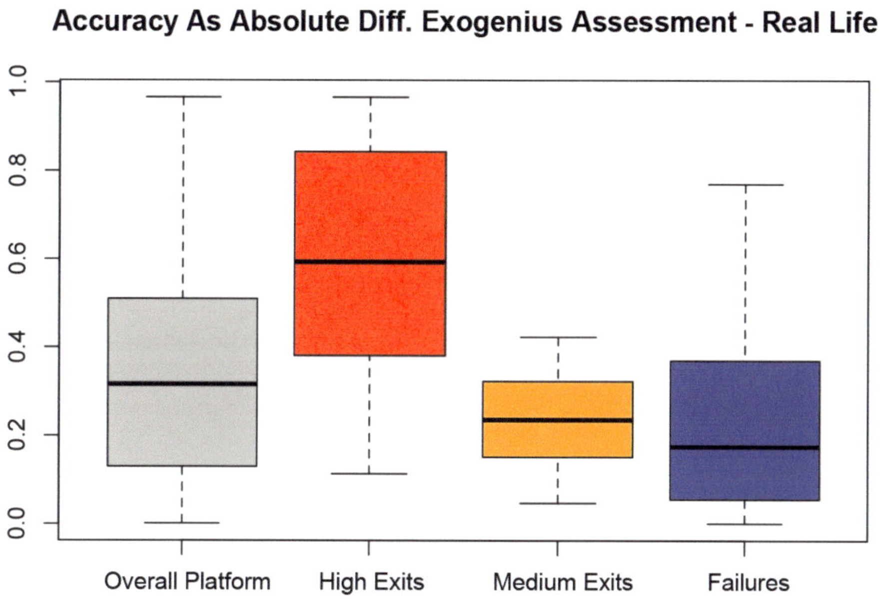

Figure 7. The overall accuracy of the Bayesian model in the expert panel experiment.

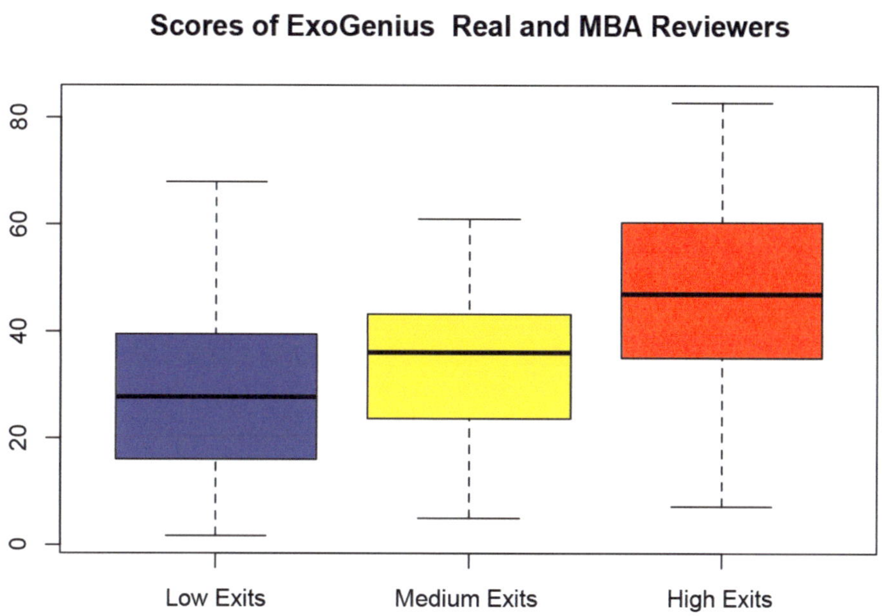

Figure 8. The overall accuracy of the Bayesian model in both experiments.

We can observe from these distributions that the "no exits" or "failures" scored low in both experiments, that the medium exits had medium scores in both experiments, and that the high exits had low, medium, and high scores in both experiments, whether we look at the final overall score or only at the exit key intermediate score (see **Figure 8**).

	Failed companies	Medium-exit companies	High-exit companies
Experiment mean scores	0.20	0.31	0.42
Experiment median scores	0.16	0.28	0.46
Accuracy	0.83	0.77	0.41

Table 1. A summary of the model accuracy based on the experimental results.

In other words, there is consistency between the two groups of reviewers with respect to each of the three groups of companies. Moreover, there is consistency in the reviewers responses and the ground truth data with respect to low-exit and medium-exit companies, but less so for high-exit companies. In other words, we can use this model to identify failures or low exits, but less so to identify high exits and, therefore, the model is designed to prune out "bad" proposals from a pool of varied investment opportunities.

Between the two experiments, we can also observe that the experts are still slightly better than MBA students at identifying low and medium exits.

The responses from the experiment for the "no exits" had a mean exit score of 20% and a median exit score of 16% and a mean and median overall score of 27% with a standard deviation of 16–17%. This means that the companies that failed in real life were reviewed with scores in the range of 16–27% in our model.

The medium exits experimental data had a mean exit score of 31%, a median of 28% and an overall mean and median of 34 and 36%, respectively, with standard deviations of 20 and 17%, respectively. This means that the companies that had medium exits (either low in capital value or took very long to exit) scored around the probabilities of 28–36% in our model.

The high exits had a mean and median exit score of 42%, an overall mean and median of 46% and a standard deviation of 28and 25%, respectively. This means that companies that were bought for more than $500 million in real life scored around 42–46% in our model (see **Table 1**).

The accuracy performance of the model was analyzed by using simple quantitative forecasting analysis. Specifically, the mean absolute deviation was used as a metric to calculate the forecasting error. The resolution value of 1 was considered for the companies with high exits, 0.5 for the medium exist, and 0 for the failed or no exit companies. The difference between these resolutions and the actual probabilities given by the reviewers was calculated as a mean absolute deviation. Based on this calculation, the overall accuracy of the model is situated at 75%, the accuracy for the no exits is valued at 83% and the accuracy for the medium and high exits is 77 and 41%, respectively (see **Table 1**).

5. Conclusions

In this research, a probabilistic model that assesses the potential for exit and overall performance of new ventures (start-ups) is presented, from building it based on practice and published statistical data, to its implementation in a readily available online platform that can be used by entrepreneurs and investors alike. The model is designed to assess quantitatively the potential of business while they are still at the very initial stages. The model is well informed with facts that we know from previous academic literature on entrepreneurship and high-growth companies, as well as informed in detail with venture capital experience and practices by working closely with them during the development phase of the model.

The model is validated using two anonymized experiments with experts in the field and MBA students and is currently translated into a commercial product. The results of these experiments and the details of the model are being presented in this chapter as both a validation method and as a viable metric or indicator that can detect ahead of time the future failures and "bad investments." This model can thus be also used by entrepreneurs to self-assess and identify points of weakness in their proposals and current seed ventures. Therefore, this research is presenting a tool for investment decision that can be easily automated and scaled up for the use of any potential investor, either angel or venture or any entrepreneur.

At the same time, these research efforts are also a good pathway to shed more transparency in the investment road map.

Acknowledgements

The author would like to thank Marco Rubin for his professional expertise and professor David Kirsch and his MBA students at the Smith School of Business for help with conducting the experiments and very useful comments.

Author details

Anamaria Berea[1]* and Daniel Maxwell[2]

*Address all correspondence to: aberea@rhsmith.umd.edu

1 Center for Complexity in Business, University of Maryland, College Park, MD, USA

2 KaDSci, VA, USA

References

[1] Tetlock PC. Superforecasting: The Art and Science of Prediction. Random House; 2015

[2] Wiltbank R, Boeker W. Returns to Angel Investors in Groups [Internet]. 2007. Available from: https://ssrn.com/abstract=1028592 or http://dx.doi.org/10.2139/ssrn.1028592

[3] Bodily S. Reducing risk and improving incentives in funding entrepreneurs. Decision Analysis. 2015;**13**(2):101–116

[4] Pearl J. Probabilistic Reasoning in Intelligent Systems: Network of Plausible Inference. Morgan Kaufmann; 1987

[5] Berea A. Essays in high-impact companies and high-impact entrepreneurship [thesis]. George Mason University; 2012

[6] Zoltan J. Acs. Foundations of High Impact Entrepreneurship. Boston, MA; 2008

[7] Shane SA. The Illusions of Entrepreneurship: The Costly Myths that Entrepreneurs, Investors and Policy Makers Live By. Yale University; 2008

[8] Arthur BW. The Nature of Technology. Free Press; 2009

[9] Auerswald P, Kauffman S, Lobo J, Shell K. The production recipes approach to modeling technological innovation: An application to learning by doing. CAE Working paper. 1998:98–10

[10] Marmer MH, Bjoern L, Dogrultan E, Berman R. Startup Genome Report. Technical report. Berkley University and Stanford University; 2011

[11] Botazzi G, Cefis E, Dosi G. Corporate growth and industrial structures: Some evidence from the italian manufacturing industry. Industrial and Corporate Change. 2002;**11**(4):705–723

[12] Wolley JL. Studying the emergence of new organizations: Entrepreneurship research design. New Perspectives on Entrepreneurship Research. 2011;**1**(1)

[13] Carvalho RN, Onishi MS, Ladeira M. Development of the Java version of the UnBBayes Framework for probabilistic reasoning. In: Congresso de Iniciacao Cientifica da UnB. Brasilia, DF, Brazil: University of Brasil; 2002

[14] GeNIe and SMILE, software developed at the Decision Systems Laboratory, School of Information Sciences, University of Pittsburgh

Chapter 17

Recent Advances in Nonlinear Filtering with a Financial Application to Derivatives Hedging under Incomplete Information

Claudia Ceci and Katia Colaneri

Additional information is available at the end of the chapter

http://dx.doi.org/10.5772/intechopen.70060

Abstract

In this chapter, we present some recent results about nonlinear filtering for jump diffusion signal and observation driven by correlated Brownian motions having common jump times. We provide the Kushner-Stratonovich and the Zakai equation for the normalized and the unnormalized filter, respectively. Moreover, we give conditions under which pathwise uniqueness for the solutions of both equations holds. Finally, we study an application of nonlinear filtering to the financial problem of derivatives hedging in an incomplete market with partial observation. Precisely, we consider the risk-minimizing hedging approach. In this framework, we compute the optimal hedging strategy for an informed investor and a partially informed one and compare the total expected squared costs of the strategies.

Keywords: nonlinear filtering, jump diffusions, risk minimization, Galtchouk-Kunita-Watanabe decomposition, partial information

1. Introduction

Bayesian inference and stochastic filtering are strictly related, since in both approaches, one wants to estimate quantities which are not directly observable. However, while in Bayesian inference, all uncertainty sources are considered as random variables, stochastic filtering refers to stochastic processes. It also covers many situations, from linear to nonlinear case, with various types of noises.

The objective of this chapter is to present nonlinear filtering results for Markovian partially observable systems where the state and the observation processes are described by jump diffusions with correlated Brownian motions and common jump times. We also aim at applying this theory to the financial problem of derivatives hedging for a trader who has limitative information on the market.

A filtering model is characterized by a signal process, denoted by X, which cannot be observed directly, and an observation process denoted by Y whose dynamics depends on X. The natural filtration of Y, $\mathbb{F}^Y = \{\mathcal{F}_t^Y, t \in [0, T]\}$, represents the available information. The goal of solving a filtering problem is to determine the best estimation of the signal X_t from the knowledge of $\mathcal{F}_t^Y$. Similar to optimal Bayesian filtering, we seek for the best estimation of the signal according to the minimum mean-squared error criterion, which corresponds to compute the posterior distribution of X_t given the available observations up to time t.

Historically, the first example of continuous-time filtering problem is the well-known Kalman-Bucy filter which concerns the case where Y gives the observation of X in additional Gaussian noise and both processes X and Y are modeled by linear stochastic differential equations. In this case, one ends up with a filter having finite-dimensional realization. Since then, the problem has been extended in many directions. To start, a number of authors including Refs. [1–3] studied the nonlinear case in the setting of additional Gaussian noise. Other references in a similar framework are given, for instance, by Refs. [4–8]. Subsequently also the case of counting process or marked point process observation has been considered (see Refs. [9–14] and reference therein). A more recent literature contains the case of mixed-type observations (marked point processes and diffusions or jump-diffusion processes), see, for, example, Refs. [15–18].

There are two major approaches to nonlinear filtering problems: the innovations method and the reference probability method. The latter is usually employed when it is possible to find an equivalent probability measure that makes the state X and the observations Y independent. This technique may appear problematic when, for instance, signal and observation are correlated and present common jump times. Therefore, in this chapter, we use the innovations approach which allows circumventing the technical issues arising in the reference probability method. By characterizing the innovation process and applying a martingale representation theorem, we can derive the dynamics of the filter as the solution of the Kushner-Stratonovich equation, which is a nonlinear stochastic partial integral differential equation. By considering the unnormalized version of the filter, it is possible to simplify this equation and make it at least linear. The resulting equation is called the Zakai equation, and due to its linear nature, it is of particular interest in many applications. We also compute the dynamics of the unnormalized filter, and we investigate pathwise uniqueness for the solutions of both equations. Normalized and unnormalized filters are probability measure and finite measure-valued processes, respectively, and therefore in general infinite-dimensional. Due to this, various recursive algorithms for statistical inference have come in to address this intractability, such as extended Kalman filter, statistical linearization, or particle filters. These algorithms intend to estimate both state and parameters. For the parameter estimation, we also mention the expectation maximization (EM) algorithm which enables to estimate parameters in models with incomplete data, see, for example, Ref. [19].

The success of the filtering theory over the years is due to its use in a great variety of problems arising from many disciplines such as engineering, informational sciences and mathematical finance. Specifically, in this chapter, we have a financial application in view. In real financial markets, it is reasonable that investors cannot fully know all the stochastic factors that may influence the prices of negotiated assets, since these factors are usually associated with economic quantities which are hard to observe. Filtering theory represents a way to measure, in some sense, this uncertainty. A consistent part of the literature over the last years has considered stochastic factor models under partial information for analyzing various financial problems, as, for example, pricing and hedging of derivatives, optimal investment, credit risk, and insurance modeling. A list, definitely nonexhaustive, is given by Refs. [15, 16, 20–26]).

In the following, we consider the problem of a trader who wants to determine the hedging strategy for a European-type contingent claim with maturity T in an incomplete financial market where the investment possibilities are given by a riskless asset, assumed to be the numéraire, and a risky asset with price dynamics given by a geometric jump diffusion, modeled by the process Y. We assume that the drift, as well as the intensity and the jump size distribution of the price process, is influenced by an unobservable stochastic factor X, modeled as a correlated jump diffusion with common jump times. By common jump times, we intend to take into account catastrophic events which affect both the asset price and the hidden state variable driving its dynamics. The agent knows the asset prices, since they are publicly available, and trades on the market by using the available information $\mathbb{F}^Y$.

Partial information easily leads to incomplete financial markets as clearly the number of random sources is larger than the number of tradeable risky asset. Therefore, the existence of a self-financing strategy that replicates the payoff of the given contingent claim at maturity is not guaranteed. Here, we assume that the risky asset price is modeled under a martingale measure, and we choose the risk-minimization approach as hedging criterion, see, for example, Refs. [27, 28].

According to this method, the optimal hedging strategy is the one that perfectly replicates the claim at maturity and has minimum cost in the mean-square sense. Equivalently, we say that it minimizes the associated risk defined as the conditional expected value of the squared future costs, given the available information (see Refs. [28, 29] and references therein).

The risk-minimizing hedging strategy under restricted information is strictly related to Galtchouk-Kunita-Watanabe decomposition of the random variable representing the payoff of the contingent claim in a partial information setting. Here, we provide a characterization of the risk-minimizing strategy under partial information via this orthogonal decomposition and obtain a representation in terms of the corresponding risk-minimizing hedging strategy under full information (see, e.g., Refs. [29, 30]) via predictable projections on the available information flow by means of the filter. Finally, we investigate the difference of expected total risks associated with the optimal hedging strategies under full and partial information.

The chapter has the following structure. In Section 2, we introduce the general framework. In Section 3, we study the filtering equations. In particular, we derive the dynamics for both normalized and unnormalized filters, and we investigate uniqueness of the solutions of the Kushner-Stratonovich and the Zakai equation. In Section 4, we analyze a financial application to risk minimization by computing the optimal hedging strategies for a European-type contingent claim under full and partial information and providing a comparison between the corresponding expected squared total costs.

2. The setting

We consider a pair of stochastic processes (X, Y), with values on $\mathbb{R} \times \mathbb{R}$ and càdlàg trajectories, on a complete filtered probability space $(\Omega, \mathcal{F}, \mathbb{F}, P)$, where $\mathbb{F} = \{\mathcal{F}_t, t \in [0, T]\}$ is a filtration satisfying the usual condition of right continuity and completeness, and T is a fixed time horizon. The pair (X, Y) represents a partially observable system, where X is a signal process that describes a phenomenon which is not directly observable and Y gives the observation of X, and it is modeled by a process correlated with the signal, having possibly common jump times.

Remark 1. *In view of the financial application discussed in Section 4, Y represents the price of some risky asset, while X is an unknown stochastic factor, which may describe the activity of other markets, macroeconomic factors or microstructure rules that influences the dynamics of the stock price process.*

We define the *observed history* as the natural filtration of the observation process Y, that is, $\mathbb{F}^Y = \{\mathcal{F}_t^Y\}_{t \in [0,T]}$, where $\mathcal{F}_t^Y := \sigma(Y_s, 0 \le s \le t)$. The σ-algebra $\mathcal{F}_t^Y$ can be interpreted as the information available from observations up to time t. We aim to compute the best estimate of the signal X from the available information, in the quadratic sense. In other terms, this corresponds to determine the filter which furnishes the conditional distribution of X_t given $\mathcal{F}_t^Y$, for every $t \in [0, T]$.

Let $\mathcal{M}(\mathbb{R})$ be the space of finite measures over $\mathbb{R}$ and $\mathcal{P}(\mathbb{R})$ the subspace of the probability measures over $\mathbb{R}$. Given $\mu \in \mathcal{M}(\mathbb{R})$, for any bounded measurable function f, we write

$$\mu(f) = \int_{\mathbb{R}} f(x)\mu(\mathrm{d}x). \tag{1}$$

Definition 2. *The filter is the $\mathbb{F}^Y$-càdlàg process π taking values in $\mathcal{P}(\mathbb{R})$ defined by*

$$\pi_t(f) := \mathbb{E}\left[f(t, X_t) | \mathcal{F}_t^Y\right] = \int_{\mathbb{R}} f(t, x)\pi_t(\mathrm{d}x), \tag{2}$$

for all bounded and measurable functions $f(t, x)$ on $[0, T] \times \mathbb{R}$.

In the sequel, we denote by π_{t-} the left version of the filter and for all functions $F(t, x, y)$ such that $\mathbb{E}|F(t, X_t, Y_t)| < \infty$ (resp. $\mathbb{E}|F(t, X_{t-}, Y_{t-})| < \infty$) for every $t \in [0,T]$, we use the notation $\pi_t(F) := \pi_t(F(t, \cdot, Y_t))$ (resp. $\pi_{t-}(F) := \pi_{t-}(F(t, \cdot, Y_{t-}))$).

In this paper, we wish to consider the filtering problem for a partially observable system (X, Y) described by the following pair of stochastic differential equations:

$$\begin{cases} \mathrm{d}X_t = b_0(t, X_t)\mathrm{d}t + \sigma_0(t, X_t)\mathrm{d}W_t^0 + \int_Z K_0(t, X_{t-}; \zeta)N(\mathrm{d}t, \mathrm{d}\zeta); X_0 = x_0 \in \mathbb{R} \\ \mathrm{d}Y_t = b_1(t, X_t, Y_t)\mathrm{d}t + \sigma_1(t, Y_t)\mathrm{d}W_t^1 + \int_Z K_1(t, X_{t-}, Y_{t-}; \zeta)N(\mathrm{d}t, \mathrm{d}\zeta); Y_0 = y_0 \in \mathbb{R} \end{cases} \tag{3}$$

where W^0 and W^1 are correlated $(\mathbb{F}, \mathbf{P})$-Brownian motions with correlation coefficient $\rho \in [-1,1]$ and $N(dt, d\zeta)$ is a Poisson random measure on $\mathbb{R}^+ \times Z$ whose intensity $\nu(d\zeta)dt$ is a σ – finite measure on a measurable space $(Z, \mathcal{Z})$. Here, $b_0, b_1, \sigma_0, \sigma_1, K_0$, and K_1 are $\mathbb{R}$-valued and measurable functions of their arguments. In particular, $\sigma_0(t, x)$ and $\sigma_1(t, x, y)$ are strictly positive for every $(t, x, y) \in [0, T] \times \mathbb{R}^2$.

For the rest of the paper, we assume that strong existence and uniqueness for system Eq. (3) holds. Sufficient conditions are collected, for instance, in Ref. [18, Appendix]. These assumptions also imply Markovianity for the pair (X, Y).

Remark 3. *Note that the quadratic variation process of* Y *defined by*

$$[Y]_t = Y_t^2 - 2\int_0^t Y_{u^-} dY_u, \quad t \in [0, T], \tag{4}$$

is $\mathbb{F}^Y$-adapted and $[Y]_t = \int_0^t \sigma_1^2(u, Y_u)du + \sum_{u \le t}(\Delta Y_u)^2$, where $\Delta Y_t := Y_t - Y_{t-}$. Therefore, it is natural to assume that the signal X does not affect the diffusion coefficient in the dynamics of Y. If Y describes the price of a risky asset, this implies that the volatility of the stock price does not depend on the stochastic factor X.

The jump component of Y can be described in terms of the following integer-valued random measure on $[0, T] \times \mathbb{R}$:

$$m(dt, dz) = \sum_{s:\Delta Y_s \neq 0} \delta_{\{s, \Delta Y_s\}}(dt, dz), \tag{5}$$

where δ_a denotes the Dirac measure at point a. Note that the following equality holds:

$$\int_0^t \int_{\mathbb{R}} zm(ds, dz) = \int_0^t \int_Z K_1(s, X_{s^-}, Y_{s^-}; \zeta)N(ds, d\zeta). \tag{6}$$

For all $t \in [0, T]$, for all $A \in \mathcal{B}(\mathbb{R})$, we define the following sets:

$$d^0(t, x) := \{\zeta \in Z : K_0(t, x; \zeta) \neq 0\}, \quad d^1(t, x, y) := \{\zeta \in Z : K_1(t, x, y; \zeta) \neq 0\}, \tag{7}$$

$$d^A(t, x, y) := \{\zeta \in Z : K_1(t, x, y; \zeta) \in A \setminus \{0\}\} \subseteq d^1(t, x, y), \tag{8}$$

$$D_t^A := d^A(t, X_{t-}, Y_{t-}) \subseteq D_t := d^1(t, X_{t-}, Y_{t-}), \quad D_t^0 := d^0(t, X_{t-}). \tag{9}$$

Typically, we have $D_t^0 \cap D_t \neq \emptyset$ $\mathbf{P}$ – *a.s.*, which means that state and observation may have common jump times. This characteristic is particularly meaningful in financial applications to model catastrophic events that produce jumps in both the stock price and the underlying stochastic factor that influences its dynamics.

To ensure existence of the first moment for the pair (X, Y) and non-explosiveness for the jump process governing the dynamics of X and Y, we make the following assumption:

Assumption 4.

$$\mathbb{E}\left[\int_0^T |b_0(t, X_t)| + \sigma_0^2(t, X_t) + \int_Z |K_0(t, X_{t-}; \zeta)| \nu(d\zeta) dt\right] < \infty, \tag{10}$$

$$\mathbb{E}\left[\int_0^T |b_1(t, X_t, Y_t)| + \sigma_1^2(t, Y_t) + \int_Z |K_1(t, X_{t-}, Y_{t-}; \zeta)| \nu(d\zeta) dt\right] < \infty, \tag{11}$$

$$\mathbb{E}\left[\int_0^T \nu(D_t^0 \cup D_t) dt\right] < \infty. \tag{12}$$

Denote by $\eta^{\mathbf{P}}(dt, dz)$ the $(\mathbb{F}, \mathbf{P})$ compensator of $m(dt, dz)$ (see, e.g., Refs. [9, 31] for the definition).

Then, in Ref. [14, Proposition 2.2], it is proved that

$$\eta^{\mathbf{P}}(dt, dz) = \lambda(t, X_{t-}, Y_{t-}) \phi(t, X_{t-}, Y_{t-}, dz) dt, \tag{13}$$

where

$$\lambda(t, x, y) \phi(t, x, y, dz) = \int_{d^1(t,x,y)} \delta_{K^1(t,x,y;\zeta)}(dz) \nu(d\zeta) \tag{14}$$

and in particular $\lambda(t, x, y) = \nu(d^1(t, x, y))$.

Remark 5. *Let us observe that both the local jump characteristics* $(\lambda(t, X_{t-}, Y_{t-}), \phi(t, X_{t-}, Y_{t-}, dz))$ *depend on* X *and, for all* $A \in \mathcal{B}(\mathbb{R})$, $\lambda(t, X_{t-}, Y_{t-}) \phi(t, X_{t-}, Y_{t-}, A) = \nu(D_t^A)$ *provides the* $(\mathbb{F}, \mathbf{P})$ *-intensity of the point process* $N_t(A) := m((0, t] \times A)$. *According to this, the process* $\lambda(t, X_{t-}, Y_{t-}) = \nu(D_t)$ *is the* $(\mathbb{F}, \mathbf{P})$ *-intensity of the point process* $N_t(\mathbb{R})$ *which counts the total number of jumps of* Y *until time* t.

2.1. The innovation process

To derive the filtering equation, we use the innovations approach. This method requires to introduce a pair (I, m^π), called the *innovation process*, consisting of the $(\mathbb{F}^Y, \mathbf{P})$-Brownian motion and the $(\mathbb{F}^Y, \mathbf{P})$-compensated jump measure that drive the dynamics of the filter. The innovation also represents the building block of $(\mathbb{F}^Y, \mathbf{P})$ -martingales.

To introduce the first component of the innovation process, we assume that

$$\mathbb{E}\left[\exp\left\{\frac{1}{2}\int_0^T \left(\frac{b_1(t, X_t, Y_t)}{\sigma_1(t, Y_t)}\right)^2 dt\right\}\right] < \infty, \tag{15}$$

and define

$$I_t := W_t^1 + \int_0^t \left(\frac{b_1(s, X_s, Y_s)}{\sigma_1(s, Y_s)} - \frac{\pi_s(b_1)}{\sigma_1(s, Y_s)}\right) ds, \quad t \in [0, T]. \tag{16}$$

The process I is an $(\mathbb{F}^Y, \mathbf{P})$-Brownian motion (see, e.g., Ref. [4]) and the $(\mathbb{F}^Y, \mathbf{P})$-compensated jump martingale measure is given by

http://dx.doi.org/10.5772/intechopen.70060

$$m^{\pi}(\mathrm{d}t, \mathrm{d}z) = m^{\pi}(\mathrm{d}t, \mathrm{d}z) - \pi_{t-}(\lambda\phi(\mathrm{d}z))\mathrm{d}t, \tag{17}$$

See, e.g. Ref. [14]. The following theorem provides a characterization of the $(\mathbb{F}^Y, \mathbf{P})$-martingale in terms of the innovation process.

Theorem 6 (A martingale representation theorem). *Under Assumption 4 and the integrability condition Eq. (15), every $(\mathbb{F}^Y, \mathbf{P})$-local martingale M admits the following decomposition:*

$$M_t = M_0 + \int_0^t \int_{\mathbb{R}} w_s(z) m^{\pi}(\mathrm{d}s, \mathrm{d}z) + \int_0^t h_s \mathrm{d}I_s, \quad t \in [0, T], \tag{18}$$

where $w(z) = \{w_t(z),\ t \in [0, T]\}$ is an $\mathbb{F}^Y$-predictable process indexed by z, and $h = \{h_t,\ t \in [0, T]\}$ is an $\mathbb{F}^Y$-adapted process such that

$$\int_0^T \int_{\mathbb{R}} |w_t(z)| \pi_{t-}(\lambda\phi(\mathrm{d}z))\mathrm{d}t < \infty, \quad \int_0^T h_t^2 \mathrm{d}t < \infty \quad \mathbf{P}-a.s.. \tag{19}$$

Proof. The proof is given in Ref. [17, Proposition 2.4]. Note that here condition (15) implies that $\mathbb{E}\left[\int_0^T \left(\frac{b_1(t, X_t, Y_t)}{\sigma_1(t, Y_t)}\right)^2 \mathrm{d}t\right] < \infty$, and also that the process L defined by

$$L_t = \exp\left(-\int_0^t \frac{b_1(s, X_s, Y_s)}{\sigma_1(s, Y_s)} \mathrm{d}W_s^1 - \frac{1}{2}\int_0^t \left(\frac{b_1(s, X_s, Y_s)}{\sigma_1(s, Y_s)}\right)^2 \mathrm{d}s\right), \tag{20}$$

for every $t \in [0, T]$, is an $(\mathbb{F}, \mathbf{P})$-martingale.

3. The filtering equations

Theorem 7 (The Kushner-Stratonovich equation). *Under Assumptions 4 and condition (15), the filter π solves the following Kushner-Stratonovich equation, that is, for every $f \in C_b^{1,2}([0, T] \times \mathbb{R})$:*

$$\pi_t(f) = f(0, x_0) + \int_0^t \pi_s(\mathcal{L}^X f)\mathrm{d}s + \int_0^t \int_{\mathbb{R}} w_s^{\pi}(f, z) m^{\pi}(\mathrm{d}s, \mathrm{d}z) + \int_0^t h_s^{\pi}(f)\mathrm{d}I_s, \quad t \in [0, T] \tag{21}$$

where

$$w_t^{\pi}(f, z) = \frac{\mathrm{d}\pi_{t-}(\lambda\phi f)}{\mathrm{d}\pi_{t-}(\lambda\phi)}(z) - \pi_{t-}(f) + \frac{\mathrm{d}\pi_{t-}(\overline{\mathcal{L}}f)}{\mathrm{d}\pi_{t-}(\lambda\phi)}(z), \tag{22}$$

$$h_t^{\pi}(f) = \sigma_1^{-1}(t)[\pi_t(b_1 f) - \pi_t(b_1)\pi_t(f)] + \rho\pi_t\left(\sigma_0 \frac{\partial f}{\partial x}\right). \tag{23}$$

Here, by $\frac{\mathrm{d}\pi_{t-}(\lambda\phi f)}{\mathrm{d}\pi_{t-}(\lambda\phi)}(z)$ and $\frac{\mathrm{d}\pi_{t-}(\overline{\mathcal{L}}f)}{\mathrm{d}\pi_{t-}(\lambda\phi)}(z)$, we mean the Radon-Nikodym derivatives of the measures $\pi_{t-}(\lambda f\phi(\mathrm{d}z))$ and $\pi_{t-}(\overline{\mathcal{L}}f)(\mathrm{d}z)$, with respect to $\pi_{t-}\left(\lambda\phi(\mathrm{d}z)\right)$. Moreover, the operator $\overline{\mathcal{L}}$ defined by $\overline{\mathcal{L}}_t f(\mathrm{d}z) := \overline{\mathcal{L}}f(., Y_{t-}, \mathrm{d}z)$ is such that for every $A \in \mathcal{B}(\mathbb{R})$,

$$\overline{\mathcal{L}}f(t,x,y,A) = \int_{d^A(t,x,y)} [f(t,x+K_0(t,x;\zeta)) - f(t,x)]\nu(\mathrm{d}\zeta) \tag{24}$$

takes into account common jump times between the signal X and the observation Y.

Finally, the operator $\mathcal{L}^X$ *given by*

$$\mathcal{L}^X f(t,x) = \frac{\partial f}{\partial t} + b_0(t,x)\frac{\partial f}{\partial x} + \frac{1}{2}\sigma_0^2(t,x)\frac{\partial^2 f}{\partial x^2} + \int_Z \{f(t,x+K_0(t,x;\zeta)) - f(t,x)\}\nu(\mathrm{d}\zeta). \tag{25}$$

denotes the generator of the Markov process X.

Proof. The theorem is proved in Ref. [17, Theorem 3.1].

Example 8 (Observation dynamics driven by independent point processes with unobservable intensities). In the sequel, we provide an example where the Kushner-Stratonovich equation simplifies and the Radon-Nikodym derivatives appearing in the dynamics of $\pi(f)$ reduce to ratios. Suppose that there exists a finite set of measurable functions $K_1^i(t,y) \neq 0$ for all $(t,y) \in [0,T] \times \mathbb{R}$, for $i \in \{1,\ldots,n\}$, such that the dynamics of Y is given by

$$\mathrm{d}Y_t = b_1(t,X_t,Y_t)\mathrm{d}t + \sigma_1(t,Y_t)\mathrm{d}W_t^1 + \sum_{i=1}^n K_1^i(t,Y_{t-})\mathrm{d}N_t^i, \quad Y_0 = y_0 \in \mathbb{R}, \tag{26}$$

where N^i are independent counting processes with $(\mathbb{F},\mathbf{P})$ intensities $\lambda^i(t,X_{t-},Y_{t-})$.

For simplicity, in this example, we assume that X and Y have no common jump times. Then, the filtering Eq. (21) reads as

$$\begin{aligned}\pi_t(f) = f(0,x_0) &+ \int_0^t \pi_s(\mathcal{L}^X f)\mathrm{d}s + \int_0^t \left\{\sigma_1(s)^{-1}[\pi_s(b_1 f) - \pi_s(b_1)\pi_s(f)] + \rho\pi_s\left(\sigma_0\frac{\partial f}{\partial x}\right)\right\}\mathrm{d}I_s \\ &+ \sum_{i=1}^n \int_0^t \mathbf{1}_{\pi_{s^-}(\lambda^i)>0}\frac{\pi_{s^-}(\lambda^i f) - \pi_{s^-}(f)\pi_{s^-}(\lambda^i)}{\pi_{s^-}(\lambda^i)}\left(\mathrm{d}N_s^i - \pi_{s^-}(\lambda^i)\mathrm{d}s\right), \quad t \in [0,T].\end{aligned} \tag{27}$$

Note that Eq. (21) has an equivalent expression in terms of the operator $\mathcal{L}_0^X$, given by

$$\begin{aligned}\mathcal{L}_0^X f(t,x,y) &= \mathcal{L}^X f(t,x) - \overline{\mathcal{L}}f(t,x,y,\mathbb{R}) \\ &= \frac{\partial f}{\partial t}(t,x) + b_0(t,x)\frac{\partial f}{\partial x} + \frac{1}{2}\sigma_0^2(t,x)\frac{\partial^2 f}{\partial x^2} + \int_{d_t^1(t,x,y)^c} \{f(t,x+K_0(t,x,\zeta)) - f(t,x)\}\nu(d\zeta),\end{aligned} \tag{28}$$

where $d^1(t,x,y)^c = \{\zeta \in Z : K_1(t,x,y,\zeta) = 0\}$. Indeed, we get

$$\mathrm{d}\pi_t(f) = \{\pi_t(\mathcal{L}_0^X f) + \pi_t(f)\pi_t(\lambda) - \pi_t(\lambda f)\}\mathrm{d}t + h_t^\pi \mathrm{d}I_t + \int_{\mathbb{R}} w^\pi(t,z)m(\mathrm{d}t,\mathrm{d}z). \tag{29}$$

Moreover, the filter has a natural recursive structure. To show this, define the sequence $\{T_n, Z_n\}_{n\in\mathbb{N}}$ of jump times and jump sizes of Y, that is, $Z_n = Y_{T_n} - Y_{T_n^-}$. These are observable data. Then, between two consecutive jump times the filter is governed by a diffusion process, that is, for $t \in (T_n \wedge T, T_{n+1} \wedge T)$

$$\pi_t(f) = \pi_{T_n}(f) + \int_{T_n}^{t} \{\pi_s(\mathcal{L}_0^X f) + \pi_s(f)\pi_s(\lambda) - \pi_s(\lambda f)\}\mathrm{d}s + \int_{T_n}^{t} h_s^{\pi}(f)\mathrm{d}I_s, \tag{30}$$

and at any jump time T_n occurring before time T, it is given by

$$\pi_{T_n}(f) = \frac{\mathrm{d}\pi_{T_n^-}(\lambda\phi f)}{\mathrm{d}\pi_{T_n^-}(\lambda\phi)}(Z_n) + \frac{\mathrm{d}\pi_{T_n^-}(\overline{\mathcal{L}}f)}{\mathrm{d}\pi_{T_n^-}(\lambda\phi)}(Z_n), \tag{31}$$

which implies that $\pi_{T_n}(f)$ is completely determined by the observed data (T_n, Z_n) and the knowledge of $\pi_t(f)$ in the time interval $[T_{n-1}, T_n)$, since $\pi_{T_n^-}(f) = \lim_{t \to T_n^-} \pi_t(f)$.

Note that the Kushner-Stratonovich equation is an infinite-dimensional nonlinear stochastic differential equation. Often, it is possible to characterize the filter in terms of a simpler equation, known as the Zakai equation which provides the dynamics of the unnormalized version of the filter. Although the Zakai equation is still infinite-dimensional, it has the advantage to be linear.

The idea for getting the dynamics of the unnormalized filter consists of performing an equivalent change of probability measure defined by

$$\left.\frac{\mathrm{d}\mathbf{P}_0}{\mathrm{d}\mathbf{P}}\right|_{\mathcal{F}_t} = Z_t, \quad t \in [0, T] \tag{32}$$

for a suitable strictly positive $(\mathbb{F}, \mathbf{P})$-martingale Z, in such a way that the so-called unnormalized filter p is the $\mathcal{M}(\mathbb{R})$-valued process defined by

$$p_t(f) := \mathbb{E}^0\left[Z_t^{-1} f(t, X_t) | \mathcal{F}_t^Y\right], \quad t \in [0, T], \tag{33}$$

Remark 9. *By the Kallianpur-Striebel formula, we get that*

$$\pi_t(f) = \frac{\mathbb{E}^0\left[f(t, X_t) Z_t^{-1} | \mathcal{F}_t^Y\right]}{\mathbb{E}^0\left[Z_t^{-1} | \mathcal{F}_t^Y\right]} = \frac{p_t(f)}{p_t(1)}, \quad t \in [0, T], \tag{34}$$

where $p_t(1) := \mathbb{E}^0\left[Z_t^{-1} | \mathcal{F}_t^Y\right]$. *This provides the relation between the filter and its unnormalized version.*

In order to compute the Zakai equation, we make the following assumption.

Assumption 10. *Suppose that there exists a transition function* $\eta^0(t, y, \mathrm{d}z)$ *such that the* $(\mathbb{F}^Y, \mathbf{P})$-*predictable measure* $\eta^0(t, Y_{t-}, \mathrm{d}z)$ *is equivalent to* $\lambda(t, X_{t-}, Y_{t-})\phi(t, X_{t-}, Y_{t-}, \mathrm{d}z)$ *and*

$$\mathbb{E}\left[\int_0^T \eta^0(t, Y_{t^-}, \mathbb{R})\mathrm{d}t\right] < \infty. \tag{35}$$

Remark 11. *In Ref. [18], a weaker assumption is considered. That condition allows to introduce an equivalent probability measure on* $(\Omega, \mathcal{F}_T^Y)$ *which is not necessarily the restriction on* $\mathcal{F}_T^Y$ *of an equivalent probability measure on* $(\Omega, \mathcal{F}_T)$.

Remark 12. *In the context of Example 8, Assumption 10 is satisfied if, for instance,* $\lambda^i(t, X_{t-}, Y_{t-}) > 0$ $\mathbf{P}$*-a.s. for every* $t \in [0, T]$.

Assumption 10 equivalently means that there exists an $(\mathbb{F}^Y, \mathbf{P})$-predictable process $\Psi(t, X_{t-}, Y_{t-}, z)$ such that

$$\lambda(t, X_{t-}, Y_{t-})\phi(t, X_{t-}, Y_{t-}, dz)dt = (1 + \Psi(t, X_{t-}, Y_{t-}, z))\eta^0(t, Y_{t-}, dz)dt \tag{36}$$

and $1 + \Psi(t, X_{t^-}, Y_{t^-}, z) > 0$ $\mathbf{P}$-a.s. for every $t \in [0, T]$, $z \in \mathbb{R}$. Setting

$$U(t, z) := \frac{1}{1 + \Psi(t, X_{t^-}, Y_{t^-}, z)} - 1, \tag{37}$$

we also assume that the following integrability condition holds:

$$\mathbb{E}\left[\exp\left\{\frac{1}{2}\int_0^T \left(\frac{b_1(s, X_s, Y_s)}{\sigma_1(s, Y_s)}\right)^2 ds + \int_0^T \int_{\mathbb{R}} U^2(s, z)\lambda(s, X_{s^-}, Y_{s^-})\phi(s, X_{s^-}, Y_{s^-}, dz)ds\right\}\right] < \infty. \tag{38}$$

The subsequent proposition provides a useful version of the Girsanov Theorem that fits to our setting.

Proposition 13. *Let Assumptions 4 and 10, and condition (38) hold and define the process* $Z_t := \mathcal{E}\left(-\int_0^t \frac{b_1(s, X_s, Y_s)}{\sigma_1(s, Y_s)} dW_s^1 + \int_0^t \int_{\mathbb{R}} U(s, z)\left(m(ds, dz) - \lambda(s, X_{s^-}, Y_{s^-})\phi(s, X_{s^-}, Y_{s^-}, dz)ds\right)\right)$, *for every* $t \in [0, T]$, *where* $\mathcal{E}(M)$ *denotes the Doléans-Dade exponential of a martingale M. Then, Z is a strictly positive* $(\mathbb{F}, \mathbf{P})$ *-martingale. Let* $\mathbf{P}^0$ *be the probability measure equivalent to* $\mathbf{P}$ *given by*

$$\left.\frac{d\mathbf{P}^0}{d\mathbf{P}}\right|_{\mathcal{F}_t} = Z_t, \quad t \in [0, T]. \tag{39}$$

Then, the process

$$\widetilde{W}_t^1 := W_t^1 + \int_0^t \frac{b_1(s, X_s, Y_s)}{\sigma_1(s, Y_s)} ds, \quad t \in [0, T] \tag{40}$$

is an $(\mathbb{F}, \mathbf{P}^0)$*-Brownian motion, and the* $(\mathbb{F}, \mathbf{P}^0)$*-predictable projection of the integer-valued random measure* $m(dt, dz)$ *is given by* $\eta^0(t, Y_{t^-}, dz)dt$.

Proof. [32, Theorem 9] ensures that Z is a martingale under Assumptions 10, 4 and integrability condition Eq. (38). Then the proof follows by Ref. [31, Chapter III, Theorem 3.24].

Note that, by Eq. (16), we get that the process $\widetilde{W^1}$ can also be written as

$$\widetilde{W}_t^1 = I_t + \int_0^t \pi_s \left(\frac{b_1}{\sigma_1} \right) \mathrm{d}s, \quad t \in [0, T] \tag{41}$$

which implies that $\widetilde{W}^1$ is also an $(\mathbb{F}^Y, \mathbf{P}^0)$-Brownian motion. Moreover, since $\eta^0(t, Y_{t^-}, \mathrm{d}z)$ is $\mathbb{F}^Y$ predictable, it provides the $(\mathbb{F}^Y, \mathbf{P}^0)$-predictable projection of the measure $m(\mathrm{d}t, \mathrm{d}z)$ and the observation process Y satisfies $\mathrm{d}Y_t = \sigma_1(t, Y_t)\mathrm{d}\tilde{W}_t^1 + \int_{\mathbb{R}} zm(\mathrm{d}t, \mathrm{d}z)$. In particular, $\eta_t^0(\mathbb{R}) := \eta^0(t, Y_{t^-}, \mathbb{R})$ is the $(\mathbb{F}^Y, \mathbf{P}^0)$-intensity of the point process which counts the total jumps of Y until time t.

Theorem 14 (The Zakai equation). *Under Assumptions 4 and 10 and condition* (38), *let* $\mathbf{P}^0$ *be the probability measure defined in Proposition 13. For every* $f \in \mathcal{C}_b^{1,2}([0, T] \times \mathbb{R})$, *the unnormalized filter defined in Eq.* (33) *satisfies the equation*

$$\begin{aligned} \mathrm{d}p_t(f) =& \{p_t(\mathcal{L}_0^X f) - p_t(\lambda f) + \eta_t^0(\mathbb{R}) p_t(f)\}\mathrm{d}t + \left\{ \frac{p_t(b_1 f)}{\sigma_1(t, Y_t)} + \rho\, p_t\left(\sigma_0 \frac{\partial f}{\partial x} \right) \right\} \mathrm{d}\widetilde{W}_t^1 \\ &+ \int_{\mathbb{R}} \left\{ p_{t^-}(f\Psi)(z) + \frac{\mathrm{d}p_{t^-}(\overline{\mathcal{L}}f)}{\mathrm{d}\eta_t^0}(z) \right\} m(\mathrm{d}t, \mathrm{d}z). \end{aligned} \tag{42}$$

See Ref. [18, Theorem 3.6] for the proof.

3.1. Uniqueness of the filtering equations

In this section, we show pathwise uniqueness for the solution of the Kushner-Stratonovich and the Zakai equations. The first result provides the equivalence of uniqueness of the solutions to the filtering Eqs. (21) and (42).

Theorem 15. *Let Assumptions 4 and 10 and condition* (38) *hold.*

i. *Assume strong uniqueness for the solution to the Zakai equation, let* μ *be a* $\mathcal{P}(\mathbb{R})$*-valued process which is a strong solution of the Kushner-Stratonovich equation. Then* $\mu_t = \pi_t \mathbf{P} -$ *a.s. for all* $t \in [0, T]$.

ii. *Conversely, suppose that pathwise uniqueness for the solution of the Kushner-Stratonovich equation holds and let* ξ *be an* $\mathcal{M}(\mathbb{R})$*-valued process which is a strong solution of the Zakai equation. Then* $\xi_t = p_t \mathbf{P} -$ *a.s. for all* $t \in [0, T]$.

Proof. The proof follows by Ref. [18, Theorems 4.5 and 4.6]. Here, note that Assumption 10 implies that the measures $\mu_{t^-}(\lambda\phi(\mathrm{d}z))$ and $\pi_{t^-}(\lambda\phi(\mathrm{d}z))$ are equivalent.

Finally, strong uniqueness for the solution of both filtering equations is established in the subsequent theorems.

Theorem 16. *Let* (X, Y) *be the partially observed system defined in Eq.* (3), *and assume in addition to Assumptions 4 and 10 and condition* (15) *that*

$$\sup_{t,x,y} \int_Z \{|K_0(t, x; \zeta)| + |K_1(t, x, y; \zeta)|\} \nu(d\zeta) < \infty. \tag{43}$$

*Let μ be a strong solution of the Kushner-Stratonovich equation. Then $\mu_t = \pi_t$ **P**-a.s. for every $t \in [0, T]$.*

Proof. See Ref. [17, Theorem 3.3].

Theorem 17. *Let (X, Y) be the partially observed system in Eq. (3). Under Assumptions 4 and 10 and conditions (38) and (43), let ξ be a strong solution to the Zakai equation, then $\xi_t = p_t$ **P**-a.s. for every $t \in [0, T]$.*

Proof. The proof follows by Ref. [18, Theorem 4.7], after noticing that under Assumption 10 the measures $\xi_{t-}(\lambda\phi(\mathrm{d}z))$ and $p_{t-}(\lambda\phi(\mathrm{d}z))$ are equivalent.

4. A financial application to risk minimization

In the current section, we focus on a financial application. We consider a simple financial market where agents may invest in a risky asset whose price is described by the process Y given in Eq. (3) and a riskless asset with price process B. Without loss of generality, we assume that $B_t = 1$ for every $t \in [0, T]$. We also assume throughout the section the following dynamics for the process Y:

$$\mathrm{d}Y_t = Y_t\Big(\sigma(t, Y_t)\mathrm{d}W_t^1 + \int_Z K(t, X_{t-}, Y_{t-}; \zeta)\Big(N(\mathrm{d}t, \mathrm{d}\zeta) - \nu(\mathrm{d}\zeta)\mathrm{d}t\Big)\Big), \quad Y_0 = y_0 \in \mathbb{R}^+ \tag{44}$$

for some functions $\sigma(t, y)$ and $K(t, x, y; \zeta)$ such that $\sigma(t, y) > 0$ and $K(t, x, y; \zeta) > -1$.

This choice for the dynamics of Y has a double advantage. On one side assuming a geometric form, together with the condition that $K(t, x, y; \zeta) > -1$ guarantees nonnegativity which is desirable when talking about prices. On the other hand, we are modeling Y directly under a martingale measure, and by Assumption 18, it turns out to be a square integrable $(\mathbb{F}, \mathbf{P})$-martingale.

Considering Eq. (44) corresponds to take in system (3)

$$\begin{gathered} b_1(t, x, y) = -y\int_Z K(t, x, y; \zeta)\nu(\mathrm{d}\zeta) \\ \sigma_1(t, y) = y\sigma(t, y), \quad K_1(t, x, y; \zeta) = yK(t, x, y; \zeta). \end{gathered} \tag{45}$$

In addition, we me make the following assumption.

Assumption 18.

$$0 < c_1 < \sigma(t, y) < c_2, \quad |K(t, x, y; \zeta)| < c_3, \quad \nu(D_t) < c_4, \tag{46}$$

for every $(t, x, y) \in [0, T] \times \mathbb{R} \times \mathbb{R}^+$, $\zeta \in Z$ and for some positive constants c_1, c_2, c_3, c_4.

Remark 19. *In the sequel, it might be useful to specify the dynamics of Y also in terms of the jump measure $m(\mathrm{d}t, \mathrm{d}z)$. Recalling Eqs. (6) and (14), we have*

$$\mathrm{d}Y_t = Y_t\sigma(t, Y_t)\mathrm{d}W_t^1 + \int_{\mathbb{R}} z\Big(m(\mathrm{d}t, \mathrm{d}z) - \lambda(t, X_{t^-}, Y_{t^-})\phi(t, X_{t^-}, Y_{t^-}, \mathrm{d}z)\mathrm{d}t\Big). \tag{47}$$

The stochastic factor X which affects intensity and jump size distribution of Y may represent the state of the economy and is not directly observable by market agents. This is a typical situation arising in real financial markets.

We model by $\mathbb{F}^Y$ the available information to investors. Since Y is $\mathbb{F}^Y$ adapted, it is in particular an $(\mathbb{F}^Y, \mathbf{P})$-martingale with the following decomposition:

$$Y_t = y_0 + \int_0^t Y_s\sigma(s, Y_s)\mathrm{d}I_s + \int_0^t \int_{\mathbb{R}} z\Big(m(\mathrm{d}s, \mathrm{d}z) - \pi_{s^-}(\lambda\phi(\mathrm{d}z))\mathrm{d}s\Big), \quad t \in [0, T]. \tag{48}$$

By Eqs. (14) and (45), in this setting the first component of the innovation process I defined in Eq. (16) is given by $I_t = W_t^1 + \int_0^t \frac{1}{Y_s\sigma(s, Y_s)} \int_{\mathbb{R}} z\Big(\lambda(s, X_s, Y_s)\phi(s, X_s, Y_s, \mathrm{d}z) - \pi_s(\lambda\phi(\mathrm{d}z))\Big)\mathrm{d}s.$

Suppose that we are given a European-type contingent claim whose final payoff is a square integrable $\mathcal{F}_T^Y$-measurable random variable ξ, that is, $\xi \in L^2(\mathcal{F}_T^Y)$ where

$$L^2(\mathcal{F}_T^Y) := \{\text{random variables } \Gamma \in \mathcal{F}_T^Y : \mathbb{E}\left[\Gamma^2\right] < \infty\}. \tag{49}$$

The objective of the agent is to find the optimal hedging strategy for this derivative. Since the number of random sources exceeds the number of tradeable risky assets, the market is incomplete. It is well known that in this setting, perfect replication by self-financing strategies is not feasible. Then, we suppose that the investor intends to pursue the risk-minimization approach. Risk minimization is a quadratic hedging method that allows determining a dynamic investment strategy that replicates perfectly the claim with minimal cost. Let us properly introduce the objects of interest. We start with the following notation. For any pair of $\mathbb{F}$-adapted (respectively, $\mathbb{F}^Y$-adapted) processes Ψ^1, Ψ^2 we refer to $\langle\Psi^1, \Psi^2\rangle^{\mathbb{F}}$ for the predictable covariation computed with respect to filtration $\mathbb{F}$ (respectively, $\langle\Psi^1, \Psi^2\rangle^{\mathbb{F}^Y}$ for the predictable covariation computed with respect to filtration $\mathbb{F}^Y$). Note that

$$\begin{aligned} \langle Y\rangle_t^{\mathbb{F}} &= \int_0^t Y_s^2\Big(\sigma^2(s, Y_{s^-}) + \int_Z K^2(s, X_{s^-}, Y_{s^-}; \zeta)\nu(\mathrm{d}\zeta)\Big)\mathrm{d}s \\ &= \int_0^t \Big(Y_s^2\sigma^2(s, Y_{s^-}) + \int_{\mathbb{R}} z^2\lambda(s, X_{s^-}, Y_{s^-})\phi(s, X_{s^-}, Y_{s^-}, \mathrm{d}z)\Big)\mathrm{d}s, \quad t \in [0, T], \end{aligned} \tag{50}$$

and since Y is also $\mathbb{F}^Y$ adapted, we also have

$$\langle Y\rangle_t^{\mathbb{F}^Y} = \int_0^t \Big(Y_s^2\sigma^2(s, Y_{s^-}) + \int_{\mathbb{R}} z^2\pi_{s^-}(\lambda\phi(\mathrm{d}z))\Big)\mathrm{d}s, \quad t \in [0, T]. \tag{51}$$

We stress that, due to the presence of a jump component, the predictable quadratic variations of Y with respect to filtrations $\mathbb{F}$ and $\mathbb{F}^Y$ are different.

Now we introduce a technical definition of two spaces, $\Theta(\mathbb{F})$ and $\Theta(\mathbb{F}^Y)$

Definition 20. *The space* $\Theta(\mathbb{F}^Y)$ *(respectively,* $\Theta(\mathbb{F})$*) is the space of all* $\mathbb{F}^Y$*-predictable (respectively,* $\mathbb{F}$*-predictable) processes* θ *such that*

$$\mathbb{E}\left[\int_0^T \theta_u^2 \mathrm{d}\langle Y\rangle_u^{\mathbb{F}^Y}\right] < \infty \quad \left(\text{respectively} \quad \mathbb{E}\left[\int_0^T \theta_u^2 \mathrm{d}\langle Y\rangle_u^{\mathbb{F}}\right] < \infty\right). \tag{52}$$

We observe that for every $\theta \in \Theta(\mathbb{F}^Y)$, thanks to $\mathbb{F}^Y$-predictability, we have

$$\mathbb{E}\left[\int_0^T \theta_u^2 \mathrm{d}\langle Y\rangle_u^{\mathbb{F}}\right] = \mathbb{E}\left[\int_0^T \theta_u^2 \mathrm{d}\langle Y\rangle_u^{\mathbb{F}^Y}\right] < \infty, \tag{53}$$

which implies that $\Theta(\mathbb{F}^Y) \subseteq \Theta(\mathbb{F})$.

Since we have two different levels of information represented by the filtrations $\mathbb{F}$ and $\mathbb{F}^Y$, we may define two classes of admissible strategies.

Definition 21. *An* $\mathbb{F}^Y$*-strategy (respectively,* $\mathbb{F}$*-strategy) is a pair* $\psi = (\theta, \eta)$ *of stochastic processes, where* θ *represents the amount invested in the risky asset and* η *is the amount invested in the riskless asset, such that* $\theta \in \Theta(\mathbb{F}^Y)$ *(respectively,* $\theta \in \Theta(\mathbb{F})$*) and* η *is* $\mathbb{F}^Y$*-adapted (respectively,* $\mathbb{F}$*-adapted).*

This definition reflects the fact that investor's choices should be adapted to her/his knowledge of the market. The value of a strategy $\psi = (\theta, \eta)$ is given by

$$V_t(\psi) = \theta_t Y_t + \eta_t, \quad t \in [0, T], \tag{54}$$

and its cost is described by the process

$$C_t(\psi) = V_t(\psi) - \int_0^t \theta_u \mathrm{d}Y_u, \quad t \in [0, T]. \tag{55}$$

In other terms, the cost of a strategy is the difference between the value process and the gain process. For a self-financing strategy, the value and the gain processes coincide, up to the initial wealth V_0, and therefore the cost is constant and equal to $C_t = V_0$, for every $t \in [0, T]$. We continue by defining the risk process, in the partial information setting.

Definition 22. *Given an* $\mathbb{F}^Y$*-strategy (respectively, an* $\mathbb{F}$*-strategy)* $\psi = (\theta, \eta)$*, we denote by* $R^{\mathcal{F}^Y}(\psi)$ *(respectively,* $R^{\mathcal{F}}(\psi)$*) the associated risk process defined as*

$$R_t^{\mathcal{F}^Y}(\psi) := \mathbb{E}\left[\left(C_T(\psi) - C_t(\psi)\right)^2 | \mathcal{F}_t^Y\right], \quad \left(\text{respectively } R_t^{\mathcal{F}}(\psi) := \mathbb{E}\left[\left(C_T(\psi) - C_t(\psi)\right)^2 | \mathcal{F}_t\right]\right), \tag{56}$$

for every $t \in [0, T]$.

Then, we have the following definition of risk-minimizing strategy under partial information.

Definition 23. *An $\mathbb{F}^Y$-strategy ψ is risk minimizing if*

i. $V_T(\psi) = \xi$,

ii. *for any other $\mathbb{F}^Y$ -strategy $\tilde{\psi}$ we have $R_t^{\mathcal{F}^Y}(\psi) \leq R_t^{\mathcal{F}^Y}(\tilde{\psi})$, for every $t \in [0, T]$.*

The corresponding definitions of risk process and risk-minimizing strategy under full information can be obtained replacing $\mathbb{F}^Y$ and $R_t^{\mathcal{F}^Y}$ with $\mathbb{F}$ and $R_t^{\mathcal{F}}$ in Definition 23. To differentiate, when it is necessary, we use the terms $\mathbb{F}^Y$-risk-minimizing strategy or $\mathbb{F}$-risk-minimizing strategy. The criterion (ii) in Definition 23 can be also written as

$$\min_{\psi \in \Theta(\mathbb{F}^Y)} \mathbb{E}\left[(C_T(\psi) - C_t(\psi))^2\right], \quad t \in [0, T], \tag{57}$$

which intuitively means that a strategy is risk minimizing if it minimizes the variance of the cost. This equivalent definition allows to obtain a nice property of risk-minimizing strategies which turn out to be *self-financing on average*, that is, the cost process C is a martingale and therefore has constant expectation (see, e.g., Ref. [27, Lemma 2] or [28, Lemma 2.3]).

In the sequel, we aim to characterize the optimal hedging strategy for the contingent claim ξ under full and partial information, that is, the $\mathbb{F}$- and the $\mathbb{F}^Y$-risk-minimizing strategies. To this, we introduce two orthogonal decompositions known as the Galtchouk-Kunita-Watanabe decompositions under full and partial information (see, e.g., [30]). To understand better the relevance of these decompositions, we assume for a moment completeness of the market and full information. Then, it is well known that for every European-type contingent claim with final payoff ξ, there exists a self-financing strategy $\psi = (\theta, \eta)$ such that

$$\xi = V_0 + \int_0^T \theta_u \mathrm{d}Y_u, \quad \mathbf{P} - \text{a.s.} \tag{58}$$

that is, a replicating portfolio is uniquely determined by the initial wealth and the investment in the risky asset. When the market is incomplete, decomposition Eq. (58) does not hold in general. Intuitively, this implies that we might expect additional terms in Eq. (58), and according to the risk-minimization criterion, this additional terms need to be such that the final cost does not deviate too much from the average cost, in the quadratic sense. Specifically, we have the following decomposition of the random variable ξ:

$$\xi = V_0 + \int_0^T \theta_u \mathrm{d}Y_u + G_T, \quad \mathbf{P} - \text{a.s.} \tag{59}$$

where G_T is the value at time T of a suitable process G. The minimality criterion requires that G is a martingale orthogonal to Y. We refer the reader to Ref. [28] for a detailed survey. Under suitable hypothesis, the above decomposition takes the name of Galtchouk-Kunita-Watanabe decomposition.

Now we wish to be more formal, and we introduce the following definitions:

Consider a random variable $\xi \in L^2(\mathcal{F}_T^Y)$. Since $\mathcal{F}_T^Y \subseteq \mathcal{F}_T$, we can define the following decompositions for ξ.

Definition 24. *a. The Galtchouk-Kunita-Watanabe decomposition of $\xi \in L^2(\mathcal{F}_T^Y)$ with respect to Y and $\mathbb{F}$ is given by*

$$\xi = U_0^{\mathcal{F}} + \int_0^T \theta_u^{\mathcal{F}} \mathrm{d}Y_u + G_T^{\mathcal{F}} \quad \mathbf{P} - a.s., \tag{60}$$

where $U_0^{\mathcal{F}} \in L^2(\mathcal{F}_0)$, $\theta^{\mathcal{F}} \in \Theta(\mathbb{F})$ and $G^{\mathcal{F}}$ is a square integrable $(\mathbb{F}, \mathbf{P})$-martingale, with $G_0^{\mathcal{F}} = 0$, orthogonal to Y, that is, $\langle G^{\mathcal{F}}, Y\rangle_t^{\mathbb{F}} = 0$ for every $t \in [0, T]$.

b. The Galtchouk-Kunita-Watanabe decomposition of $\xi \in L^2(\mathcal{F}_T^Y)$ with respect to Y and $\mathbb{F}^Y$ is given by

$$\xi = U_0^{\mathcal{F}^Y} + \int_0^T \theta_u^{\mathcal{F}^Y} \mathrm{d}Y_u + G_T^{\mathcal{F}^Y} \quad \mathbf{P} - a.s., \tag{61}$$

where $U_0^{\mathcal{F}^Y} \in L^2(\mathcal{F}_0^Y)$, $\theta^{\mathcal{F}^Y} \in \Theta(\mathbb{F}^Y)$ and $G^{\mathcal{F}^Y}$ is a square integrable $(\mathbb{F}^Y, \mathbf{P})$ -martingale, With $G_0^{\mathcal{F}^Y} = 0$, strongly orthogonal to Y, that is, $\langle G^{\mathcal{F}}, Y\rangle_t^{\mathbb{F}^Y} = 0$ for every $t \in [0, T]$.

In the sequel, we refer to Eqs. (60) and (61) as the Galtchouk-Kunita-Watanabe decompositions under full information and under partial information, respectively. Since Y is a square integrable martingale with respect to both filtrations $\mathbb{F}$ and $\mathbb{F}^Y$, decompositions Eqs. (60) and (61) exist.

Next proposition provides a relation between the integrands $\theta^{\mathcal{F}}$ and $\theta^{\mathcal{F}^Y}$ of decompositions Eqs. (60) and (61) in terms of predictable projections. For any $(\mathbb{F}, \mathbf{P})$-predictable process A of finite variation, we denote by $A^{p,\mathbb{F}^Y}$ its $(\mathbb{F}^Y, \mathbf{P})$-dual-predictable projection.[1]

Proposition 25. *The integrands in decompositions Eqs. (60) and (61) satisfy the following relation:*

$$\theta_t^{\mathcal{F}^Y} = \frac{\mathrm{d}\left(\int_0^t \theta_u^{\mathcal{F}} \mathrm{d}\langle Y\rangle_u^{\mathbb{F}}\right)^{p,\mathbb{F}^Y}}{\mathrm{d}\langle Y\rangle_t^{p,\mathbb{F}^Y}}, \quad t \in [0, T]. \tag{62}$$

Here, $\langle Y\rangle^{p,\mathbb{F}^Y}$ denotes the $(\mathbb{F}^Y, \mathbf{P})$-dual-predictable projection of $\langle Y\rangle^{\mathbb{F}}$ and it is given by

[1]We call $(\mathbb{F}^Y, \mathbf{P})$- *dual predictable projection* of a process A the $\mathbb{F}^Y$-predictable finite variation process $A^{p,\mathbb{F}^Y}$ such that for any $\mathbb{F}^Y$-predictable-bounded process ϕ we have

$$\mathbb{E}\left[\int_0^T \phi_s \mathrm{d}A_s\right] = \mathbb{E}\left[\int_0^T \phi_s \mathrm{d}A_s^{p,\mathbb{F}^Y}\right]$$

$$\langle Y\rangle_t^{p,\mathbb{F}^Y} = \langle Y\rangle_t^{\mathbb{F}^Y} = \int_0^t Y_s^2\sigma^2(s, Y_{s^-})\mathrm{d}s + \int_0^t\int_{\mathbb{R}} z^2\pi_{s^-}(\lambda\phi(\mathrm{d}z))\mathrm{d}s, \quad t\in[0,T]. \tag{63}$$

Proof. First note that the $(\mathbb{F}^Y, \mathbf{P})$-dual-predictable projection of the process $\langle Y\rangle^{\mathbb{F}}$ coincides with the predictable quadratic variation of the process Y itself, computed with respect to its internal filtration, given in Eq. (51), since for any $(\mathbb{F}^Y, \mathbf{P})$-predictable-(bounded) process ϕ, we have that $\mathbb{E}\left[\int_0^T \phi_t \mathrm{d}\langle Y\rangle_t^{\mathbb{F}}\right] = \mathbb{E}\left[\int_0^T \phi_t \mathrm{d}\langle Y\rangle_t^{\mathbb{F}^Y}\right]$. This proves Eq. (63).

Let

$$\theta_t := \frac{\mathrm{d}\left(\int_0^t \theta_u^{\mathcal{F}}\mathrm{d}\langle Y\rangle_u^{\mathbb{F}}\right)^{p,\mathbb{F}^Y}}{\mathrm{d}\langle Y\rangle_t^{p,\mathbb{F}^Y}}, \quad t\in[0,T]. \tag{64}$$

By the Galtchouk-Kunita-Watanabe decomposition Eq. (60), we can write

$$\xi = U_0^{\mathcal{F}} + \int_0^T \theta_u \mathrm{d}Y_u + G_T^{\mathcal{F}} + \widetilde{G}_T \quad \mathbf{P}-\text{ a.s.}, \tag{65}$$

where $\widetilde{G}_t := \int_0^t (\theta_u^{\mathcal{F}} - \theta_u)\mathrm{d}Y_u$, for every $t\in[0,T]$. We observe that for every $\mathbb{F}^Y$-predictable process ϕ the following holds:

$$\begin{aligned}\mathbb{E}\left[\int_0^T \phi_u\theta_u \mathrm{d}\langle Y\rangle_u^{\mathbb{F}}\right] &= \mathbb{E}\left[\int_0^T \phi_u\theta_u \mathrm{d}\langle Y\rangle_u^{\mathbb{F}^Y}\right]\\ &= \mathbb{E}\left[\int_0^T \phi_u(\theta_u^{\mathcal{F}}\mathrm{d}\langle Y\rangle_u^{\mathbb{F}})^{p,\mathbb{F}^Y}\right] = \mathbb{E}\left[\int_0^T \phi_u\theta_u^{\mathcal{F}}\mathrm{d}\langle Y\rangle_u^{\mathbb{F}}\right].\end{aligned} \tag{66}$$

By choosing $\phi = \theta$ and applying the Cauchy-Schwarz inequality, we obtain

$$\mathbb{E}\left[\int_0^T (\theta_u)^2 \mathrm{d}\langle Y\rangle_u^{\mathbb{F}^Y}\right] \le \mathbb{E}\left[\int_0^T (\theta_u^{\mathcal{F}})^2 \mathrm{d}\langle Y\rangle_u^{\mathbb{F}}\right] < \infty. \tag{67}$$

This implies that $\theta\in\Theta(\mathbb{F}^Y)\subseteq\Theta(\mathbb{F})$ and that $\widetilde{G}$ is an $(\mathbb{F}, \mathbf{P})$-martingale. Taking the conditional expectation with respect to $\mathcal{F}_T^Y$ in Eq. (65) leads to

$$\xi = \mathbb{E}[U_0^{\mathcal{F}}|\mathcal{F}_T^Y] + \int_0^T \theta_u \mathrm{d}Y_u + G_T^{\mathcal{F}} + \widetilde{G}_T = \mathbb{E}[U_0^{\mathcal{F}}|\mathcal{F}_0^Y] + \int_0^T \theta_u \mathrm{d}Y_u + \widehat{G}_T^{\mathcal{F}^Y} \quad \mathbf{P}-\text{a.s.} \tag{68}$$

where

$$\widehat{G}_t^{\mathcal{F}^Y} := \mathbb{E}[U_0^{\mathcal{F}}|\mathcal{F}_t^Y] - \mathbb{E}[U_0^{\mathcal{F}}|\mathcal{F}_0^Y] + \mathbb{E}[G_T^{\mathcal{F}}|\mathcal{F}_t^Y] + \mathbb{E}\left[\widetilde{G}_T|\mathcal{F}_t^Y\right], \quad t\in[0,T], \tag{69}$$

which provides the Galtchouk-Kunita-Watanabe decomposition Eq. (61) if we can show that the $(\mathbb{F}^Y, \mathbf{P})$-martingale $\widehat{G}^{\mathcal{F}^Y}$ is strongly orthogonal to Y, that is, if for any $(\mathbb{F}^Y, \mathbf{P})$-predictable-(bounded) process ϕ the following holds:

$$\mathbb{E}\left[\widehat{G}_T^{\mathcal{F}^Y}\int_0^T \phi_u \mathrm{d}Y_u\right] = 0. \tag{70}$$

Note that orthogonality of the term $\mathbb{E}\left[U_0^{\mathcal{F}}|\mathcal{F}_t^Y\right] - \mathbb{E}\left[U_0^{\mathcal{F}}|\mathcal{F}_0^Y\right] + \mathbb{E}\left[G_T^{\mathcal{F}}|\mathcal{F}_t^Y\right]$ follows by the orthogonality of $G^{\mathcal{F}}$ and Y. Moreover, we have

$$\mathbb{E}\left[\mathbb{E}\left[\widetilde{G}_T|\mathcal{F}_T^Y\right]\int_0^T \phi_u \mathrm{d}Y_u\right] = \mathbb{E}\left[\widetilde{G}_T\int_0^T \phi_u \mathrm{d}Y_u\right] = \mathbb{E}\left[\int_0^T \phi_u(\theta_u^{\mathcal{F}} - \theta_u)\mathrm{d}\langle Y\rangle_u^{\mathbb{F}}\right], \tag{71}$$

and by Eq. (64)

$$\begin{aligned}\mathbb{E}\left[\int_0^T \phi_u\theta_u \mathrm{d}\langle Y\rangle_u^{\mathbb{F}}\right] &= \mathbb{E}\left[\int_0^T \phi_u\theta_u \mathrm{d}\langle Y\rangle_u^{\mathbb{F}^Y}\right]\\ &= \mathbb{E}\left[\int_0^T \phi_u \mathrm{d}\left(\int_0^u \theta_r^{\mathcal{F}} \mathrm{d}\langle Y\rangle_r\right)^{p,\mathbb{F}^Y}\right] = \mathbb{E}\left[\int_0^T \phi_u\theta_u^{\mathcal{F}} \mathrm{d}\langle Y\rangle_u^{\mathbb{F}}\right],\end{aligned} \tag{72}$$

which proves strong orthogonality.

Theorem 26 shows the relation between the Galtchouk-Kunita-Watanabe decompositions and the optimal strategies under full and partial information.

Theorem 26. *i. Every contingent claim $\xi \in L^2(\mathcal{F}_T^Y, \mathbf{P})$ admits a unique $\mathbb{F}$-risk-minimizing strategy $\psi^{*,\mathcal{F}} = (\theta^{*,\mathcal{F}}, \eta^{*,\mathcal{F}})$, explicitly given by*

$$\theta^{*,\mathcal{F}} = \theta^{\mathcal{F}}, \quad \eta^{*,\mathcal{F}} = V(\psi^{*,\mathcal{F}}) - \theta^{*,\mathcal{F}}Y, \tag{73}$$

where $V_t(\psi^{*,\mathcal{F}}) = \mathbb{E}[\xi|\mathcal{F}_t]$ for every $t \in [0, T]$, with minimal cost

$$C_t(\psi^{*,\mathcal{F}}) = U_0^{\mathcal{F}} + G_t^{\mathcal{F}}, \quad t \in [0, T]. \tag{74}$$

Here, $\theta^{\mathcal{F}}$, $U_0^{\mathcal{F}}$, and $G^{\mathcal{F}}$ are given in Definition 24 part a.

ii. Moreover, it also admits a unique $\mathbb{F}^Y$-risk-minimizing strategy $\psi^{*,\mathcal{F}} = (\theta^{*,\mathcal{F}}, \eta^{*,\mathcal{F}^Y})$, explicitly given by

$$\theta^{*,\mathcal{F}^Y} = \theta^{\mathcal{F}^Y}, \quad \eta^{*,\mathcal{F}^Y} = V(\psi^{*,\mathcal{F}^Y}) - \theta^{*,\mathcal{F}^Y}Y, \tag{75}$$

where $V_t(\psi^{*,\mathcal{F}^Y}) = \mathbb{E}\left[\xi|\mathcal{F}_t^Y\right]$ for every $t \in [0, T]$, with minimal cost

$$C_t(\psi^{*,\mathcal{F}^Y}) = U_0^{\mathcal{F}^Y} + G_t^{\mathcal{F}^Y}, \quad t \in [0, T], \tag{76}$$

and $\theta^{\mathcal{F}^Y}$, $U_0^{\mathcal{F}^Y}$ *and* $G^{\mathcal{F}^Y}$ *are given in Definition 24 part b.*

Proof. The proof of part *i.* is given, for example, in Ref. [28, Theorem 2.4]. For part *ii.*, note that using the martingale representation of Y with respect to its inner filtration given in Eq. (48) and the fact that $\xi \in L^2(\mathcal{F}_T^Y)$, it is possible to reduce the partial information case to full information and apply again [28, Theorem 2.4]. □

Proposition 25 helps us in the computation of the optimal strategy under partial information. Indeed, it is sufficient to compute the corresponding strategy $\theta^{*,\mathcal{F}}$ under full information and the Radon-Nikodym derivative given in Eq. (62). To get more explicit representations, we assume that the payoff of the contingent claim has the form $\xi = H(T, Y_T)$, for some function $H : [0, T] \times \mathbb{R}^+ \to \mathbb{R}$. Let $\mathcal{L}^{X,Y}$ denote the Markov generator of the pair (X, Y), that is

$$\begin{aligned}\mathcal{L}^{X,Y} f(t,x,y) = {} & \frac{\partial f}{\partial t} + b_0(t,x)\frac{\partial f}{\partial x} + b_1(t,x,y)\frac{\partial f}{\partial y} + \frac{1}{2}\sigma_0^2(t,x)\frac{\partial^2 f}{\partial x^2} + \rho y \sigma_0(t,x)\sigma(t,y)\frac{\partial^2 f}{\partial x \partial y} \\ & + \frac{1}{2} y^2 \sigma^2(t,y)\frac{\partial^2 f}{\partial y^2} + \int_Z \Delta f(t,x,y;\zeta)\nu(d\zeta)\end{aligned} \tag{77}$$

for every $f \in C_b^{1,2,2}([0, T] \times \mathbb{R} \times \mathbb{R}^+)$, where

$$\Delta f(t,x,y;\zeta) := f(t, x + K_0(t,x;\zeta), y(1 + K(t,x,y;\zeta))) - f(t,x,y). \tag{78}$$

By the Markov property, we have that for any $t \in [0, T]$ there exists a measurable function $h(t, x, y)$ such that

$$h(t, X_t, Y_t) = \mathbb{E}[H(T, Y_T)|\mathcal{F}_t]. \tag{79}$$

If the function h is sufficiently regular, for instance $h \in C_b^{1,2,2}([0, T] \times \mathbb{R} \times \mathbb{R}^+)$, we can apply Itô's formula and get that

$$h(t, X_t, Y_t) = h(0, X_0, Y_0) + \int_0^t \mathcal{L}^{X,Y} h(s, X_s, Y_s) ds + M_t^h \tag{80}$$

where M^h is the $(\mathbb{F}, \mathbf{P})$-martingale given by

$$\begin{aligned}dM_t^h = {} & \int_0^t \frac{\partial h}{\partial x}(s, X_s, Y_s)\sigma_0(s, X_s) dW_s^0 + \int_0^t \frac{\partial h}{\partial y}(s, X_s, Y_s) Y_s \sigma(s, Y_s) dW_s^1 \\ & + \int_0^t \int_Z \Delta h(s, X_{s^-}, Y_{s^-}; \zeta)\Big(N(ds, d\zeta) - \nu(d\zeta) ds\Big).\end{aligned} \tag{81}$$

By Eq. (79), the process $\{h(t, X_t, Y_t),\ t \in [0, T]\}$ is an $(\mathbb{F}, \mathbf{P})$-martingale. Then, the finite variation term vanishes, which means that the function h satisfies $\mathcal{L}^{X,Y} h(t, X_t, Y_t) = 0$, $\mathbf{P}$-a.s. and for almost every $t \in [0, T]$. The next proposition provides the risk-minimizing strategy under partial information.

Proposition 27. *Assume* $h \in C_b^{1,2,2}([0,T] \times \mathbb{R} \times \mathbb{R}^+)$. *Then the first components* $\theta^{*,\mathcal{F}}$ *and* $\theta^{*,\mathcal{F}^Y}$ *of the risk-minimizing strategies under full and partial information are given by*

$$\theta_t^{*,\mathcal{F}} = \frac{g(t, X_{t-}, Y_{t-})}{Y_{t-}^2\sigma^2(t, Y_{t-}) + \int_{\mathbb{R}} z^2\lambda(t, X_{t-}, Y_{t-})\phi(t, X_{t-}, Y_{t-}, \mathrm{d}z)}, \quad t \in [0,T] \tag{82}$$

$$\theta_t^{*,\mathcal{F}^Y} = \frac{\pi_{t-}(g)}{Y_{t-}^2\sigma(t, Y_{t-}) + \int_{\mathbb{R}} z^2\pi_{t-}(\lambda\phi(\mathrm{d}z))}, \quad t \in [0,T] \tag{83}$$

respectively, where the function $g(t, x, y)$ *is*

$$g(t,x,y) = \rho\,\sigma_0(t,x)y\sigma(t,y)\frac{\partial h}{\partial x} + y^2\sigma^2(t,y)\frac{\partial h}{\partial y} + \int_Z yK(t,x,y;\zeta)\Delta h(t,x,y;\zeta)\nu(d\zeta). \tag{84}$$

Proof. Consider decomposition Eq. (60) for $\xi = H(T, Y_T)$. Then, conditioning on $\mathcal{F}_t$ we get

$$h(t, X_t, Y_t) = U_0 + \int_0^t \theta_s^{*,\mathcal{F}}\mathrm{d}Y_s + G_t^{\mathcal{F}}. \tag{85}$$

Taking the covariation with respect to Y and $\mathbb{F}$, we obtain

$$\langle h(\cdot, X, Y), Y\rangle_t^{\mathbb{F}} = \int_0^t \theta_s^{*,\mathcal{F}}\mathrm{d}\langle Y\rangle_s^{\mathbb{F}}. \tag{86}$$

On the other hand, $h(t, X_t, Y_t) = M_t^h$, then taking Eqs. (81) and (44) into account we get that

$$\langle h(\cdot, X, Y), Y\rangle_t^{\mathbb{F}} = \int_0^t g(s, X_s, Y_s)\mathrm{d}s, \tag{87}$$

where $g(t, x, y)$ is given in Eq. (84). Hence, by Eqs. (50) and (87), we may represent $\theta^{*,\mathcal{F}}$ as

$$\theta_t^{*,\mathcal{F}} = \frac{\mathrm{d}\langle h(\cdot, X, Y), Y\rangle_t^{\mathbb{F}}}{\mathrm{d}\langle Y\rangle_t^{\mathbb{F}}} = \frac{g(t, X_{t-}, Y_{t-})}{Y_{t-}^2\sigma^2(t, Y_{t-}) + \int_{\mathbb{R}} z^2\lambda(t, X_{t-}, Y_{t-})\phi(t, X_{t-}, Y_{t-}, \mathrm{d}z)} \tag{88}$$

Note that by Eq. (51) and

$$\left(\int_0^t \theta_u^{*,\mathcal{F}}\mathrm{d}\langle Y\rangle_u^{\mathbb{F}}\right)^{p,\mathbb{F}^Y} = \left(\int_0^t g(s, X_s, Y_s)\mathrm{d}s\right)^{p,\mathbb{F}^Y} = \int_0^t \pi_s(g)\mathrm{d}s, \tag{89}$$

applying Eq. (62) we get representation Eq. (83).

Our ultimate objective in this section is to investigate on the relation between costs of the $\mathbb{F}$-optimal strategy and the $\mathbb{F}^Y$-optimal strategy, or equivalently the associated risk processes.

It clearly holds that $\theta^{*,\mathcal{F}^Y} \in \Theta(\mathbb{F})$, and then the $\mathbb{F}^Y$-risk-minimizing strategy is also an $\mathbb{F}$-strategy. Considering the corresponding risks, we have

$$\begin{aligned}\mathbb{E}\left[\left(C_T(\psi^{*,\mathcal{F}^Y}) - C_t(\psi^{*,\mathcal{F}^Y})\right)^2 |\mathcal{F}_t^Y\right] &= \mathbb{E}\left[\mathbb{E}\left[\left(C_T(\psi^{*,\mathcal{F}^Y}) - C_t(\psi^{*,\mathcal{F}^Y})\right)^2 |\mathcal{F}_t\right] |\mathcal{F}_t^Y\right] \\ \geq \mathbb{E}\left[\mathbb{E}\left[\left(C_T(\psi^{*,\mathcal{F}}) - C_t(\psi^{*,\mathcal{F}})\right)^2 |\mathcal{F}_t\right] |\mathcal{F}_t^Y\right] &= \mathbb{E}\left[\left(C_T(\psi^{*,\mathcal{F}}) - C_t(\psi^{*,\mathcal{F}})\right)^2 |\mathcal{F}_t^Y\right],\end{aligned} \tag{90}$$

and then $\mathbb{E}[R_t^{\mathcal{F}}(\psi^{*,\mathcal{F}})] \leq \mathbb{E}\left[R_t^{\mathcal{F}^Y}(\psi^{*,\mathcal{F}^Y})\right]$, for every $t \in [0, T]$. In the remaining part of the paper, we assume that $\mathcal{F}_0^Y = \mathcal{F}_0 = \{\Omega, \emptyset\}$, and we wish to measure the difference in the total risk taken by an informed investor, endowed with a filtration $\mathbb{F}$, and a partially informed investor, whose information is described by $\mathbb{F}^Y$. Precisely, we compute the difference $R_0^{\mathcal{F}^Y}(\psi^{*,\mathcal{F}^Y}) - R_0^{\mathcal{F}}(\psi^{*,\mathcal{F}})$. By decompositions Eqs. (60) and (61), we have that $C_T(\psi^{*,\mathcal{F}}) - C_0(\psi^{*,\mathcal{F}}) = G_T^{\mathcal{F}}$ and $C_T(\psi^{*,\mathcal{F}^Y}) - C_0(\psi^{*,\mathcal{F}^Y}) = G_T^{\mathcal{F}^Y}$ and also

$$G_T^{\mathcal{F}^Y} = U_0^{\mathcal{F}} - U_0^{\mathcal{F}^Y} + \int_0^T (\theta_r^{*,\mathcal{F}} - \theta_r^{*,\mathcal{F}^Y}) \mathrm{d}Y_r + G_T^{\mathcal{F}}, \tag{91}$$

since $\mathcal{F}_0^Y = \mathcal{F}_0 = \{\Omega, \emptyset\}$, $U_0^{\mathcal{F}} = U_0^{\mathcal{F}^Y}$. Then computing the square of $G_T^{\mathcal{F}^Y}$ and taking the expectation we get

$$\mathbb{E}\left[(G_T^{\mathcal{F}^Y})^2\right] = \mathbb{E}\left[(G_T^{\mathcal{F}})^2\right] + \mathbb{E}\left[\left(\int_0^T (\theta_r^{*,\mathcal{F}} - \theta_r^{*,\mathcal{F}^Y}) \mathrm{d}Y_r\right)^2\right] + 2\mathbb{E}\left[G_T^{\mathcal{F}} \int_0^T (\theta_r^{*,\mathcal{F}} - \theta_r^{*,\mathcal{F}^Y}) \mathrm{d}Y_r\right]. \tag{92}$$

It follows from Itô isometry and the fact that $G^{\mathcal{F}}$ is orthogonal to Y, that

$$\mathbb{E}\left[(G_T^{\mathcal{F}^Y})^2\right] = \mathbb{E}\left[(G_T^{\mathcal{F}})^2\right] + \mathbb{E}\left[\int_0^T (\theta_r^{*,\mathcal{F}} - \theta_r^{*,\mathcal{F}^Y})^2 \langle Y \rangle_r^{\mathbb{F}}\right]. \tag{93}$$

Then the difference that we want to evaluate becomes

$$\begin{aligned}R_0^{\mathcal{F}^Y}(\psi^{*,\mathcal{F}^Y}) - R_0^{\mathcal{F}}(\psi^{*,\mathcal{F}}) &= \mathbb{E}\left[(G_T^{\mathcal{F}^Y})^2\right] - \mathbb{E}\left[(G_T^{\mathcal{F}})^2\right] = \mathbb{E}\left[\int_0^T (\theta_r^{*,\mathcal{F}} - \theta_r^{*,\mathcal{F}^Y})^2 \mathrm{d}\langle Y \rangle_r^{\mathbb{F}}\right] \\ &= \mathbb{E}\left[\int_0^T (\theta_r^{*,\mathcal{F}})^2 \mathrm{d}\langle Y \rangle_r^{\mathbb{F}}\right] + \mathbb{E}\left[\int_0^T (\theta_r^{*,\mathcal{F}^Y})^2 \mathrm{d}\langle Y \rangle_r^{\mathbb{F}}\right] - 2\mathbb{E}\left[\int_0^T \theta_r^{*,\mathcal{F}} \theta_r^{*,\mathcal{F}^Y} \mathrm{d}\langle Y \rangle_r^{\mathbb{F}}\right].\end{aligned} \tag{94}$$

Using Eq. (62) and the definition of $\mathbb{F}^Y$-dual-predictable projections, we have that

$$\mathbb{E}\left[\int_0^t \theta_r^{*,\mathcal{F}^Y} \theta_r^{*,\mathcal{F}} \mathrm{d}\langle Y \rangle_r^{\mathbb{F}}\right] = \mathbb{E}\left[\int_0^t (\theta_r^{*,\mathcal{F}^Y})^2 \mathrm{d}\langle Y \rangle_r^{\mathbb{F}^Y}\right] = \mathbb{E}\left[\int_0^t (\theta_r^{*,\mathcal{F}^Y})^2 \mathrm{d}\langle Y \rangle_r^{\mathbb{F}}\right], \tag{95}$$

which implies

$$R_0^{\mathcal{F}^Y}(\psi^{*,\mathcal{F}^Y}) - R_0^{\mathcal{F}}(\psi^{*,\mathcal{F}}) = \mathbb{E}\left[\int_0^T (\theta_r^{*,\mathcal{F}})^2 \mathrm{d}\langle Y\rangle_r^{\mathbb{F}}\right] - \mathbb{E}\left[\int_0^T (\theta_r^{*,\mathcal{F}^Y})^2 \mathrm{d}\langle Y\rangle_r^{\mathbb{F}^Y}\right]. \tag{96}$$

Plugging in the expressions for the optimal strategies given in Eqs. (82) and (83), respectively, and denoting $\Sigma(t, X_t, Y_t) := Y_t^2\left(\sigma^2(t, Y_t) + \int_Z z^2\lambda(t, X_{t-}, Y_{t-})\phi(t, X_{t-}, Y_{t-}, \mathrm{d}z)\right)$, we have

$$\begin{aligned} R_0^{\mathcal{F}^Y}(\psi^{*,\mathcal{F}^Y}) - R_0^{\mathcal{F}}(\psi^{*,\mathcal{F}}) &= \mathbb{E}\left[\int_0^T \left(\frac{g^2(t, X_t, Y_t)}{\Sigma(t, X_t, Y_t)} - \frac{\pi_t^2(g)}{\pi_t(\Sigma)}\right)\mathrm{d}t\right] \\ &\leq C\mathbb{E}\left[\int_0^T \left(g^2(t, X_t, Y_t) - \pi_t^2(g)\right)\mathrm{d}t\right] = C\mathbb{E}\left[\int_0^T \left(g(t, X_t, Y_t) - \pi_t(g)\right)^2 \mathrm{d}t\right] \end{aligned} \tag{97}$$

for some $C > 0$, where the inequality follows by Assumption 18, and in the last equality, we used $\mathbb{E}\left[\int_0^T 2g(t, X_t, S_t)\pi_t(g)\mathrm{d}t\right] = \mathbb{E}\left[\int_0^T 2\pi_t(g)^2\mathrm{d}t\right]$.

We can conclude by saying that we found an upper bound for the expected difference between the total risks taken by an informed investor and a partially informed one which is directly proportional to the mean-squared error between the process $\{g(t, X_t, S_t),\ t \in [0, T]\}$ and its filtered estimate $\pi(g) = \{\pi_t(g),\ t \in [0, T]\}$.

Author details

Claudia Ceci[1]* and Katia Colaneri[2]

*Address all correspondence to: c.ceci@unich.it

1 Department of Economics, University of Chieti-Pescara, Pescara, Italy

2 Department of Economics, University of Perugia, Perugia, Italy

References

[1] Kushner H. On the differential equations satisfied by conditional probability densities of Markov processes, with applications. Journal of the Society for Industrial and Applied Mathematics, Series A: Control. 1964;**2**(1):106-119

[2] Kushner H. Dynamical equations for optimal nonlinear filtering. Journal of Differential Equations. 1967;**3**(2):179-190

[3] Zakai M. On the optimal filtering of diffusion processes. Probability Theory and Related Fields. 1969;**11**(3):230-243

[4] Lipster RS, Shiryaev A. Statistics of Random Processes I. Springer-Verlag; Berlin Heidelberg, 1977

[5] Kallianpur G. Stochastic Filtering Theory. Springer; Springer Verlag New York, 1980

[6] Elliott RJ. Stochastic Calculus and Applications. Springer; Berlin Heidelberg New York, 1982

[7] Kurtz TG, Ocone D. Unique characterization of condition distribution in nonlinear filtering. Annals of Probability. 1988;**16**:80-107

[8] Bhatt AG, Kallianpur G, Karandikar RL. Uniqueness and robustness of solution of measure-valued equations of nonlinear filtering. The Annals of Probability. 1995;**23**(4): 1895-1938

[9] Brémaud P. Point Processes and Queues. Springer-Verlag; New York, 1980

[10] Kliemann WH, Koch G, Marchetti F. On the unnormalized solution of the filtering problem with counting process observations. IETIT. 1990;**36**:1415-1425

[11] Ceci C, Gerardi A. Filtering of a Markov jump process with counting observations. Applied Mathematics & Optimization. 2000;**42**:1-18

[12] Frey R, Runggaldier W. A nonlinear filtering approach to volatility estimation with a view towards high frequency data. International Journal of Theoretical and Applied Finance. 2001;**4**(2):199-210

[13] Ceci C, Gerardi A. A model for high frequency data under partial information: A filtering approach. International Journal of Theoretical and Applied Finance. 2006;**9**(4):1-22

[14] Ceci C. Risk minimizing hedging for a partially observed high frequency data model. Stochastics: An International Journal of Probability and Stochastic Processes. 2006;**78**(1): 13-31

[15] Frey R, Runggaldier W. Pricing credit derivatives under incomplete information: A nonlinear-filtering approach. Finance and Stochastics. 2010;**14**:495-526

[16] Frey R, Schimdt T. Pricing and hedging of credit derivatives via the innovation approach to nonlinear filtering. Finance and Stochastics. 2011;**16**(1):105-133

[17] Ceci C, Colaneri K. Nonlinear filtering for jump diffusion observations. Advances in Applied Probability. 2012;**44**(3):678-701

[18] Ceci C, Colaneri K. The Zakai equation of nonlinear filtering for jump-diffusion observations: Existence and uniqueness. Applied Mathematics & Optimization. 2014;**69**(1):47-82

[19] Elliott R, Malcolm W. Discrete-time expectation maximization algorithms for Markov-modulated Poisson processes. IEEE Transactions on Automatic Control. 2008;**53**(1):247-256

[20] Björk T, Davis M, Landén C. Optimal investment under partial information. Mathematical Methods of Operations Research. 2010;**71**(2):371-399

[21] Ceci C, Colaneri K, Cretarola A. Local risk-minimization under restricted information to asset prices. Electronic Journal of Probability. 2015;**20**(96):1-30

[22] Ceci C, Colaneri K, Cretarola A. Hedging of unit-linked life insurance contracts with unobservable mortality hazard rate via local risk-minimization. Insurance: Mathematics and Economics. 2015;**60**:47-60.

[23] Ceci C, Gerardi A. Pricing for geometric marked point processes under partial information: entropy approach. International Journal of Theoretical and Applied Finance. 2009;**12**:179-207

[24] Frey R. Risk minimization with incomplete information in a model for high-frequency data. Mathematical Finance. 2000;**10**(2):215-222

[25] Nagai H, Peng S. Risk-sensitive dynamic portfolio optimization with partial information on infinite time horizon. Annals of Applied Probability. 2000;**12**:173-195

[26] Bäuerle N, Rieder U. Portfolio optimization with jumps and unobservable intensity process. Mathematical Finance. 2007;**17**(2):205-224

[27] Föllmer H, Sondermann D. Hedging of non redundant contingent claims. In: Hildenbrand W, Mas-Colell A, editors. Contribution to Mathematical Economics. North Holland, Amsterdam New York Oxford Tokyo; 1986. pp. 205-223

[28] Schweizer M. A guided tour through quadratic hedging approaches. In: Jouini E, Cvitanic J, Musiela M, editors. Option Pricing, Interest Rate and Risk Management. Cambridge University Press; Cambridge, 2001. pp. 538-574

[29] Schweizer M. Risk minimizing hedging strategies under partial information. Mathematical Finance. 1994;**4**:327-342

[30] Ceci C, Cretarola A, Russo F. GKW representation theorem under restricted information. An application to risk-minimization. Stochastics and Dynamics. 2014;**14**(2):1350019 (p. 23)

[31] Jacod J, Shiryaev A. Limit Theorems for Stochastic Processes. 2nd ed. Berlin: Springer; 2003

[32] Protter P, Shimbo K, Ethier SN, Feng J, Stockbridge RH eds, No arbitrage and general semimartingales. In: Markov Processes and Related Topics: A Festschrift for Thomas G. Kurtz. Institute of Mathematical Statistics; Beachwood, Ohio, USA, 2008. pp. 267-283

Chapter 18

Airlines Content Recommendations Based on Passengers' Choice Using Bayesian Belief Networks

Sien Chen, Wenqiang Huang, Mengxi Chen,
Junjiang Zhong and Jie Cheng

Additional information is available at the end of the chapter

http://dx.doi.org/10.5772/intechopen.70131

Abstract

Faced with the increasingly fierce competition in the aviation market, the strategy of consumer choice has gained increasing significance in both academia and practice. As ever-increasing travel choices and growing consumer heterogeneity, how do airline companies satisfy passengers' needs? With a vast amount of data, how do airline managers combine information to excavate the relationship between independent variables to gain insight about passengers' choices and value system as well as determining best personalized contents to them? Using the real case of China Southern Airlines, this paper illustrates how Bayesian belief network (BBN) can enable airlines dynamically recommend relevant contents based on predicting passengers' choice to optimize the loyalty. The findings of this study provide airline companies useful insights to better understand the passengers' choices and develop effective strategies for growing customer relationship.

Keywords: consumer choice, Bayesian belief network, recommendation system

1. Introduction

In a world of increasingly global competition, companies have to compete on the effectiveness and efficiency of their marketing strategies to capture new opportunities to satisfy customers' needs. In other words, having the greatest product at the lowest price is not competitive enough. Choice behavior is affected by a consumer's own preference for entire product categories and particular brands, allowing companies to collect market and industry data, learn about consumer preference, and change sales tactics. In general, companies must consider consumer choice and offer their customers varieties of differentiated products and different types of choices to meet consumer demand when they formulate revenue decisions and marketing strategies. For instance, most airlines have different fare classes (e.g., economy class versus first class) that differ in the level of services and facilities available for customers. Companies have to understand the choices that consumers make when facing such a product assortment and provide appropriate contents for each consumer. Once individual choice has been modeled, the choice prediction would be of great value to managers for the estimation of the impact of a change in product formulation [1, 2].

What one cares most is choice, the selection of a suitable content from a set of available alternatives. Given the growing diversity of the purchasing channels and information media, companies are increasingly interested in modeling and understanding an actual process through which consumers choose products, in addition to measure consumers' future choices. Better understanding of consumer choices and predicting preference is important for enterprises to introduce new products and implement target marketing. Preference prediction could also be used more extensively by companies to guide decision optimization [3]. The researchers and managers are mostly interested in knowing human choice behavior, particularly the underlying choice mechanisms, and reveal and investigate fundamental reasons behind it. Choice behavior is complex and yet rational, as a result, decision makers seek to simplify the formulation of choice process. Capturing the consumers' choice decision, a method that estimates direct and indirect effects, the situation-specific variables and clear causal relationship could offer better representing choice behavior mechanisms. Based on choice mechanism, companies need to measure preference and predict consumer decision making to conduct market research to design products. In this chapter, we aim to investigate these concerns:

- How can we infer consumers' choice for content in the future?
- How to design a model using only current period choices to infer consumers' inter-temporal preferences?
- Is the process dynamic and it allows the researchers to analyze influences of effect changes?

With increasing awareness of transportation competition, if airline companies hope to survive and make profit, they have to realize that customer resource is the most valuable competitive advantage and try their best to satisfy the needs of their customers. Besides, Mowen (1988) emphasized that managers can reset promotional strategies to satisfy different types of consumers' desire most effectively, through the best channels [7]. Based on the above analysis, we decided to introduce a personalized content recommendation system to satisfy China air passengers' desire. As we all know, good relationship with customers is crucial for airline companies to keep advantage in competition and furthermore make profit in the long run. Using real history data from China Southern Airline and Bayesian network can fix best personalized contents to each individual passenger.

2. Consumer choice behavior and Bayesian belief network

This chapter introduces Bayesian belief networks (BBNs) for predicting air passengers' choice. On the basis of these choices, airlines can recommend best relevant content to passengers, including products, service, tips, notices, feature introductions, and information sharing to improve their travel experience, satisfaction, and loyalty. The remainder of this paper is organized as follows. Section 2 briefly discusses a review of consumer choice behavior, provides some definitions, and illustrates advantages of Bayesian belief networks. In the next section, we establish BBN models by using the case of China Southern Airlines with real transaction data, including passengers' basic information, history decision options, and purchase characteristics to predict the possible contents which the consumer will choose, followed by model results and discussion.

2.1. Consumer choice behavior

When consumers face multiple alternative products, brands, and services, they tend to repeat the same choices that proved satisfactory in similar situations [4]. Information integration theory offers a specific mechanism to describe how individuals integrate separate pieces of available information into an overall index of preferences [5, 6]. The theory proposes that in situations where information about the products and brands are available in the marketplace, consumers tend to value and weight product attributes more often at the time of making a purchase decision. We formulated a comprehensive evaluation by combing consumers' values and weights under certain rules. Marketing managers should carefully study these decision-making processes and results to understand where consumers can collect relevant information, how consumers form beliefs, and what criteria consumers use to make product or service choices. As a result, companies can develop products that emphasize the appropriate attributes, and managers can reset promotional strategies to satisfy different types of consumers' desire most effectively, through the best channels [7]. Another interest issue to academia is in determining whether there are systematic differences in consumers' choice behavior. Identifying and understanding these differences are important for developing or formulating effective marketing strategies.

Consumer choice behavior has been mainly conceptualized as a combination of some sociodemographics and the attributes of alternatives [8]. Constructs, such as utility, attitude, or cognition, are used to map the attributes into one of the choice behavior. However, little research has been conducted about the choice process models considering the socio-demographic characteristics and the attributes of their decision alternatives in a recent study.

2.1.1. Consumer choice behavior in airline industry

In recent years, the airline industry faces the economic challenges, which are coupled with volatile fuel prices and pressure of environment protection. In addition, with increasing awareness of competition caused by the development of other transportation alternatives, if airline companies hope to survive and make profit, they have to realize that customer resource is the most valuable competitive advantage and try their best to satisfy the needs of their customers. Thus, good relationship with customers is crucial for airline companies to keep advantage in competition and furthermore make profit in the long run. Domestic and international airline companies have long shifted their attention to customer relationship management [9].

Relationship with passengers has been taken as one of the most important goals for every airline company to maximize passengers' loyalty and revenues. Besides good performance in airline operation and business management, another important key for success is leveraging power of customer relationship to attain superb performance. Those airline companies, who could correctly estimate trends and risks in the airline market and take necessary actions to satisfy their customers, could be much more successful in the industry.

Although understanding changing needs of passengers is of great importance for airlines, passengers' decision-making processes have received relatively little managerial attention. Therefore, further understanding of those decision-making processes is crucial for airline companies to improve their operations and business models [10].

With today's ever-increasing travel choices and growing consumer heterogeneity, numerous factors affect passengers' choices, for example, their socio-demographic status, decision-making patterns, cultural background, ticket cost, travel objectives, time schedule, and so on. The role of each factor is difficult to define, let alone the interaction between different factors. Leisure travelers are becoming more and more supply-oriented selecting airlines with most convenient schedules and best service experience. A change from selecting the travel destination and seeking for appropriate transportations to one, which a desirable airline service is initially set and the trip is arranged around in, will very likely alter the dominant social ideological trend of travel behavior.

2.2. Bayesian belief network

Belief networks are probabilistic graphical representations of models that capture relationship between different variables. Belief networks use either directed or undirected graphs to represent a dependency model. The directed acyclic graph (DAG) is more flexible and expressive. It is also able to investigate a wide range of probabilistic interdependency than undirected graphs. For example, induced and transitive dependency cannot be modeled accurately by undirected graphs, but can be easily represented in DAGs.

A causal belief network is made up of various types of nodes. An arc between two nodes represents a causal relation, an originating node of the arc is a parent node and the others are child nodes. A root node has no parents while a leaf node has no children. Each node has an underlying conditional probability table (CPT) that describes the option distribution for specific nodes associated with each possible combination of the parent nodes. Bayesian belief network (BBN) is a specific type of causal belief network, consisting of a set of nodes, where each node represents a variable in the dependency model and the connecting arcs represent the causal relationship among variables. **Figure 1** shows a simple BBN example about heart disease and heartache patients. The CPT's of the nodes is also illustrated in **Figure 1**. As for any causal belief network, the nodes represent stochastic variables and the arcs identify direct causal influences among the linked variables [11]. Each node or variable may take one of a number of possible states. The certainty of these states is determined from the belief in each possible state of all the nodes. The belief in each state of a node is updated whenever the belief

http://dx.doi.org/10.5772/intechopen.70131

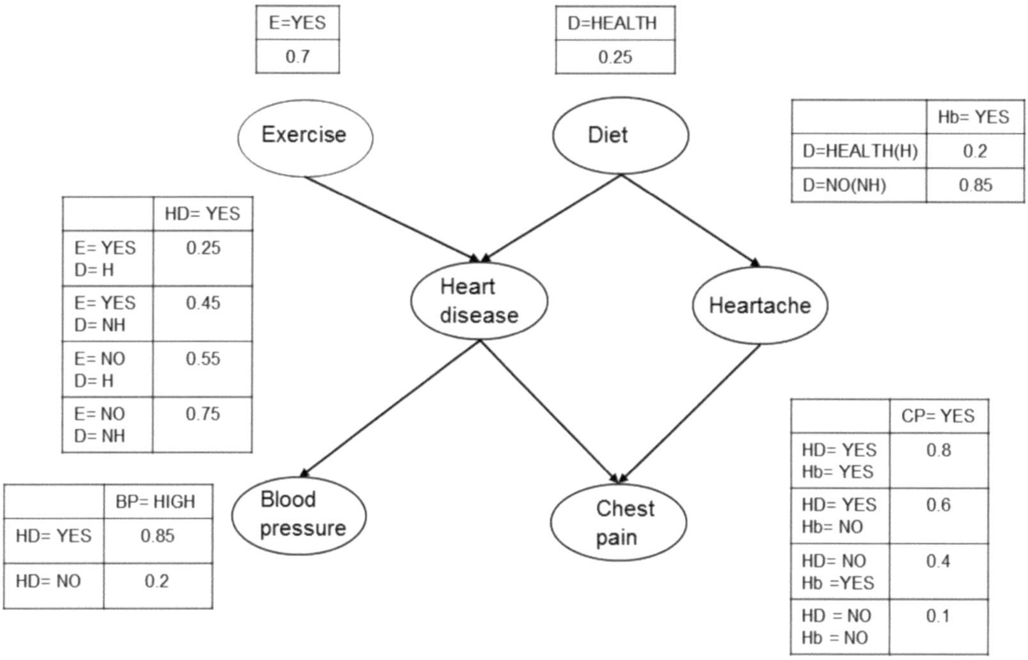

Figure 1. A Bayesian belief network depicting relationship among heart disease and heartache patients.

in each state of any directly connected node changes. The difference between Bayesian belief networks and other causal belief networks is that BBNs use Bayesian calculus to process the state probabilities of each node from the predetermined conditional and prior probabilities. The belief network is dynamic and their probabilities are subject to changes.

A Bayesian belief network is a graphical representation of a Bayesian probabilistic dependency within a knowledge domain [12], particularly appropriate for target recognition problems, where the category, identity, and class of target groups are to be recognized [13]. Bayesian belief networks have proven to be very useful, befitting to small and incomplete data collections. A Bayesian network can be, for example, used to save a considerable amount of space, explicit treatment of uncertainty, and support for decision analysis, casual relationship, and fast responses. Bayesian network is also suited to structural learning applications, and a combination of different sources of the preferred knowledge [14]. Besides, Bayesian approach finds the inclusion optimal model structure from data constructed by the a priori knowledge, and a constraint-based approach finds the optimal model structure from conditional dependences in each pair of variables. Given the ascertained information, Bayesian belief networks are used to determine or infer the posterior probability distributions for the variables of interest [11]. As such, they do not include decisions or utilities that typify the preferences of the users, but the user make decisions based on these probability distributions [11]. The causal relationships in Bayesian belief networks allow the correlation between variables to be modeled and predictions to be made. Comparing to classical statistical approaches, Bayesian belief networks have a distinct advantage [15]. BBN becomes not only a powerful tool for knowledge representation but reasoning under conditions of uncertainty [16], frequently dealing with real-world problems such as building medical diagnostic systems, forecasting, and manufacturing process control for several decades [17]. Nowadays BBN has been extended to other applications including software risk management [18], ecosystem and environmental management [19], and transportation [20]. There is a great impact of key events on long-term transport mode choice decisions using Bayesian belief network, precisely Bayesian decision network, for the exploration of the suggested formalism in measuring, analyzing, and predicting dynamic travel mode choice in relation to key events and critical incidents [21]. However, seldom researches are found using BBN as an application in airlines marketing management. This paper introduces Bayesian belief networks using relative and contextual variables to estimate a logic relationship and test the causal mechanisms of current passengers' choice and predict their future preference.

3. Case study: BBN in China Southern Airlines

Air passengers make their choices using prior information available as well as information they obtain from the internal and external environments. Passengers integrate all the information actually available to them (including prior information and any information affect them) and turn them into preferences of a product. The basic aim is to support airline decision makers in their analysis of the impact of variables on passenger demand in the future. Prediction of passenger choice for the distant future is critical to guide managers in the specification of marketing strategies to be used. Such distant future predictions necessitate large-scale models of passenger choice but that pressing need contrasts sharply with the capabilities of traditional forecasting and modeling techniques. In this study, both qualitative and quantitative approaches are studied. Developed as such, the BBN is expected to guide airline managers in their future product decisions, facilitating analysis of specific decisions based on predicting the choice modes of passengers; highlighting the causal relationships among variables in the process and finally showing the impact of changes. To represent the dynamic nature of the causal relationship and to draw inferences based on the uncertainty concerning the states of the variables; this part constructs a Bayesian belief network for airline content recommendation mode using a case study of Chinese airline. A basic assumption of BBN is that when the conditional probabilities for each variable are multiplied, the joint probability distributions for all variables in the network are then calculated [22]. The structure is determined based on experts' judgments on content recommendation mode and a logic relationship.

Three components of a belief network are important: the nodes representing variables, the links among nodes, and states representing the expected utilities or probabilities. Therefore, the first step of the process is the development of a casual network. For this purpose, relevant variables and the logic relationship of this network should be determined. In the next stage, belief networks explore how the changes of states of variables (nodes) influence consumers' future choices and the needs of contents. Therefore, the static causal model is transformed into a dynamic one through the calculation of the Bayesian belief network. The resulting network is subjected to scenario analysis to help airline decision makers in their analysis on future product designs.

3.1. Determination of the basic variables and casual relations

To obtain a mutually, selectively exhaustive list of basic variables of the airline companies, interviews are conducted with airline domain experts, who are encouraged to identify the variables that might be relevant to the research. Thereafter, 35 variables are generated based on the situation of China and with weights of the expert judgments and estimation. The decision variables are classified into four groups:

1. Personal characteristics
2. Experience and behavior characteristics
3. Preference characteristics
4. Individuals' perceptions

Personal characteristics include airline passengers' demographic status and member information related to air travel. Experience and behavioral characteristics include passenger purchase behavior, decisions in choosing products, and attributes of particular experience. Preference characteristics include consumer preference and travel patterns. Individual perception describes the evaluation of passengers' loyalty, satisfaction, and comfort. After the identification of variables, the next step is the determination of the causal relations among all the variables. The use of this network is proposed to capture the knowledge and assumptions and to understand the mechanism of consumer choice processes`. The whole network is built up using Netica. The changes exist in the network are subjected to field tests using real world data from Chinese data sources.

3.2. Implementation of the BBN

The content recommendation is a new attempt in airline companies' new marketing strategies. After obtaining and integrating consumers' choice behavior, airline companies forecast and measure passengers' preference to predict intertemporal choices in the future. Based on the predicted choices, airline decision makers formulate relevant content and recommend it to target consumer groups. Passengers can get information about what they want to know which can improve their loyalty, satisfaction, and comfort with airlines. Better customer relationship, more market share in the fierce competition. Content recommendations include products, services, tips, notices, introductions, and information sharing. Products include popular routes; international and domestic hotels; duty free gifts, etc. Services include special assistance, baggage inquiry, online check-in, pre-paid luggages, and so on. Tips include travel guide, entertainment activities, lounge locations, and flight delays. News and promotions, mileage redemption, and offers are also included in notices. Introductions involve frequent flyer program, activities, flight and hotel, boarding and arrival procedure, and so on. Information sharing is a new measure applied to web search with the popularity of social media. Airline industry starts to realize this platform can further improve service experience. Passengers can link the data of Weibo (China popular social media) or WeChat to flight reservation process that is easy for them to know who are in the same flight [23]. The network in **Figure 2** shows airline content recommendations for given choices of passengers.

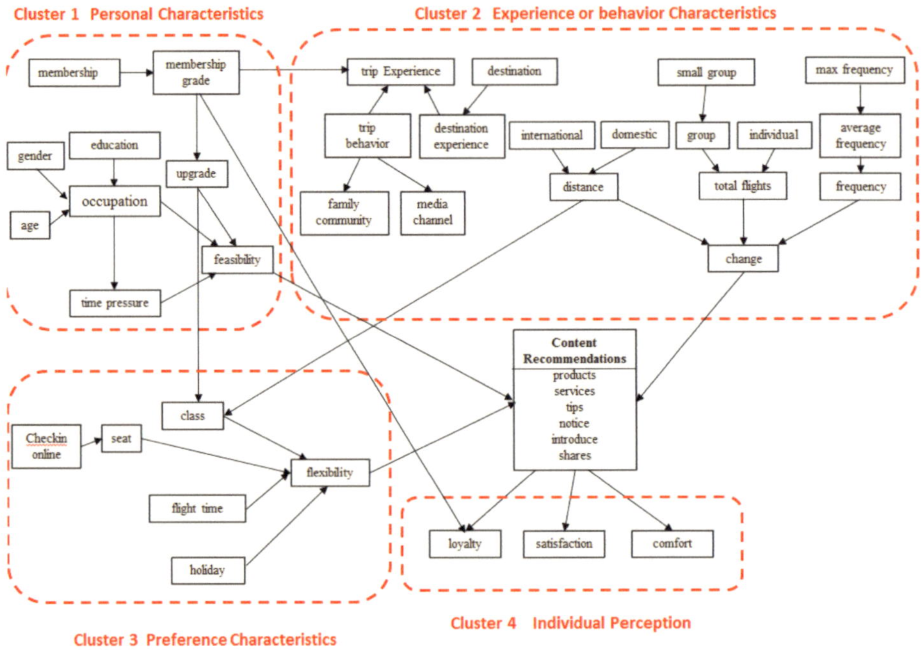

Figure 2. Network for content recommendation mode.

The first cluster illustrates personal information. Demographic data elements include gender, age, and education. By using these three attributes, one can speculate individual occupation and time pressure. Distinguishing leisure travelers and business travelers depends on time sensitivities. The node 'feasibility' indicates the air travel feasibility of each node combination, upgrade (yes or no), travel mode (leisure or business), and time pressure (yes or no). The second cluster depicts experience and behavior characteristic. The experience characteristic describes passengers' trip and destination experience. The third cluster represents passenger's preference such as fare class preference, seat preference, flight time preference, and holiday preference. It is worth to emphasize that distance and upgrade may lead passenger to change their class selections. Passengers will choose more comfortable classes when they take longer range flights. When it comes to membership upgrades, passengers are more likely to choose traveling first class to accumulate qualified miles. The fourth cluster describes individuals' perception evaluating the effect of variable changes on passengers' loyalty, satisfaction, and degree of comfort. This cluster refers to benefit variables that intend to cover the most significant perception. One of the benefit variables, namely, loyalty is affected by membership class. The higher the membership class, the higher level of stickiness to an Airline company. In this aspect, the outcome should take the weights of the benefit nodes into account.

3.3. Results and discussion

The data from airlines are used to complete all CPTs of the nature nodes. After completing all tables, we use Netica software to compile the network and determine the probabilities of six contents. **Figure 3** shows the compiled decision network.

http://dx.doi.org/10.5772/intechopen.70131

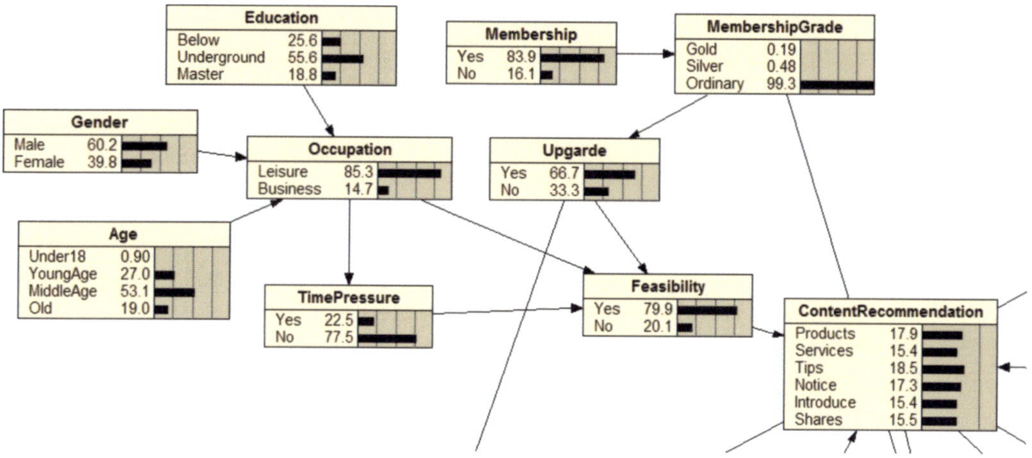

Figure 3. Compiled decision network (Cluster 1).

From **Figure 3**, the probability of tips is the highest, reaching 18.5% in total among all the contents decision options. Due to tips containing travel guide, entertainment activities, lounge caution, and delay calling, different kinds of hints remind passengers to have considerable experience. The probabilities of products and notice are around 17%. The other three contents are similar under the average level just over 15.4%. The beliefs and probabilities will be updated when evidence for certain nature nodes change. We will discuss some examples below.

After entering the evidence 'Yes' for the node 'Feasibility', which is colored gray in **Figure 4**, the belief for the decision nodes are auto-updated and recalculated. We find that only the probability of 'Introduce' option changed according to this new evidence. When air traveling is totally feasible for passengers no matter his or her travel purpose is holiday or business, the 'Introduce' is less useful to provide them flight information, boarding, and arrival procedure what they already know. They concern more about the services, delay caution, popular routes, holiday destination, and so on.

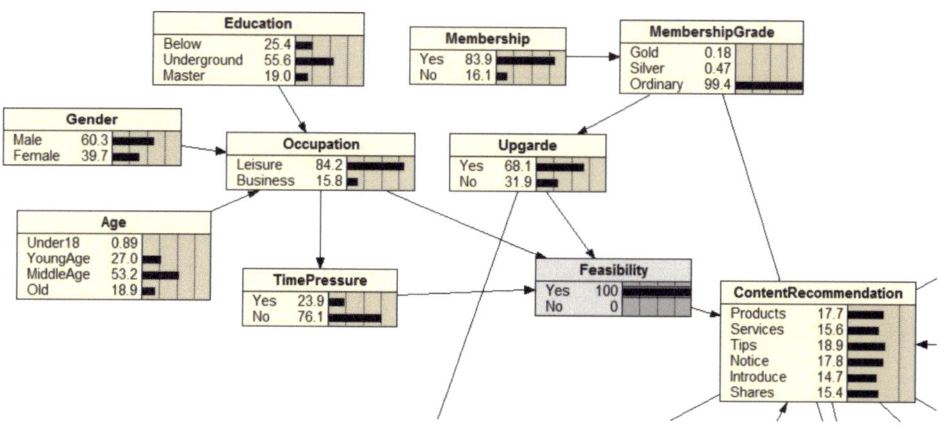

Figure 4. Compiled decision network ('Feasibility' = 'yes').

Figure 5 represents the influence of the evidence 'Yes' for the nature node 'Change' on the decision options mode. The beliefs and probabilities updated automatically. The results illustrates if passengers change their purchase behavior or trip modes, for example, taking high-speed train, the effective method to retain their customers, airline managers could recommend relative tips and notice to them and give them more comfortable services.

We compiled network with the 'Long' for 'Distance', 'High' for 'TotalFlight', and 'High' for 'Frequency', respectively. The consequences are represented on **Figure 6 (a–c)**. The trends of three results are similar. The probabilities of 'Products', 'Introduce', and 'Shares' rise outstanding. However, what surprised us is that the probabilities of 'Services' decrease sharply. This result gives decision makers a good suggestion that passengers who have high frequency traveling behavior need products recommendation, destination introduce, web link to share when they experience long range journey. In the same way, the service is not as important as other aspects.

As each passenger has own preference. The information about preference is too diverse, so that we introduce a nature node 'Flexibility' to describe the overall variation of consumers' preferences. Controlling states of these nodes, including seat, class, flight time and holiday, have no obvious effects on decision option modes. Therefore, we set 'Flexibility' to 'Yes' for sure in **Figure 7**. 'Shares' has the biggest change in the entire content recommendation options mode that means 'Share' is the most useful method to address flexibility problems whose regular pattern is hard to capture. Facing this situation, airline managers share links to their passengers on social media to release a service "meeting & sitting in the same flight" [23]. As the results shown in **Figure 8**, the membership class has significant influence on customers' loyalty. Members with highest qualification are stickier to their choices of airlines; the probability of high loyalty ascends from 35.7 to 85.2% when we set 'Yes' for state 'Gold'. Moreover, more than half of silver card owners remain loyal to their airline companies. For airlines, managers should better service loyal customers, reduce the loss of customers and mining new customers.

In the first part, we come up with three questions: How can we infer consumers' choice for content in the future? How to design a model using only current period choices to infer consumers' inter-temporal preferences? Is the process dynamic and it allows the researchers to analyze influences of effect changes?

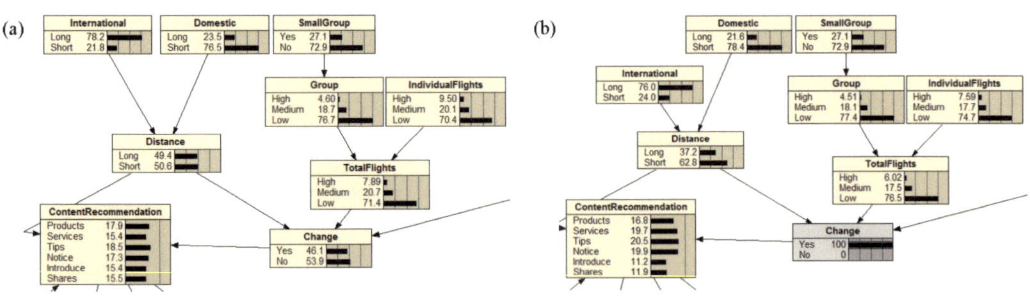

Figure 5. (a) Compiled decision network (Cluster 2). (b) Compiled decision network ('Change' = 'Yes').

http://dx.doi.org/10.5772/intechopen.70131

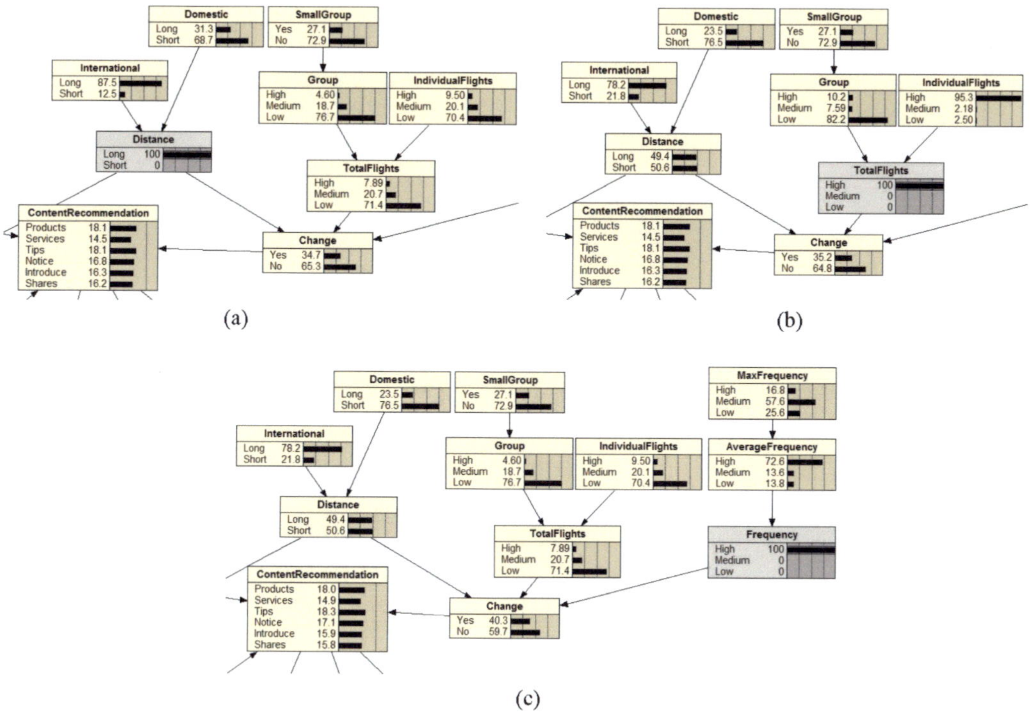

Figure 6. (a) Compiled decision network ('Distance' = 'Long'). (b) Compiled decision network ('TotalFlights'='High'). (c) Compiled decision network ('Frequency' = 'High').

This paper uses automatic updating process to explain the dynamics of belief networks. BBN model represents a complex network that constructs and model consumer choice process. From the examples above, we investigate clearly how the evidence of one state of a node change affects the probability of decision options. Based on China Southern Airline historical real data, we predict the passengers' choice and help airline managers recommend relative contents to satisfy passengers' needs.

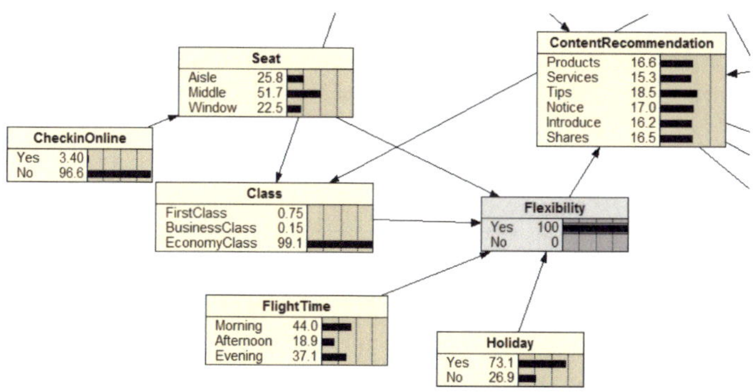

Figure 7. Compiled decision network ('Flexibility' = 'Yes').

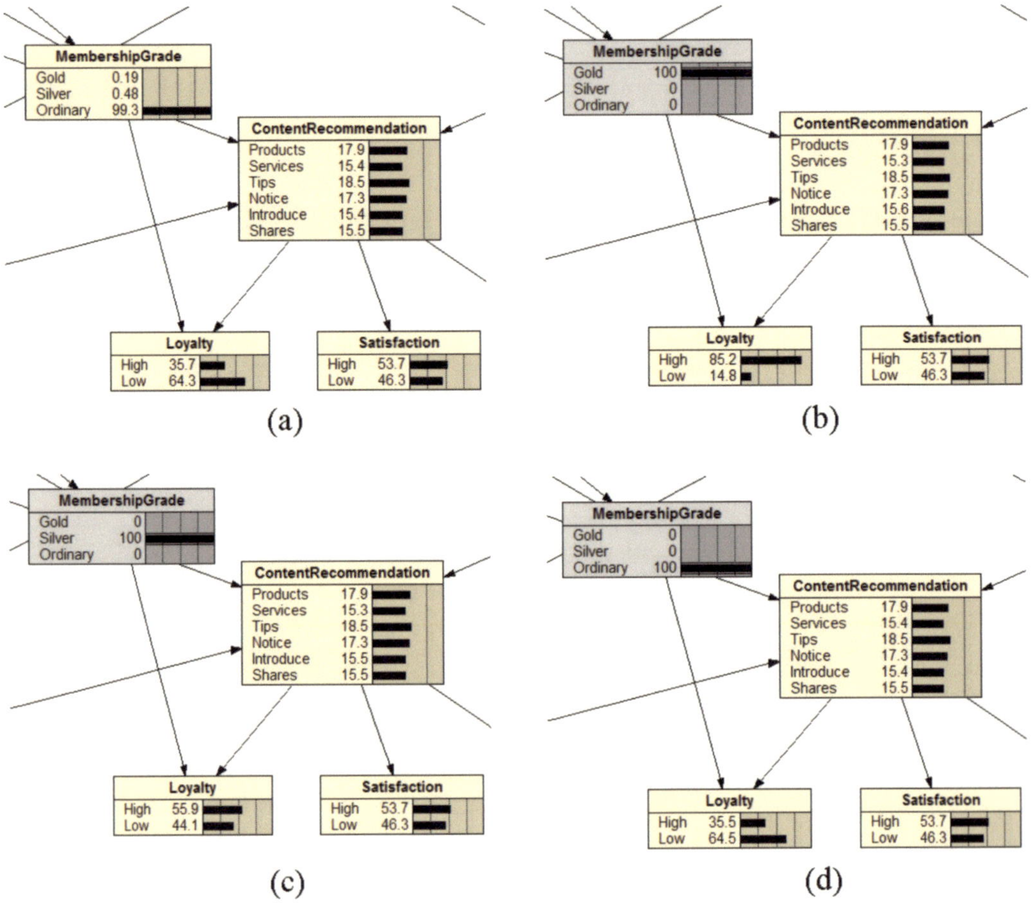

Figure 8. (a–d) Compiled decision network of Cluster 4.

4. Conclusion and implications

This article measures air passengers' preference and predicts their choices in the future based on current choice behavior using Bayesian belief network. This network can represent complex choice behavior and causal relationship among different variables, and the use of the probability of options can capture passengers' dynamic decision-making processes. The most powerful of the Bayesian network is that the probability of getting results from each stage is a reflection of mathematics and science. In other words, the network will infer reasonable results if we obtain enough information based on statistical knowledge.

We illustrate it by conducting a detailed empirical study of a data set from a Chinese Airline company. Our research demonstrates that understanding the extent to which the consumer choice behavior is beneficial for airline mangers strategic decision making.

To help with formulating better marketing strategies, the airline companies may consider adoption of the following procedures.

1. To track detailed consumer behavior: inquiry of products, reservation, payment, ticket issue, check-in, waiting, cabin service, luggage claim, mileage accumulation.
2. To analyze consumer behavior: purchase behavior, tour experience, choice behavior, preference.
3. To set up high-level products: high-level customization; customized design and products design, relevant product support.
4. To use social media: share web link, extract information from social media and social network.

A good strategy should analyze passengers' trip behavior and preferences to conduct cross selling, filter unnecessary information, and to present consumer recommendations and offer the most valuable product portfolio to customers.

We expect that, together with the need for the more specific features; BBN combined with Artificial Intelligence and deep learning are of great value to addressing uncertainty problems and consumer choice behavior in the future.

Author details

Sien Chen[1,2]*, Wenqiang Huang[3], Mengxi Chen[4], Junjiang Zhong[5] and Jie Cheng[6]

*Address all correspondence to: sien.chen@postgrad.manchester.ac.uk

1 Alliance Manchester Business School, University of Manchester, Manchester, UK

2 Antai College of Economics and Management, Shanghai Jiao Tong University, Shanghai, China

3 China Southern Airlines, Guangzhou, China

4 Shanghai Jiao Tong University, Shanghai, China

5 Xiamen University of Technology, Xiamen, China

6 Gausscode Technology Inc, CA, USA

References

[1] Bucklin RE, Srinivasan V. Determining interbrand substitutability through survey measurement of consumer preference structures. Journal of Marketing Research. 1991;**28** (February):58-71

[2] Wittink DR, Philippe C. Commercial use of conjoint analysis: An update. Journal of Marketing. 1989;**53**(July):91-6

[3] Rao VR. Applied Conjoint Analysis. New York: Springer; 2014

[4] Hansen F. Consumer choice behavior: An experimental approach. Journal of Marketing Research. 1969;**6**(4):436-443

[5] Anderson NH. Contributions to Information Integration Theory Volume II: Social. Lawrence Erlbaum Associates. Psychology Press, New York; 1991

[6] Bettman J, Capon N, Lutz RJ. Cognitive algebra in multi-attribute attitude models. Journal of Marketing Research. 1975;**12**(May):151-164

[7] Mowen JC. Beyond consumer decision making. Journal of Consumer Marketing. 1988;**5**(1):15-25.

[8] Wierenga B, van Raaj WF. Consumentengedrag. Leiden; Stenfert Kroese BV; 1987

[9] Davenport TH. At the Big Data Crossroads: Turning Towards a Smarter Travel Experience. 2013. Available from: http://www.bigdata.amadeus.com/assets/pdf/Amadeus_Big_Data.pdf (Accessed: March 2, 2018, 14:50)

[10] Buhalis D, Law R. Progress in information technology and tourism management: 20 years on and 10 years after the Internet—The state of eTourism research. Tourism Management. 2008;**28**(4):587-590

[11] Suermondt HJ. Explanation in Bayesian belief networks. PhD thesis, Palo Alto, California: Medical Information Sciences, Stanford University; March 1992

[12] Jensen FV. An Introduction to Bayesian Networks. London UK: UCL Press; 1996

[13] Stewart L, McCarty Jr P. The use of Bayesian belief networks to fuse continuous and discrete information for target recognition, tracking and situation assessment. Proceedings of the SPIE, 1992;**1699**:177-185

[14] Uusitalo L. Advantages and challenges of Bayesian networks in environmental modelling. Ecological Modelling. 2007;**203**(3/4):312-318

[15] Heckerman D. A Tutorial on Learning with Bayesian Networks, Technical Report MSR-TR-95-06, Redmond, WA: Microsoft Corporation; 1996

[16] Cheng J, Greiner R, Kelly J, Kelly J, Bell D, Liu W. Learning Bayesian networks from data: An information-theory based approach. Artificial Intelligence. 2002;**137**(1/2):43-90

[17] Heckerman D, Mamdani A, Wellman MP. Real-world applications of Bayesian networks. Communications of the ACM. 1995;**38**(3):24-26

[18] Fan C, Yu Y. BBN-based software project risk management. The Journal of Systems and Software. 2004;**73**(2):193-203

[19] Uusitalo, L. Advantages and challenges of Bayesian networks in environmental modelling. Ecological Modelling. 2007;**203**(3/4):312-318

[20] Ulegine F, Onsel S, Topcu YI, Aktas E, Kabak O. An integrated transportation decision support system for transportation policy decisions: The case of Turkey. Transportation Research Part A, Policy and Practice. 2007;**41**(1):40-97

[21] Verhoeven M, Arente TA, Timmermans HJP, van der Waerden PJHJ. Modeling the impact of key events on long-term transport mode choice decisions: A decision network approach using event history data, Transportation Research Record. 2005;**1926**:106-114. DOI: 10.3141/1926-13

[22] Fusun U, Sule O, Iker Topcu Y, Emel A, Ozgur K. An integrated transportation decision support system for transportation policy decisions: The case of Turkey. Transportation Research Part A. 2007;**41**:80-97. DOI: 10.1016/j.tra.2006.05.010

[23] Peveto A. KLM surprise: How a little research earned 1,000,000 impressions on Twitter. 2011. Available from:. http://www.digett.com/2011/01/11/klm-surprise-how-little-research-earned-1000000-impressions-twitter (Accessed: March 5, 2018, 9:20)